ENGINEERING PSYCHOPHYSIOLOGY

Issues and Applications

ENGINEERING PSYCHOPHYSIOLOGY

Issues and Applications

Edited by

Richard W.Backs
Central Michigan University
Wolfram Boucsein
University of Wuppertal, Germany

CRC Press
Taylor & Francis Group
Boca Raton London New York

CRC Press is an imprint of the
Taylor & Francis Group, an **informa** business

Contents

PART II: APPLICATIONS

Preface

The purpose of this volume is to promote engineering psychophysiology as a discipline and to demonstrate its value to a new audience who, we hope, will consist of ergonomists, human factors psychologists, and engineers. We use a rather broad definition of what constitutes engineering, including all aspects of the fields known as human engineering, industrial engineering, and safety and systems engineering. We had two goals for this volume that are reflected in its subtitle: *Issues and Applications.*

The goal for the Issues section is to introduce the components critical for the successful application of psychophysiological methods to problems in engineering. In particular, these chapters are intended to provide an introduction for the reader who is unfamiliar with psychophysiology. They are not comprehensive reviews, nor are they tutorials. Instead, their purpose is to provide the newcomer to the discipline with an overview of the basic theoretical, measurement, instrumentation, and experimental design questions inherent in the use of psychophysiological methods. Chapter 1 provides a very brief historical context of engineering psychophysiology as a discipline, introduces some theoretical constructs (e.g., arousal and stress-strain) and provides a taxonomic bibliography of recent research using the predominant measurement techniques in the field. Chapter 2 reviews the major theoretical approaches (i.e., mental workload and stress-strain) that form the foundation for much of the research of the discipline. Chapter 3 critiques the classical approach to experimental design in psychophysiology and suggests ways in which research in engineering psychophysiology can and should move beyond the classical approach. Chapter 4 reviews bioelectrical signal processing and illustrates how psychophysiological research in complex tasks differs from typical laboratory psychophysiology. Finally, chapter 5 reviews advantages and disadvantages of ambulatory assessment of psychophysiological measures. The newcomer is intended to use these chapters as guides to the more in-depth treatments of each issue that can be found in the chapter references.

The primary goal for the Applications section is to illustrate the many ways that psychophysiological methods are already being used in engineering applications. We also use a broad definition of *application,* one that includes laboratory and simulation research as well as field studies. However, what all the chapters have in common is that they address questions that are relevant for applying psychophysiological methods in the field. Our intent is to stimulate investigators to use these methods in new problem areas; therefore, the content of these chapters varies widely. Some chapters review specific psychophysiological measures (e.g., chap. 13). Other chapters review work performed on specific engineering problems (e.g., chaps. 11, 14, and 15), including problems in industrial and safety and systems engineering (e.g., chaps. 6, 8, 9, 10, and 16). Several chapters focus on the use of psychophysiological measures to assess mental workload (e.g., chaps. 7 and 12). Finally, there is a report on the status of the discipline in Japan (chap. 17).

The secondary goal for the Applications section is to have a broad sampling of measures and measurement techniques that have been used in engineering psychophysiology. In addition to the more typical psychophysiological measures of the central, autonomic, and peripheral nervous systems (i.e., electroencephalographic, cardiovascular, respiratory, electrodermal, oculomotor, and electromyographic), we also have included research using endocrine measures. Some chapters focus on a single response or response system, whereas others use multiple measures. The latter approach is increasingly common in psychophysiological research and has much to recommend it (see chap. 3).

We believe that the timing is right for this volume. The cost of instrumentation needed for psychophysiological research continues to fall. Simultaneously, the capability of the instruments continues to increase. Another important development is the rapid advancement of ambulatory technology enabling the transition of psychophysiological methods from the laboratory to the natural environment (see chap. 5). As a result, the interest in psychophysiological methods has never been greater.

An indication of this increased interest is the recent founding of the Psychophysiology in Ergonomics (PIE) society that is also a technical group of the International Ergonomics Association and an interest group of the Society for Psychophysiological Research. This group has more than 180 members representing 15 countries. PIE has sponsored two international conferences and several special issues of major journals such as *Ergonomics* and *Biological Psychology* (more information on PIE can be obtained at:

http://www.uni-wuppertal.de/FB3/psychologie/physio/pie.htm).

ACKNOWLEDGMENTS

We recognize and thank some of the many people who made this project possible. First, of course, are the contributors; the volume would not have been possible without their good will throughout the revision process. Second, we thank the reviewers, especially those contributors who also volunteered as reviewers: Mike Bonnet, Alex Gundel, Rob Henning, Dick Jennings, Art Kramer, Petra Netter, Raja Parasuraman, Bill Ray, John Stern, and Larry Walrath.

—Richard W.Backs
—Wolfram Boucsein

I

ISSUES

Chapter 1

Engineering Psychophysiology as a Discipline: Historical and Theoretical Aspects

Wolfram Boucsein
University of Wuppertal, Germany
Richard W.Backs
Central Michigan University

A BRIEF HISTORY OF ENGINEERING PSYCHOPHYSIOLOGY

Engineering psychophysiology is a term we have chosen to describe research that applies psychophysiological methods to the traditional problems addressed within the discipline of engineering psychology. The name is somewhat arbitrary and reflects our preferences to identify with the scientific study of human interactions with technology and to use the term engineering psychology to describe this endeavor. Unfortunately, a multiplicity of terms other than engineering psychology have also been used for this purpose, which has been problematic for the field ever since it was identified as a separate discipline.

Münsterberg (1913, 1914a, 1914b) may have been the first to promote the application of psychological science to improve everyday life. Münsterberg advocated the use of basic laboratory data to solve applied problems and the use of those problems to direct basic laboratory research. Collectively, he referred to the application of experimental psychology to address problems in the fields of education, medicine, law, business, and industry as *psychotechnics*. Of most relevance to the present chapter is Münsterberg's work in what he called "economic psychology," which anticipated modern industrial/organizational and engineering psychology.

However, the roots of modern engineering psychology are firmly planted in the contributions of psychologists in the United States and the United Kingdom to the design of weapons systems and personnel protection equipment during World War II. Psychologists had been involved in personnel testing and selection for many years, but their involvement in engineering and design was something new. Many of their early activities involved the application of the knowledge base and methodologies of experimental psychology to the design and evaluation of weapons systems. Natural descriptors of these activities that appeared in the United States were terms such as *applied experimental psychology, engineering psychology, human factors engineering,* or *human engineering,* along with the term *ergonomics* in the United Kingdom. All of these terms are still in use, and the distinctions among them have largely disappeared such that they are often used interchangeably.

However, a distinction was made in the earliest efforts to define the discipline. *Engineering psychology* was defined as the study of human behavior in human-machine (or more broadly human-environment) systems. Thus, the engineering psychologist generated the knowledge base that the human factors engineer could apply to the design of new systems (F.V.Taylor, 195 7). Early on, the scientist and the practitioner were often the same person and they tended to be more generalists with regard to the type of problem areas on which they worked. Today, the scientist and the practitioner are seldom the same person and both tend to be much more specialized with limited areas of expertise.

The specialty of the contributors to this volume is defined not so much by the nature of the problems on which they work, but instead by the type of measurement techniques they bring to bear on the problems. We refer to these techniques collectively *as psychophysiological methodology* because they emphasize the noninvasive assessment of physiological functioning. The use of these techniques has increased dramatically since the 1980s, along with the technological advancement and reduced cost of the equipment needed for collection and analysis of these types of data. However, psychophysiological measures have been used in engineering psychology since the earliest days of the discipline.

Perhaps the beginning of engineering psychophysiological research was in 1939 (although the report was not published until after the war) with a study of psychophysiological "tension" in flight using what may have been the first airborne polygraph (Williams, Macmillan, & Jenkins, 1946, cited in Roscoe, Corl, & LaRoche, 1997). Early psychophysiological research was limited by the technology and zeitgeist of the period and tended to emphasize muscular (including eye movement) measures during skilled psychomotor tasks. Examples of early work that we claim as engineering psychophysiology include the series of studies by R.C.Davis and colleagues using an electromyogram (EMG) to determine how muscular tension affected flight control (e.g., Davis, 1956) and the classic study by Fitts, Jones, and Milton (1950) of pilots' eye movements during instrument landings that helped to determine the arrangement of the basic six instruments (airspeed, attitude, altitude, rate-of-turn, heading, rate-of-climb) according to what is called a standard "T" arrangement that is still generally used today. Other early examples are the studies of eye movements during vigilance tasks by Mackworth, Kaplan, and Metlay (1964) and Baker (1960) that found subjects could fail to detect rare signals even when fixated on them.

Engineering psychophysiology flourished in the United States in the 1970s, because of technological advancements in data acquisition and analysis equipment and because of the "biocybernetic" research funding from the Department of Defense (e.g., Gomer, 1980). The technological advances meant that a wider range of physiological responses could be meaningfully examined in applied contexts, especially electroencephalogram (EEG) measures and the recently discovered event-related brain potentials. The level of support produced a wealth of research from government, university, and private research laboratories during this period. One representative example of the productivity of this period is the work of Donchin and his colleagues at the University of Illinois (see Donchin, Kramer, & Wickens, 1986, for a summary).

Unfortunately, engineering psychophysiological research in the United States waned during the 1980s. The expectations generated by the biocybernetic movement were unrealistically high, which may have led to some disillusion with psychophysiological methods in general. Furthermore, the difficulties of transitioning psychophysiological methodologies from the laboratory to operational environments, or even to full-fidelity simulators, became clear.

Whereas engineering psychophysiological research in the United States continued its traditional focus on defense, especially aviation, applications during the 1970s and 1980s, research in Europe had a different focus. European research during this period was concerned with the physiological effects of increasing workplace automation (see Rohmert, 1987, for a review). This type of research adopted a stress-strain theoretical approach (see the section on the three arousal model), unlike the information-processing approach of most of the engineering psychophysiology in the United States.

More recently, the ongoing computerization of workplaces has shifted the attention of European researchers from physical stress to issues of mental workload (see chap. 2, this volume). Although this development has been recognized by psychophysiologists, neither a deliberate theoretical background nor specific methods have been provided for that particular field of application (e.g., Gale & Christie, 1987). Nevertheless, psychophysiological research continues to gain influence on the organization of work settings (see chap. 14, this volume), as well as on the identification and prevention of stress related disorders (see chap. 16, this volume).

A THEORETICAL MODEL AND AN EMPIRICAL FRAMEWORK FOR ENGINEERING PSYCHOPHYSIOLOGY

Although the short and selective history given in the previous section documents the technological progress of engineering psychophysiological research, there is one area in which only limited progress has been made, the development of psychophysiological theory. Fitts, in his 1963 review, said, "The greatest gap in the physiological area, from the viewpoint of engineering psychology, is the lack of theory or empirical data relating physiological variables to performance data" (p. 924). The gap has certainly closed with regard to empirical data since then. However, psychophysiologists, with a few exceptions, have not contributed to a theoretical approach backing up their measures (e.g., Boucsein, 1991; Kahneman, 1973; Pribram & McGuinness, 1975; Sanders, 1983).

One of the reasons for this lack of progress is the nontheoretical multivariate shotgun approach frequently used in psychophysiological research. In addition, psychophysiologists have always been occupied with the development of refined techniques for data acquisition and evaluation. As a consequence, rather simple theoretical frameworks like the one-dimensional arousal concept (Poulton, 1970) are still used in the field. We foster a three-arousal model first proposed by Pribram and McGuinness (1975) to be used as a theoretical framework as a possibility for generating hypotheses concerning the specific sensitivity of the psychophysiological variables.

A Three-Arousal Model

Figure 1.1 depicts a simplified model of the three arousals and their neurophysiological backup (for more details, see Boucsein, 1992, 1993; McGuinness & Pribram, 1980). The right panel of Fig. 1.1 provides examples of typical physiological measures that may be used as indicators for the respective arousal systems.

Arousal System 1 is centered around the amygdala and is labeled the *affect arousal* system. It is responsible for focusing attention and generating hypothalamic reaction patterns such as orienting responses.

Arousal System 2 is centered around the hippocampus and is labeled the *effort* system. It has the ability to connect or disconnect input and output. The physiological patterns generated by this system can be regarded as concomitants of central information processing.

Arousal System 3 is centered around the basal ganglia and is labeled the *preparatory activation* system. Its activation results in an increased readiness of motor brain areas.

Arousal System 1 is responsible for the elicitation of orienting responses, but will also focus attention. It corresponds to the lower level system, whereas Arousal

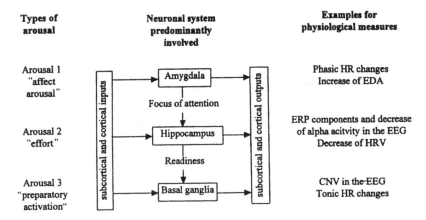

| Types of arousal | Neuronal system predominantly involved | Examples for physiological measures |

FIG. 1.1. A three-arousal model for engineering psychophysiology.
CNV=contingent negative variation; EEG=electroencephalogram;
EDA=electrodermal activity; ERP=event-related potentials;
HR=heart rate; HRV=heart rate variability.

System 2 corresponds to Broadbent's (1971) higher level system. Arousal System 2 is related to the basic hippocampal circuit involved in the behavioral inhibition system proposed by Gray (1982), whereas Arousal System 3 is related to the behavioral activation system described by Fowles (1980). If a situation changes or certain stimulation occurs, "affect arousal" will show up in such measures as phasic heart rate (HR) or the amplitude of the electrodermal response (EDR). The requirement for central processing capacity will appear in specific components of the event-related potential (ERP). Readiness for reaction will increase brain negativity and may be indicated as a response contingent negative variation (CNV) in the EEG (McGuinness & Pribram, 1980), accompanied by an increase in tonic HR (Fowles, 1980) as an expression of the "preparatory activation system."

   This rather straightforward chain of situation-reaction relationship can be modified by characteristics of the situation such as additional emotional load caused by danger or time pressure. The "effort" system has the property to disconnect Arousal Systems 1 and 3 to prevent immediate action and facilitate deliberate analysis performed by certain cortical-subcortical brain circuits (Gray, 1982). However, Arousal System 1 facilitates the focusing of attention to the situational demands. In this case, an increase of spontaneous, nonstimulus-specific EDRs will indicate an emotional load that comes with the situation (Fowles, 1980). The ongoing central information processing will be reflected both by the dominance of the beta frequency band over the alpha band in the EEG and a decrease in heart rate variability (HRV).

Although far from covering all possible psychophysiological relationships in engineering psychology, the model proposed here may help provide a framework for generating refined hypotheses regarding the action of different arousal processes on physiological outcomes. It is crucial for such a neurophysiologically based approach to be supported with empirical results from the field. We are going to exemplify this with respect to workload, which is a topic that has always been in the focus of using psychophysiology in the field of ergonomics (see chaps. 2 and 6, this volume).

From a psychological point of view, the issue of workload may be embedded in the broader concept of stress-strain processes. Although not very familiar to most North Americans, the use of the term *strain* for the impact of stress is gradually spreading from Europe (Weiner, 1982). A common distinction has been made between physical and mental workload or strain. Following a proposal by Klimmer and Rutenfranz (1983), mental load may be further split into mental and emotional strain. *Mental strain* refers to the concept of mental effort (see chap. 6, this volume). *Emotional strain* refers to the excess mental effort that comes from anxiety evoking cognitive aspects of the task, such as the imposition of deadlines, worries about one's capacity to perform the job, and so on (Boucsein, 1991). Table 1.1 provides hypotheses concerning the sensitivity of physiological variables for measuring different kinds of strain as derived from the literature, representative examples of which are given in Tables 1.2 through 1.7. The literature cited in the tables is limited to field and simulator studies published since 1985. Laboratory studies are substituted when no field or simulator study was found for the particular variable or type of sensitivity. The types of studies are indicated in the tables.

The rows in Table 1.1 are grouped according to the various physiological systems. If there is evidence from the literature that a particular physiological measure is associated with a specific strain (i.e., physical, mental, or emotional), an upward- or downward-pointing arrow is inserted in the appropriate column to indicate that the measure increases or decreases with an increase in that kind of strain. Two arrows indicate greater evidence for the relation.

Measures of Central Nervous System Activity (Table 1.2)

Spontaneous changes in the EEG are observed to changing mental demands such as a decrease of alpha activity (8–12 Hz) and concomitant increase of beta activity (> 13 Hz) that occur with increased mental load. They may also be used to indicate different states of vigilance or attention. Increased theta band activity (4–7 Hz) has been seen during periods of increased mental load. As a more refined measure of stimulus-directed central processing capacity, an ERP (see Fig. 1.1), the most frequently discussed component of which is the P3 or P300, can be taken from the EEG by averaging procedures. Averaging procedures also apply to the CNV or the *Bereitschaftspotential* used for determining reaction-directed

TABLE 1.1
Psychophysiological Parameters of Different Categories of Strain

Physiological Measure	Physical	Mental	Emotional
EEG alpha activity (8–12 Hz)		↓↓	
EEG theta activity (4–7 Hz)		↑↑	
P3 amplitude		↑↑	
P3 latency		↑	
CNV amplitude		↑	
Heart rate	↑↑	↑	↑
0.1 Hz component		↓↓	
Respiratory sinus arrhythmia		↓↓	
Additional heart rate		↑	
Respiration rate	↑	↑	
Finger pulse volume amplitude		↓	↓
Systolic blood pressure	↑↑	↑	
Diastolic blood pressure	↑	↑	
EDR amplitude		↑	↑
EDR recovery time		↑	
Spontaneous EDR frequency		↑	↑↑
Eyeblink rate		↑↑	↑
Saccadic eye movements		↑	
Pupillary diameter		↑	↑
Electromyogram	↑	↑	↑
Muscle tremor	↑↑		
Core temperature	↑↑		
Finger temperature			↑
Epinephrine		↑↑	↑
Norepinephrine	↑↑		↑
Cortisol		↑	↑↑

Note. The variables are grouped according to their respective physiological systems. "↑" means that the values of the parameter in question increase with increasing strain, and "↓" that they decrease. More evidence is provided for double arrows than for single ones.

TABLE 1.2
Demonstrated Sensitivity of Central Nervous System Measures

Measure	Sensitivity	Study	Type
EEG alpha activity ↓	Workload in real flight	Sterman and Mann (1995)	F
EEG alpha activity ↓	Complexity of attached work	Brookings et al. (1996)	S
EEG alpha rhythm de-synchronization ↑	Sleepiness in pilots during consecutive night flights	Gundel et al. (1995)	F
EEG alpha asymmetry	Alpha suppression in the right hemisphere during fast pacing	Marquis et al. (1984)	L
EEG alpha activity ↑	During tactical defensive exercise after sleep deprivation	Haslam (1982)	F
EEG alpha activity ↑ EEG theta activity ↑	Sleepiness during train driving at night	Torsvall and Åkerstedt (1983, 1987)	F
EEG alpha and theta burst activity ↑	Sleepiness in long-haul truck drivers	Kecklund and Åkerstedt (1993)	F
EEG theta activity ↑	During flight segments with high cognitive demands	Hankins and Wilson (1998)	F
EEG theta activity ↑	With increased memory load during driving	Lorenz et al. (1992)	F
EEG frontal-midline theta duration ↑	High mental demands in HCI	Yamamoto et al. (1989)	L
EEG frontal-midline theta duration ↑	Increasing memory load in HCI	Yamada (1998)	L
P3 amplitude ↑ P3 latency ↑ N2 amplitude ↑	Increased threat by missiles in military simulated flight	Lindholm and Sisson (1985)	S
P3 amplitude ↑	Capture of attention in radar operators	Kramer et al. (1995)	S
P3 amplitude ↑ CNV amplitude ↑	During low error rates in a reaction time task	Hohnsbein et al. (1998)	L

(continued)

TABLE 1.2 (continued)

Measure	Sensitivity	Study	Type
CNV amplitude ↑	Uncertainty of an aversive event	Backs and Grings (1985)	L
P3 latency ↑	Mental workload imposed by additional tasks in simulated flight	Fowler (1994)	S
P3 latency ↑ P3 latency ↓	Mental workload; Mental effort	Hohnsbein et al. (1995)	L
P3 amplitude ↓	With increased task demands in simulated flight	Kramer et al. (1987)	S
P3 amplitude ↓	With increased magnitude of communications in simulated helicopter flight	Sirevaag et al. (1993)	S
P3 amplitude ↓	During HCI compared to a paper-and-pencil task	Trimmel and Huber (1998)	L

Note.   HCI = human–computer interaction; ATC = air traffic control; Type = type of study; F = field; S = simulation; L = laboratory.

central processing (see chap. 12, this volume). Additional event-averaged EEG procedures that may be used in ergonomics are the lateralized readiness potential (LRP; Hohnsbein, Falkenstein, & Hoormann, 1998) and the eye-fixation related potential (EFRP; see chap. 17, this volume). However, all these procedures require a highly controlled work situation with access to reactions to single stimuli. Furthermore, due to their artifact-proneness, EEG variables in general have not been used much in field studies (Gevins et al., 1995). Brain imaging has not been used in the field for technical reasons (Kodama, Yoshida, Yamauchi, Takahashi, & Echigo, 1997).

Measures of Cardiorespiratory Activity (Table 1.3)

One of the most frequently applied measures in the field is HR. It is used as an indicator of physical as well as mental load (see chap. 6, this volume). More refined analyses of evaluating cardiac activity use measures of HRV, such as the power of the 0.1 Hz component and respiratory sinus arrhythmia (RSA), as

TABLE 1.3
Demonstrated Sensitivity of Cardiorespiratory Measures

Measure	Sensitivity	Study	Type
HR ↓	Physical fatigue during sleep deprivation	Myles and Romet (1987)	F
HR ↓	During prolonged city bus driving	Milosevic (1997)	F
HR ↓	During night shift in attached specialists;	Costa (1993)	F
HR ↑	Numbers of aircraft to handle		
HR ↑	Difficult compared to easy city bus driving	Felnémeti and Boon-Heckl (1985)	F
HR ↑	Physical load in factory employees	Green et al. (1986)	F
HR ↑	Physical strain in steel workers	Vitalis et al. (1994)	F
HR ↑	Physical and environmental stress in firefighters	Romet and Frim (1987)	S
HR ↓	Mechanized tasks in forest harvesting;	Smith et al. (1985)	F
HR ↑	Ambient temperature		
HR ↑	Ambient temperature in laundry workers	Brabant et al. (1989)	F
HR ↑	Physical strain in tree planting	Trites et al. (1993)	F
HR ↑	Physical load induced by ambient temperature change	Kristal-Boneh et al. (1997)	F
HR ↑	During high physical and mental load in truck drivers	Vivoli et al. (1993)	F
HR ↑	Workload in ATC specialists	Lille and Burnod (1983)	F
HR ↑	Short- and long-term workload in simulated flight	Lindholm and Cheatham (1983)	S
HR ↑	Workload in simulated commercial transport flight	Metalis (1991)	S
HR ↑	Workload and effort in flight	Hart and Hauser (1987)	F

(continued)

TABLE 1.3 (continued)

Measure	Sensitivity	Study	Type
HR ↑	Workload in flight, especially during takeoff and landing	Roscoe (1987, 1993)	F
HR ↑	Mental workload in real flight; ↑ in command, ↓ with practice	Kakimoto et al. (1988)	F
HR ↑	During takeoff and landing	Hankins and Wilson (1998)	F
HR ↑	Touch panel compared to other inputs in HCI	Haider et al. (1982)	F
HR ↑	High demands in simulated power line operations	Rau (1996)	S
HR ↑ Respiration rate ↑	Increased threat by missiles in military simulated flight	Lindholm and Sisson (1985)	S
HR ↑ Respiration rate ↑ End-tidal PCO_2 ↓	During computer based data entry work	Schleifer and Ley (1994)	L
Respiration rate ↑	Increased metabolic demands	Brookings et al. (1996)	S
Respiration rate ↑	Mental load in a continuous memory task	Backs and Seljos (1994)	L
HR ↓ HRV ↑ (mean squared successive differences of IBIs)	Monotony in train drivers' mental load during fast train driving	Myrtek et al. (1994)	F
HR ↑ RPI ↓ (interval between R-wave and pulse at the ear)	Workload and task demand; Speech associated reactivity in ATC specialists	Henderson et al. (1990)	F
HR ↑ RSA ↓	Physiological demands imposed by trailer length and fatigue in trucks	Apparies et al. (1998)	F
HR ↑ HRV ↑ (variation coefficient of IBIs)	Mentally demanding tasks during short-haul bus driving	Göbel et al. (1998)	F
HRV ↑ (variation coefficient of IBIs)	Mental load induced by speeding and swerving in car driving	De Waard et al. (1995)	F
HRV ↑ (variation coefficient of IBIs)	Mental effort in patent examiners	Boucsein and Thum (1997)	F

(continued)

TABLE 1.3 (continued)

Measure	Sensitivity	Study	Type
HR ↑ 0.1 Hz component ↓ Respiration rate ↑ Breath amplitudes ↓	During increased mental and physical load in fighter pilots and weapon systems officers	Wilson (1993)	F
HR ↑ 0.1 Hz component ↓	During takeoff and landing; During mentally demanding problem solving in flight engineers	Tattersall and Hockey (1995)	S
HR ↑ 0.1 Hz component ↓	Mental workload in pilots in simulated combat; High information load	Svensson et al. (1997)	S
HR ↑ 0.1 Hz component ↓	Difficulty of curvature in driving; Speedy driving in moderate curves	Richter et al. (1998)	F
RSA ↑	Due to respiratory activity during verbal compared to nonverbal communication in simulated helicopter flight	Sirevaag et al. (1993)	S
HR ↑ 0.1 Hz component ↓	During simulated shiftwork under noise	Boucsein and Ottmann (1996)	S
0.1 Hz component ↓	Mental workload in pilots during abnormal situations	Itoh et al. (1990)	S
0.1 Hz component ↓	Large changes in mental work-load during simulated flight	Veltman and Gaillard (1993, 1998)	S
0.1 Hz component ↓	Mental load in bus drivers	Mulders et al. (1982)	F
Additional HR ↑	Mental load (arithmetic)	Carroll et al. (1986, 1987)	L
Additional HR ↑	Mental load (psychological challenge)	Turner et al. (1985, 1988)	L
Finger pulse volume ↓	During exposure to loud noise	Millar and Steels (1990)	L
Finger pulse volume ↓	During a stressful interview	Smith et al. (1984)	L
Systolic BP ↑	Before and after parachute jumps	Sharma et al. (1994)	F
HR ↑ Systolic BP ↑	During a data query task with system response times	Emurian (1991)	L

(continued)

TABLE 1.3 (continued)

Measure	Sensitivity	Study	Type
HR ↑ Systolic BP ↑ Diastolic BP ↑	In ATC specialists with increasing number of planes to handle	Rose and Fogg (1993)	F
Systolic BP ↑	During various mental stress tests	Seibt et al. (1998)	L
Diastolic BP ↑	In complicated compared to simple HCI	Gao et al. (1990)	L
HR ↓ Systolic BP ↓ Diastolic BP ↓	During the last hours of long-haul bus driving as sign of fatigue	Raggatt and Morrissey (1997)	F

Note. For more details on HR and pilot workload, see Roscoe (1992), and for the use of HRV as a measure of workload in real and simulated flight, see Jorna (1993). IBI = interbeat interval.

very sensitive indicators of mental strain. Additional HR can be calculated as the difference of the actual HR to a standard situation such as ergometer performance. Althoguh tonic HR changes may indicate the need for energy supply in response preparation, phasic HR patterns can be used to determine stimulus-directed central processing activity (see Fig. 1.1). Recording of respiration without wearing a mask, which may interfere with performance, is possible by using a thermistor in the nostrils or a strain gauge around the chest. However, these devices do not provide more than respiration rate, which is mainly used to control for respiratory artifacts in various psychophysiological measures, but may also be used as a measure itself (Wientjes, 1992). A decrease in finger pulse volume is a very sensitive measure of defensiveness and, therefore, mainly indicates both mental and emotional strain. Blood pressure (BP) is also frequently recorded as an indicator of physical or mental strain, although problems with its continuous recording still persist (Wesseling, Settels, & de Wit, 1986).

Measures of Electrodermal Activity (Table 1.4)

Compared to other biosignals taken from the skin, electrodermal activity (EDA) can be regarded as the most convenient measure for workload. Phasic EDA is measured as skin conductance response (SCR) or skin resistance response (SRR). SCR amplitudes may reflect the amount of affective or emotional arousal elicited by a stimulus or situation (see Fig. 1.1). Both amplitude and recovery of an EDR have been demonstrated to be sensitive for certain aspects of central information processing (Boucsein, 1992), and may be used in a manner similar to ERPs as indicators of mental strain. The frequency of spontaneous electrodermal changes (called frequency of nonspecific SCRs or NS.SCR frequency) is a valid indicator

TABLE 1.4
Demonstrated Sensitivity of the Electrodermal System

Measure	Sensitivity	Study	Type
SCR amplitude (during 5-second intervals) ↑	Short-term workload during approach in simulated flight	Lindholm and Cheatham (1983)	S
SRL ↓	Before and after tower jump of parachutists	Sharma et al. (1994)	F
NS.SCR frequency ↑	Emotional strain during prolonged interruptions of patent examiners	Boucsein and Thum (1997)	F
NS.SCR frequency ↑	Emotional load by prolonged SRTs and time pressure	Schaefer et al. (1986)	L
NS.SCR frequency ↑	Emotional load by prolonged SRTs and time pressure	Kuhmann et al. (1987)	L
NS.SCR frequency ↑	Perceived difficulty of road curvature during car driving	Richter et al. (1998)	F
SCR amplitude ↑	Increased emotional load during car driving	Heino et al. (1990)	F
SCR amplitude ↑	Probability of an aversive event	Backs and Grings (1985)	L
SCR amplitude ↑	User-hostile task structure in HCI	Muter et al. (1993)	L
Integrated SCR amplitude ↑	Developing emotional strain	Kuhmann (1989)	L
NS.SRR frequency ↓	Decreasing emotional load under long system response times in HCI when time pressure is absent	Kuhmann et al. (1990)	L
NS.SCR frequency ↓	After prolonged simulated night shift work and prolonged work under noise; also as aftereffect of night shift work	Boucsein and Ottmann (1996)	S

Note. HCI = human–computer interaction; SCR = skin conductance response; SRL = skin resistance level; SRR = skin resistance response; SRT = system response time; NS.SCR = nonspecific skin conductance response.

of emotional strain and has shown specific sensitivity during computerized work (see chap. 14, this volume).

Measures of Somatomotor Activity (Table 1.5)

Eye movements and eyeblink rate are sensitive indicators of mental, and sometimes emotional, strain. Eyeblink rate has also been used to detect fatigue in various fields of application (Stern, Boyer, & Schroeder, 1994). Pupillary dilation is a sensitive measure of both mental and emotional strain. Its application to field situations poses a problem principally because of difficulty in maintaining the eye in the field of view. Changes in light intensity pose no major problem as long as the amount of light impinging on the eye is monitored and corrections made. The EMG can be used to evaluate the strain of certain muscle groups caused by physical work as well as to evaluate mental and emotional strain. Especially in computerized work, it may be used to quantify specific kinds of strain such as neck muscle tension (see chap. 14, this volume). Muscle tremor is used as a specific indicator of physical strain and fatigue.

TABLE 1.5
Demonstrated Sensitivity of Somatomotor Measures

Measure	Sensitivity	Study	Type
Eyeblink rate ↑	Mediational load (reasoning, decision making, judgment)	Wierwille et al. (1985)	S
Eyeblink rate ↑	Increasing memory load in HCI	Yamada (1998)	L
Eyeblink rate ↓	During car driving in complicated road sections	Richter et al. (1998)	F
Eyeblink rate ↓ Horizontal eye movements ↑	During visual demands of VFR and IFR flight; Inside–outside check	Hankins and Wilson (1998)	F
Eyeblink rate ↓ Eye closure duration ↓	High demands on the visual system of fighter pilots and weapon systems officers	Wilson (1993)	F
Eye closure duration ↑	With increased communication load in simulated helicopter flight	Sirevaag et al. (1993)	S
Eyeblink amplitude ↓ Eye closure duration ↑ Long eye closure rate ↑	With increased error rate resulting from pilot fatigue	Morris and Miller (1996)	S

(continued)

TABLE 1.5 (continued)

Measure	Sensitivity	Study	Type
Saccade duration ↑ Saccadic velocity ↓	During time on task in simulated ATC work; During eyeblink occurrence	McGregor and Stern (1996)	S
Saccadic velocity ↑	Increasing task demands in a choice reaction time task	App and Debus (1998)	L
Glissadic saccades ↑	Vigilance decrement in a signal detection task	Wang (1998)	L
Slow eye movements ↑	Sleepiness during train driving at night	Torsvall and Åkerstedt (1983, 1987)	F
Oculographic activity ↓ EMG power ↑	Light pen compared to other inputs in HCI	Haider et al. (1982)	F
EMG power ↑	In task specific muscles (neck and shoulder in HCI) with mental workload	Hanson et al. (1993)	L
EMG power ↑	In task specific muscles in cashiers	Lannersten and Harms-Ringdahl (1990)	F
EMG power shift from higher to lower band Muscle tremor ↑	Localized muscular fatigue; Result of fatigue (after work)	Gomer et al. (1987)	F
Pupillary diameter ↓	After visual workload in HCI	Saito and Taptagaporn (1991)	L
Pupillary diameter ↑	Increased information-processing demands in HCI	Backs and Walrath (1992)	L
Pupillary diameter ↑	With mental load due to increased task difficulty	Rößger et al. (1993)	L

Note. HCI = human computer interaction; VFR = visual flight rules; IFR = instrument flight rules; ATC = air traffic control.

Measures of Body Temperature (Table 1.6)

Core temperature, rectal or oral, is frequently used in addition to heart rate, especially in field studies on night and shift work, since it displays marked circa' dian variations and can be recorded easily. Body temperature is also increased during physical strain. However, it is not suitable for short-term changes. Finger

TABLE 1.6
Demonstrated Sensitivity of Body Temperature Measures

Measure	Sensitivity	Study	Type
Rectal temperature appropriately phased	Amplitude possibly related to "commitment" in shift workers	Minors and Waterhouse (1983)	F
Rectal temperature adjustment highly variable	During eastward compared to westward flight in pilots	Gander et al. (1989)	F
Rectal temperature ↑	Adjustment to prolonged night work during early morning hours	Minors and Waterhouse (1985)	F
Rectal temperature ↑	Physical and environmental stress in firefighters	Romet and Frim (1987)	S
Oral temperature ↑	In morning shift following physical training	Härmä et al. (1988)	F
Oral temperature ↑	During night shift in ATC specialists	Costa (1993)	F
Oral temperature ↓	Shift-related performance decrement	Daniel and Potasova (1989)	F
Otic temperature ↓	During night shift in power plant operators	Milosevic and Cabarkapa (1985)	F

Note.   ATC = air traffic control.

temperature measured by thermistors may be used as an indicator for emotional strain (Levenson, Ekman, & Friesen, 1990).

Endocrine Measures (Table 1.7)

The measurement of stress-related hormonal changes has been used in various ergonomic studies (see chap 16, this volume). Catecholamines (epinephrine and norepinephrine) and cortisol are the most frequently used stress hormones. Because blood samples are difficult to obtain during work, analyses of urine fractions are preferred. However, these cannot indicate effects of short-term strain. Epinephrine indicates long-lasting mental, and sometimes emotional, strain, whereas norepinephrine acts as an indicator of physical strain. Cortisol can now be recorded, with sufficient reliability, from saliva probes.

TABLE 1.7
Demonstrated Sensitivity of Endocrine Measures

Measure	Sensitivity	Study	Type
Catecholamines ↑ Epinephrine ↑	Stress in emergency ambulance; Additional psychological stress	Lehmann et al. (1983)	F
Epinephrine ↑	During and after highly demanding computer work	Johansson and Aronsson (1984)	F
Epinephrine ↑	Workload in bus drivers	Mulders et al. (1982)	F
Epinephrine ↑	With increased work density in driving examiners	Meijman et al. (1992)	F
Epinephrine ↑	In assembly line work compared to a flexible work setting	Melin et al. (1999)	F
Catecholamines ↑ Epinephrine ↑ Epinephrine and Cortisol ↑	In assembly line workers; During and after HCI; Mental effort in plant operators	Frankenhaeuser and Johansson (1986)	F
Catecholamines ↑ Epinephrine ↑	In truck drivers; In stressful traffic conditions	Vivoli et al. (1993)	F
Norepinephrine ↑ (not significant in epinephrine)	Increased difficulty of reading during HCI	Tanaka et al. (1988)	L
Catecholamines ↓ Epinephrine ↑	In simulated nightshift work; Under noise in night shift	Boucsein and Ottmann (1996)	S
Catecholamines ↑ 17-keto-steroids ↑	Before and after parachute jumps	Sharma et al. (1994)	F
Catecholamines ↑ Cortisol ↑	Time pressure, work demands; Irritation, tenseness, tiredness	Lundberg et al. (1989)	F
Cortisol ↑	In ATC specialists with increasing number of planes	Rose and Fogg (1993)	F
Salivary cortisol ↑	Mental workload in real flight parallel to HR ↑	Kakimoto et al. (1988)	F
Salivary cortisol ↑	During simulated helicopter ditching	Hytten et al. (1989)	S
Salivary cortisol ↑	Low level of performance in simulated flight	Veltman and Gaillard (1993)	S

(continued)

TABLE 1.7 (continued)

Measure	Sensitivity	Study	Type
Salivary cortisol ↑	Actual and perceived workload in ATC specialists	Zeier (1994) Zeier et al. (1996)	F
Salivary cortisol ↓	During consecutive nocturnal bomber missions	French et al. (1994)	S
Salivary cortisol ↓	Adaptation to night shift work	Hennig et al. (1998)	F
Norepinephrine metabolites ↑ Vanillyl mandelic acid ↑	Examination stress in high school students	Frankenhaeuser et al. (1986)	F
Vanillyl mandelic acid ↑	At the end of night shift in ATC specialists	Costa (1993)	F
Prolactin ↑	During jet fighter mission (because of dopamine ↓?)	Leino et al. (1995)	F

Note.   HCI= human–computer interaction; HR = heart rate; ATC = air traffic control.

Researchers in engineering psychophysiology may use this empirical framework to determine the specific sensitivity of designated physiological measures. Such an inductive approach should, however, be complemented by deductive considerations resulting from neurophysiological modeling as shown in Fig. 1.1. Stimulus-related emotional strain will mainly go with Arousal System 1 (i.e., "affect arousal"), whereas mental strain will very likely be a result of Arousal System 2 (i.e., "effort") or Arousal System 3 (i.e., "preparatory activation"). If emotional load is going to persist, certain measures related to the Arousal System 1, such as spontaneous EDA, will be affected. Thus, if the situation under investigation is thought to exert some emotional load on the subject, care should be taken to include measures that reflect both the "affect arousal" and the "effort" system. If the physiological variables are chosen accordingly, the sensitivity and diagnostic ability of the psychophysiological approach can be considerably improved and hypotheses about what kind of changes might be expected can be formed more accurately.

In addition to the physiological measures, subjective and behavioral measures constitute a genuine part of the psychophysiological approach. Subjective arousal, well-being, ratings of bodily symptoms, and other stress-related variables can be obtained discontinuously during working pauses by means of rating scales. As behavioral indicators, various performance measures such as speed, accuracy, and amount of work can be continuously obtained from the task itself. In addition, behavior, including mimic and gross body movements, may be videotaped. However, the analysis of such data is time consuming and is, therefore, not often performed by engineering psychologists.

OUTLOOK TO THE FUTURE
OF ENGINEERING PSYCHOPHYSIOLOGY

We believe that engineering psychophysiology is at the beginning of a renewed and exciting phase of research activity. Although defense applications are often still responsible for pushing the envelope of technological development, the microcomputerization and substantially reduced cost of commercially available data acquisition and analysis equipment has reached the stage where engineering psychophysiologists can leave the laboratory. Psychophysiological data can now be validly collected in the field for many application domains.

Physiological recordings can be taken continuously without interrupting the work flow and have the further advantage of not being subjected to "faking" such as psychological scales. However, not all of the various bioelectrical and biochemical measures used in psychophysiology are easy to obtain at real work places. Their use is limited by recording artifacts caused by body movements or electromagnetic fields. Furthermore, the number of recording channels of equipment for field studies at real work places is more restricted. (For a more detailed discussion, see chap. 5, this volume).

Field studies limit the number of dependent variables to be recorded. In addition, changes in working conditions cannot be controlled as easily as in the laboratory, which limits the isolation of independent variables. Therefore, psychophysiological methods should be tested in simulated workplaces before being applied in the field. This can be done easily for most automated workplaces where the human-machine interaction takes place using a computer. With such a combined laboratory-field approach, hypotheses from field observations can be tested under highly controlled laboratory conditions. Subsequent studies at real workplaces may be performed only with psychophysiological variables that have been shown to be relevant measures at simulated workplaces.

There is, however, no single physiological system that can cover all needs for gathering data on questions raised in the field. Therefore, we strongly recommend using a multivariate approach that includes measures for at least each of the three kinds of arousal mentioned in Fig. 1.1 and, if applicable, for the different kinds of strain mentioned in Table 1.1. Proper advances in engineering psychophysiology as a not-so-new scientific discipline can only be made if applications accompany improvements in the theoretical framework.

Although the contributors to this volume may represent only a small sample of the various psychophysiological methods and domains, their expertise in using those methods in the field qualifies them goal setters in engineering psychophysiology.

REFERENCES

App, E., & Debus, G. (1998). Saccadic velocity and activation: Development of a diagnostic tool for assessing energy regulation. *Ergonomics, 41,* 689–697.

Apparies, R.J., Riniolo, T.C., & Porges, S.W. (1998). A psychophysiological investigation of the effects of driving longer-combination vehicles. *Ergonomics, 41,* 581–592.

Backs, R.W., & Grings, W.W. (1985). Effects of UCS probability on the contingent negative variation and electrodermal response during long ISI conditioning. *Psychophysiology, 22,* 268–275.

Backs, R.W., & Seljos, K.A. (1994). Metabolic and cardiorespiratory measures of mental effort: the effects of level of difficulty in a working memory task. *International Journal of Psychophysiology, 16,* 57–68.

Backs, R.W., & Walrath, L.C. (1992). Eye movement and pupillary response indices of men-tal workload during visual search of symbolic displays. *Applied Ergonomics, 23,* 243–254.

Baker, C.H. (1960). Observing behavior in a vigilance task. *Science, 132,* 674–675.

Boucsein, W. (1991). Arbeitspsychologische Beanspruchungsforschung heute—eine Herausforderung an die Psychophysiologie [Research on stress-strain processes in today's industrial psychology—A challenge for psychophysiology]. *Psychologische Rundschau, 42,* 129–144.

Boucsein, W. (1992). *Electrodermal activity.* New York: Plenum Press.

Boucsein, W. (1993). Psychophysiology in the computer workplace—goals and methods. In P.Ullsperger (Ed.), *Psychophysiology of mental workload* (pp. 35–42). Berlin: Schriftenreihe der Bundesanstalt für Arbeitsmedizin, Sonderschrift 2.

Boucsein, W., & Ottmann, W. (1996). Psychophysiological stress effects from the combination of night-shift work and noise. *Biological Psychology, 42,* 301–322.

Boucsein, W., & Thum, M. (1997). Design of work/rest schedules for computer work based on psychophysiological recovery measures. *International Journal of Industrial Ergonomics, 20,* 51–57.

Brabant, C., Bédard, S., & Mergler, D. (1989). Cardiac strain among women workers in an industrial laundry. *Ergonomics, 32,* 615–628.

Broadbent, D.E. (1971). *Decision and stress.* London: Academic Press.

Brookings, J.B., Wilson, G.F., & Swain, C.R. (1996). Psychophysiological responses to changes in workload during simulated air traffic control. *Biological Psychology, 42,* 361–377.

Carroll, D., Turner, J.R., & Hellawell, J.C. (1986). Heart rate and oxygen consumption during active psychological challenge: The effects of level of difficulty. *Psychophysiology, 23,* 174–181.

Carroll, D., Turner, J.R., & Rogers, S. (1987). Heart rate and oxygen consumption during mental arithmetic, a video game, and graded static exercise. *Psychophysiology, 24,* 112–118.

Costa, G. (1993). Evaluation of workload in air traffic controllers. *Ergonomics, 36,* 1111–1120.

Daniel, J., & Potasova, A. (1989). Oral temperature and performance in 8 h and 12 h shifts. *Ergonomics, 32,* 689–696.

Davis, R.C. (1956). *Electromyographic factors in aircraft control: The relation of muscular tension to performance* (AF SAM Rept. 55–122). Randolph Air Force Base, TX: U.S. Air Force School of Aerospace Medicine.

DeWaard, D., Jessurun, M., & Steyvers, J.J.M. (1995). Effect of road layout and road environment on driving performance, drivers' physiology and road appreciation. *Ergonomics, 38,* 1395–1407.

Donchin, E., Kramer, A.F., & Wickens, C. (1986). Applications of brain event-related potentials to problems in engineering psychology. In M.G.H.Coles, E.Donchin, & S.W. Porges (Eds.), *Psychophysiology: Systems, processes, and applications* (pp. 702–718). New York: Guilford.

Emurian, H.H. (1991). Physiological responses during data retrieval: Comparison of constant and variable system response times. *Computers in Human Behavior, 7,* 291–310.

Felnémeti, A., & Boon-Heckl, U. (1985). Belastungsuntersuchung an Salzburger Busfahrern [Investigating stress in Salzburg's bus drivers]. *Zeitschrift für Verkehrssicherheit, 31,* 16–21.

Fitts, P.M. (1963). Engineering psychology. In S.Koch (Ed.), *Psychology: A study of science* (Vol. 5, pp. 908–933). New York: McGraw-Hill.

Fitts, P.M., Jones, R.E., & Milton, J.L. (1950). Eye movements of aircraft pilots during instrument landing approaches. *Aeronautical Engineering Review, 9,* 1–16.

Fowler, B. (1994). P300 as a measure of workload during a simulated aircraft landing task. *Human Factors, 336,* 670–683.

Fowles, D.C. (1980). The three arousal model: Implications of Gray's two-factor learning theory for heart rate, electrodermal activity, and psychopathy. *Psychophysiology, 17,* 87–104.

Frankenhaeuser, M., & Johansson, G. (1986). Stress at work: Psychobiological and psychosocial aspects. *International Review of Applied Psychology, 35,* 287–299.

Frankenhaeuser, M., Lundberg, U., von Wright, M,R., von Wright, J., & Sedvall, G. (1986). Urinary monoamine metabolites as indices of mental stress in healthy males and females. *Pharmacology, Biochemistry & Behavior, 24,* 1521–1525.

French, J., Bisson, R.U., Neville, K.J., Mitcha, J., & Storm, W.F. (1994, May). Crew fatigue during simulated, long duration B-1B bomber missions. *Aviation, Space, and Environmental Medicine,* A1–A6.

Gale, A., & Christie, B. (Eds.). (1987). *Psychophysiology and the electronic workplace.* Chichester: Wiley.

Gander, P.H., Myhre, G., Graeber, C.R., Andersen, H.T., & Lauber, J.K. (1989). Adjustment of sleep and the circadian temperature rhythm after flights across nine time zones. *Aviation, Space, and Environmental Medicine, 60,* 733–743.

Gao, C., Lu, D., & She, Q. (1990). The effects of VDT data entry work on operators. *Ergonomics, 33,* 917–924.

Gevins, A., Leong, H., Du, R., Smith, M.E., Le, J., DuRousseau, D., Zhang, J., & Libove, J. (1995). Towards measurement of brain function in operational environments. *Biological Psychology, 40,* 169–186.

Göbel, M., Springer, J., & Scherff, J. (1998). Stress and strain of short haul bus drivers—psychophysiology as a design oriented method for analysis. *Ergonomics, 41,* 563–580.

Gomer, F.E. (1980). *Biocybernetic applications for military systems* (MDC E2191). St. Louis, MO: McDonnell Douglas Corporation.

Gomer, F.E., Silverstein, L.D., Berg, W.K., & Lassiter, D.L. (1987). Changes in electromyographic activity associated with occupational stress and poor performance in the workplace. *Human Factors, 292,* 131–143.

Gray, J.A. (1982). *The neuropsychology of anxiety: An inquiry into the functions of the septohippocampal system.* Oxford, England: Clarendon Press.

Green, M.S., Luz, Y., Jucha, E., Cocos, M., & Rosenberg, N. (1986). Factors affecting ambulatory heart rate in industrial workers. *Ergonomics, 29,* 1017–1027.

Gundel, A., Drescher, J., Maaß, H., Samel, A., & Vejvdova, M. (1995). Sleepiness of civil airline pilots during two consecutive night flights of extended duration. *Biological Psychology, 40,* 131–141.

Haider, E., Luczak, H., & Rohmert, W. (1982). Ergonomics investigations of workplaces in a police command-control centre equipped with TV displays. *Applied Ergonomics, 13,* 163–170.

Hankins, T.C., & Wilson, G.F. (1998). A comparison of heart rate, eye activity, EEG and subjective measures of pilot mental workload during flight. *Aviation, Space, and Environmental Medicine, 69,* 360–367.

Hanson, E.K.S., Schellekens, J.M.H., Veldman, J.B.P., & Mulder, L.J.M. (1993). Psychomotor and cardiovascular consequences of mental effort and noise. *Human Move-ment Science, 12,* 607–626.

Härmä, M.I., Ilermarinen, J., Knauth, P., Rutenfranz, J., & Hänninen, O. (1988). Physical training intervention in female shift workers: I. The effects of intervention of fitness, fatigue, sleep, and psychosomatic symptoms. *Ergonomics, 31,* 39–50.

Hart, S.G., & Hauser, J.R. (1987). Inflight application of three pilot workload measurement techniques. *Aviation, Space and Environmental Medicine, 58,* 402–410.

Haslam, D.R. (1982). Sleep loss, recovery sleep, and military performance. *Ergonomics, 25,* 163–178.

Heino, A., van der Molen, H.H., & Wilde, G.J.S. (1990). *Risk-homeostatic processes in car following behaviour: Electrodermal responses and verbal risk estimates as indicators of the perceived level of risk during a car-driving task.* Groningen, The Netherlands: Reports from the University's Traffic Research Center.

Henderson, P.R., Bakal, D.A., & Dunn, B.E. (1990). Cardiovascular response patterns and speech: A study of air traffic controllers. *Psychosamatic Medicine, 52,* 27–41.

Hennig, J., Kieferdorf, P., Moritz, C., Huwe, S., & Netter, P. (1998). Changes in cortisol secretion during shiftwork: implications for tolerance to shiftwork? *Ergonomics, 41,* 610–621.

Hohnsbein, J., Falkenstein, M., & Hoormann, J. (1995). Effects of attention and time-pressure on P300 subcomponents and implications for mental workload research. *Biological Psychology, 40,* 73–81.

Hohnsbein, J., Falkenstein, M., & Hoormann, J. (1998). Performance differences in reaction tasks are reflected in event-related brain potentials (ERPs). *Ergonomics, 41,* 622–634.

Hytten, K., Jensen, A., & Vaernes, R. (1989). Effects of underwater escape training: A psychophysiological study. *Aviation, Space, and Environmental Medicine, 60,* 460–464.

Itoh, Y., & Hayashi, Y. (1990). The ergonomic evaluation of eye movement and mental workload in aircraft pilots. *Ergonomics, 33,* 719–733.

Johansson, G., & Aronsson, G. (1984). Stress reactions in computerized administrative work. *Journal of Occupational Behaviour, 5,* 159–181.

Jorna, P.G. (1993). Heart rate and workload variations in actual and simulated flight. *Ergonomics, 36,* 1043–1054.

Kahneman, D. (1973). *Attention and effort*. Englewood Cliffs, NJ: Prentice-Hall.

Kakimoto, Y., Nakamura, A., Tarui, H., Nagasawa, Y., & Yagura, S. (1988). Crew workload in JASDF C-1 transport flights: I. Change in heart rate and salivary cortisol. *Aviation, Space, and Environmental Medicine, 59,* 511–516.

Kecklund, G., & Åkerstedt, T. (1993). Sleepiness in long distance truck driving: an ambulatory EEG study of night driving. *Ergonomics, 36,* 1007–1017.

Klimmer, F., & Rutenfranz, J. (1983). Folgen mentaler und emotionaler Belastungen [Consequences of mental and emotional stress]. In W.Rohmert & J.Rutenfranz (Eds.), *Praktische Arbeitsphysiologie* (pp. 135–141). Stuttgart: Thieme.

Kodama, H., Yoshida, T., Yamauchi, Y., Takahashi, A., & Echigo, J. (1997). Application of functional MRI to ergonomics. *Proceedings of the 13th Triennial Congress of the International Ergonomics Association* (Vol 7, pp. 246–248). Helsinki, Finland: Finnish Institute of Occupational Health.

Kramer, A.F., Sirevaag, E.J., & Braune, R. (1987). A psychophysiological assessment of operator workload during simulated flight missions. Special Issue: Cognitive psychophysiology. *Human Factors, 29,* 145–160.

Kramer, A.F., Trejo, L.J., & Humphrey, D. (1995). Assessment of mental workload with task-irrelevant auditory probes. *Biological Psychology, 40,* 83–100.

Kristal-Boneh, E., Harari, G., & Green, M.S. (1997). Heart rate response to industrial work at different outdoor temperatures with or without control systems at the plant. *Ergonomics, 40,* 729–736.

Kuhmann, W. (1989). Experimental investigation of stress-inducing properties of system response times. *Ergonomics, 32,* 271–280.

Kuhmann, W., Boucsein, W., Schaefer, F., & Alexander, J. (1987). Experimental investigation of psychophysiological stress-reactions induced by different system response times in human-computer interaction. *Ergonomics, 30,* 933–943.

Kuhmann, W., Schaefer, F., & Boucsein, W. (1990). Effekte von Wartezeiten innerhalb einfacher Aufgaben: Eine Analogie zu Wartezeiten in der Mensch-Computer-Interaktion [Effects of waiting times within sample problems: An analogy to waiting times in hu-man-computer interaction]. *Zeitschrift für Experimentelle und Angewandte Psychologie, 37,* 242–265.

Lannersten, L., &. Harms-Ringdahl, K. (1990). Neck and shoulder muscle activity during work with different cash register systems. *Ergonomics, 33,* 49–65.

Lehmann, M., Dörges, V., Huber, G., Zöllner, G., Spöri, U., & Keul, J. (1983). Zum Verhalten der freien Katecholamine im Blut und Harn bei Sanitätern und ärzten während des Einsatzes [Behavior of free plasma and urine catecholamines of ambulance men and physicians during medical service]. *International Archives of Occupational and Environmental Health, 51,* 209–222.

Leino, T., Leppäluoto, J., Huttunen, P., Ruokonen, A., & Kuronen, P. (1995). Neuroendocrine responses to real and simulated BA Hawk MK 51 flight. *Aviation, Space, and Environmental Medicine, 66,* 108–113.

Levenson, R.W., Ekman, P., & Friesen, W.V. (1990). Voluntary facial action generates emotion-specific autonomic nervous system activity. *Psychophysiology, 27,* 363–384.

Lille, F., & Burnod, Y. (1983). Professional activity and physiological rhythms. *Advances in Biological Psychiatry, 11,* 64–71.

Lindholm, E., & Cheatham, C.M. (1983). Autonomic activity and workload during learning of a simulated aircraft carrier landing task. *Aviation, Space, and Environmental Medicine, 54,* 435–439.

Lindholm, E., & Sisson, N. (1985). Physiological assessment of pilot workload in simulated and actual flight environments. *Behavior Research Methods, Instruments, and Computers, 17,* 191–194.

Lorenz, J., Lorenz, B., &. Heineke, M. (1992). Effect of mental task load on fronto-central theta activity in a deep saturation dive to 450 msw. *Undersea Biomedical Research, 19,* 243–262.

Lundberg, U., Hedman, M., Melin, B., &. Frankenhaeuser, M. (1989). Type A behavior in healthy males and females as related to physiological reactivity and blood lipids. *Psychosomatic Medicine, 51,* 113–122.

Mackworth, N.M., Kaplan, I.T., & Metlay, W. (1964). Eye movements during vigilance. *Perceptual and Motor Skills, 18,* 397–402.

Marquis, F.A., Glass, A., & Corlett, E.N. (1984). Speed of work and EEG asymmetry. *Biological Psychology, 19,* 205–211.

McGregor, D.K., & Stern, J.A. (1996). Time on task and blink effects on saccade duration. *Ergonomics, 39o* 649–660.

McGuinness, D., & Pribram, K. (1980). The neuropsychology of attention: Emotional and motivational controls. In M.C. Wittrock (Ed.), *The brain and psychology* (pp. 95–139). New York: Academic Press.

Meijman, T.F., Mulder, G., van Dormolen, M., & Cremer, R. (1992). Workload of driving examiners: A psychophysiological field study. In H. Kragt (Ed.), *Enhancing industrial performance* (pp. 245–260). London: Taylor & Francis.

Melin, B., Lundberg, U., Söderlund, J., & Granquist, M. (1999). Stress reactions in assembly work: Comparison between two contrasting work organizations as related to sex. *Journal of Organizational Behaviour, 20,* 47–61.

Metalis, S.A. (1991). Heart period as a useful index of pilot workload in commercial transport aircraft. *The International Journal of Aviation Psychology, 1,* 107–116.

Millar, K., & Steels, M.J. (1990). Sustained peripheral vasoconstriction while working in continuous intense noise. *Aviation, Space, and Environmental Medicine, 61,* 695–698.

Milosevic, S. (1997). Drivers' fatigue studies. *Ergonomics, 40,* 381–389.

Milosevic, S., & Cabarkapa, M. (1985). Variation in body temperature and performance during weekly rotating shift system. *Studia Psychologica, 27,* 225–233.

Minors, D.S., & Waterhouse, J. M. (1983). Circadian rhythm amplitude—is it related to rhythm adjustment and/or worker motivation? *Ergonomics, 26,* 229–241.

Minors, D.S., & Waterhouse, J.M. (1985). Circadian rhythms in deep body temperature, urinary excretion and alertness in nurses on night work. *Ergonomics, 28,* 1523–1530.

Morris, T.L., & Miller, J.C. (1996). Electrooculographic and performance indices of fatigue during simulated flight. *Biological Psychology, 42,* 343–360.

Münsterberg, H. (1913). *Psychology and industrial efficiency.* Boston: Houghton Mifflin.

Münsterberg, H. (1914a). *Grundzüge der Psychotechnik* [Fundamentals of psychotechnics]. Leipzig: Barth.

Münsterberg, H. (1914b). *Psychology, general and applied.* New York: Appleton.

Mulders, H.P.G., Meijman, T.F., O'Hanlon, J.F., & Mulder, G. (1982). Differential psychophysiological reactivity of city bus drivers. *Ergonomics, 25,* 1003–1011.

Muter, P., Furedy, J.J., Vincent, A., & Pelcowitz, T. (1993). User-hostile systems and patterns of psychophysiological activity. *Computers in Human Behavior, 9,* 105–111.

Myles, W.S., & Romet, T.T. (1987). Self-paced work in sleep deprived subjects. *Ergonomics, 30,* 1175–1184.

Myrtek, M., Deutschmann-Janicke, E., Strohmaier, H., Zimmermann, W., Lawerenz, S., Brügner, G., & Müller, W. (1994). Physical, mental, emotional, and subjective workload components in train drivers. *Ergonomics, 37,* 1195–1203.

Poulton, E.C. (1970). *Environment and human efficiency.* Springfield, IL: C.C.Thomas.

Pribram, K.H., & McGuinness, D. (1975). Arousal, activation, and effort in the control of attention. *Psychological Review, 82,* 116–149.

Raggatt, P.T.F., & Morrissey, S.A. (1997). A field study of stress and fatigue in long-distance bus drivers. *Behavioral Medicine, 23,* 122–129.

Rau, R. (1996). Psychophysiological assessment of human reliability in a simulated complex system. *Biological Psychology, 42,* 287–300.

Richter, P., Wagner, T., Heger, R., & Weise, G. (1998). Psychophysiological analysis of mental load during driving on rural roads—a quasi-experimental field study. *Ergonomics, 41,* 593–609.

Rößger, R., Rötting, M., & Unema, P. (1993). Experimentelle Untersuchung zum Einfluß von Leuchtdichteveränderungen und mentaler Beanspruchung auf den Pupillendurchmesser [Experimental studies of the influence of changes of luminence and mental load on pupil diameter]. *Zeitschrift für Arbeitswissenschaft, 47,* 141–147.

Rohmert, W. (1987). Physiological and psychological work load measurement and analysis. In G.Salvendy (Ed.), *Handbook of human factors* (pp. 402–428). New York: Wiley-Interscience.

Romet, T.T., & Frim, J. (1987). Physiological responses to fire fighting activities. *European Journal of Applied Physiology, 56,* 633–638.

Roscoe, A.H. (1987). Pilot arousal during the approach and landing. *Aviation Medicine Quarterly, 1,* 31–36

Roscoe, A.H. (1992). Assessing pilot workload. Why measure heart rate, HRV and respiration? *Biological Psychology, 34,* 259–287.

Roscoe, A.H. (1993). Heart rate as a psychophysiological measure for in-flight workload assessment. *Ergonomics, 36,* 1055–1062.

Roscoe, S.N., Corl, L., & LaRoche, J. (1997). *Predicting human performance.* Pierrefonds, Quebec: Helio Press.

Rose, R.M., & Fogg, L.F. (1993). Definition of a responder: Analysis of behavioral, cardiovascular, and endocrine responses to varied workload in air traffic controllers. *Psychosomatic Medicine, 55,* 325–338.

Saito, S., & Taptagaporn, S. (1991). Pupillary reflexes and accommodation as physiological indices of visual fatigue due to VDT operation. In H.J.Bullinger (Ed.), *Human aspects in computing: design and use of interactive systems and work with terminals* (pp. 233–237). Amsterdam: Elsevier.

Sanders, A.F. (1983). Towards a model of stress and human performance. *Acta Psychologica, 53,* 61–97.

Schaefer, F, Kuhmann, W., Boucsein, W., & Alexander, J. (1986). Beanspruchung durch Bildschirmtätigkeit bei experimentell variierten Systemresponsezeiten [Stress effects of experimentally varied system response times at visual display units]. *Zeitschrift für Arbeitswissenschaft, 40*(12 NF), 31–38.

Schleifer, L.M., & Ley, R. (1994). End-tidal PCO_2 as an index of psychophysiological activity during VDT data-entry work and relaxation. *Ergonomics, 37,* 245–254.

Seibt, R., Boucsein, W., & Scheuch, K. (1998). Effects of different stress settings on cardiovascular parameters and their relationship to daily life blood pressure in normotensives, borderline hypertensives and hypertensives. *Ergonomics, 41,* 634–648.

Sharma, V.M., Sridharan, K., Selvamurthy, W., Mukherjee, A.K., Kumaria, M.M.L., Upadhyay, T.N., Ray, U.S., Hegde, K.S., Raju, V.R.K., Panwar, M.R., Asnani, V. & Dimri, G.P. (1994). Personality traits and performance of military parachutist trainees. *Ergonamics, 37,* 1145–1155.

Sirevaag, E.J., Kramer, A.F., Wickens, C.D., Reisweber, M., Strayer, D.L., & Grenell, J.F. (1993). Assessment of pilot performance and mental workload in rotary wing aircraft. *Ergonomics, 36,* 1121–1140.

Smith, L.A., Wilson, G.D., & Sirois, D.L. (1985). Heart-rate response to forest harvesting work in the south-eastern United States during summer. *Ergonomics, 28,* 655–664.

Smith, T.W., Houston, B.K., & Zurawski, R.M. (1984). Finger pulse volume as a measure of anxiety in response to evaluative threat. *Psychophysiology, 21,* 260–264.

Stern, J., Boyer, D.J., & Schroeder, D. (1994). Blink rate: A possible measure of fatigue. *Human Factors, 36,* 285–297.

Sterman, M.B., & Mann, C.A. (1995). Concepts and applications of EEG analysis in aviation performance evaluation. *Biological Psychology, 40,* 115–130.

Svensson, E., Angelborg-Thanderz, M., Sjöberg, L., & Olsson, S. (1997). Information complexity—mental workload and performance in combat aircraft. *Ergonomics, 40,* 362–380.

Tanaka, T., Fukumoto, T., Yamamoto, S., & Noro, K. (1988). The effects of VDT work on urinary excretion of catecholamines. *Ergonomics, 31,* 1753–1763.

Tattersall, A.J., & Hockey, G.R. J. (1995). Level of operator control and changes in heart rate variability during simulated flight maintenance. *Human Factors, 37,* 682–698.

Taylor, F.V. (1957). Psychology and the design of machines. *American Psychologist, 12,* 249–258.

Torsvall, L., & Åkerstedt, T. (1983). Sleepiness and irregular work hours. *Sleep Research, 12,* 376.

Torsvall, L., & Åkerstedt, T. (1987). Sleepiness on the job: Continuously measured EEG changes in train drivers. *Electroencephalography and Clinical Neurophysiology, 66,* 502–511.

Trimmel, M., & Huber, R. (1998). After-effects of human-computer interaction indicated by P300 of the event-related brain potential. *Ergonomics, 41,* 649–655.

Trites, D.G., Robinson, D.G., & Banister, E.W. (1993). Cardiovascular and muscular strain during a tree planting season among British Columbia silviculture workers. *Ergonomics, 36,* 935–949.

Turner, J.R., & Carroll, D. (1985). Heart rate and oxygen consumption during mental arithmetic, a video game, and graded exercise: Further evidence of metabolically-exaggerated cardiac adjustments? *Psychophysiology, 22,* 261–267.

Turner, J.R.,Carroll, D., Hanson, J., & Sims, J. (1988). A comparison of additional heart rates during active psychological challenge calculated from upper body and lower body dynamic exercise. *Psychophysiology, 25,* 209–216.

Veltman, J.A., & Gaillard, A.W.K. (1993). Indices of mental workload in a complex task environment. *Neuropsychobiology, 28,* 72–75.

Veltman, J.A., & Gaillard, A.W.K. (1998). Physiological workload reactions to increasing levels of task difficulty. *Ergonomics, 41,* 656–669.

Vitalis, A., Pournaras, N.D., Jeffrey, G.B., Tsagarakis, G., Monastiriotis, G., & Kavvadias, S. (1994). Heart rate strain indices in greek steelworkers. *Ergonomics, 37,* 845–850.

Vivoli, G., Bergomi, M., Rovesti, S., Carrozzi, G., & Vezzosi, A. (1993). Biochemical and haemodynamic indicators of stress in truck drivers. *Ergonomics, 36,* 1089–1097.

Wang, L. (1998). Glissadic saccades: a possible measurement of vigilance. *Ergonomics, 41,* 721–732.

Weiner, J.S. (1982). The Ergonomics Society—The society's lecture 1982: The measurement of human workload. *Ergonomics, 25,* 953–965.

Wesseling, K.H., Settels, J.J., & de Wit, B. (1986). The measurement of continuous finger arterial pressure non-invasively in stationary subjects. In T.H.Schmidt, T.M.Dembroski, & G.Blümchen (Eds.), *Biological and psychological factors in cardiovascular disease* (pp. 355–375). Berlin: Springer.

Wierwille, W.W., Rahimi, M., &. Casali, J.G. (1985). Evaluation of 16 measures of mental workload using a simulated flight task emphasizing mediational activity. *Human Factors, 27,* 489–502.

Wientjes, C.J. E. (1992). Respiration in psychology: Methods and applications. *Biological Psychology, 34,* 179–204.

Williams, A.C., Jr., Macmillan, J.W., & Jenkins, J.G. (1946). *Preliminary experimental investigation of "tension" as a determinant of performance in flight training* (Rep. No. 54, Publication Bulletin L-503–25). Washington, DC: Civil Aeronautics Administration, Division of Research.

Wilson, G. (1993). Air-to-ground training missions: a psychophysiological workload analysis. *Ergonomics, 36,* 1071–1087.

Yamada, F. (1998). Frontal midline theta rhythm and eyeblinking activity during a VDT task and a video game: useful tools for psychophysiology in ergonomics. *Ergonomics, 41,* 678–688.

Yamamoto, S., Matsuoka, S., & Ishikawa, T. (1989). Variations in EEG activities during VDT operation. In M.J.Smith & G.Salvendy (Eds.), *Work with computers: Organizational, management, stress and health aspects* (pp. 225–232). Amsterdam: Elsevier.

Zeier, H. (1994). Workload and psychophysiological stress reactions in air traffic controllers. Ergonomics, 37, 525–539.

Zeier, H., Brauchli, P., & Joller-Jemelka, H.I. (1996). Effects of work demands on immunoglobulin A and cortisol in air traffic controllers. *Biological Psychology, 42,* 413–423.

Chapter 2

Theoretical and Methodological Issues in Psychophysiological Research

Anthony W.K.Gaillard
TNO Human Factors Research Institute, The Netherlands
Arthur F.Kramer
University of Illinois, Urbana-Champaign

There is renewed interest in using psychophysiological measures in the workplace (see, e.g., Boucsein, Luczak, Stern, & Yagi, 1996; Caldwell et al., 1994; Gaillard, Boucsein, & Stern, 1996). This interest has been stimulated by a host of factors ranging from new measurement technologies to the development of new theoretical approaches. Attempts at using physiological indices before the 1980s were hampered by technical problems: the nonportability of the apparatus and the loss of data due to recording artifacts. Ambulatory equipment and data analysis have progressed considerably since then. The equipment available in 1999 is small, relatively inexpensive, and user-friendly.

In the 1970s the research efforts of psychophysiologists were focused on finding the "golden yardstick" with which workload could be measured. Early ideas of a direct relationship between workload and physiological measures have changed. Investigators no longer adhere to a one-dimensional concept of arousal in which the difficulty of the task is associated with increases in the level of physiological activity. The strategy of recording many psychophysiological variables to measure workload has been replaced by more sophisticated ap-proaches. A selected set of variables is chosen to investigate specific effects of workload and of the work environment on the state, efficiency, and well-being of the employee. Different reactivity patterns may be identified on the basis of psychophysiological measures, in response to the demands in the (work) environment. Some reactivity patterns may be associated with efficient behavior and improved performance, as opposed to other patterns that appear to trigger pathophysiological processes that increase health risks.

The examination of physiological reactivity in laboratory tasks, simulated environments, or real-life settings, has become a goal in itself. For example, heart rate (HR) is measured not only as an index of workload, but also because of the interest in the dynamics of cardiovascular reactivity in demanding conditions. This interest is based on the growing evidence that the working environment has a strong impact on well-being and health. Psychophysiological measures may play an important role not only in specifying the conditions that promote efficient and healthy behavior in the work environment, but also in revealing the mechanisms that mediate between work demands and health.

In this chapter, we address the following questions: Why should we measure physiological variables, when seemingly simple indices, such as performance measures and subjective ratings, can do the job? What types of results can be expected from psychophysiological studies? What is the relevance of this information to the understanding of workload or work-health relationships? How can this information be used to solve problems or to improve working conditions?

In the next section, we outline the psychophysiological approach by presenting a general framework into which many studies in this area can be placed. This section also discusses the advantages and disadvantages of the psychophysiological approach, and an overview is given of the issues that may be addressed by this approach. We then discuss the relationship between physiological states and psychological processes. At the end of this discussion, two general approaches are distinguished that are further specified in the next two sections, which provide an overview of the psychophysiology of workload and present research on the work-health relationship.

THE PSYCHOPHYSIOLOGICAL APPROACH

Psychophysiology provides unique and unobtrusive techniques to study the human being (e.g., Andreassi, 1995). In both laboratory and applied settings, psychophysiological measures can provide information that is not readily available from performance measures and subjective ratings. Psychophysiology attempts to understand psychological processes by examining concurrent activity in the brain, heart, muscles, and so on. In applied settings, the psychophysiological method may be used to study reactions of employees to job demands and environmental stressors. Physiological measures are assumed to provide information on the state of the brain and body that is difficult to obtain otherwise. A state should be regarded as the result of many physiological and psychological processes that regulate brain and body in an attempt to put an individual in an optimal condition to meet the demands of the work environment.

A General Framework

This section provides a global overview of the theories and approaches used by investigators applying psychophysiological methods. Table 2.1 schematically illustrates the position of physiological measures in relation to other reaction levels of the employee and to the different types of demands, processes, and outcomes in the long run. Table 2.1 describes how demands in the work environment lead to acute reactions via different processes, which may result in different long-term outcomes.

Demands. In a work situation, the subject not only reacts to the specific requirements of the task on a cognitive level, but also to other demands, such as the emotional (e.g., daily hassles, job insecurity) or physical aspects (e.g., noise) of the environment.

Processes. Work demands may initiate a variety of processes that may occur independently and in parallel. For example, when a task has to be done under time pressure, not only is task-related information processed but the possibility of not being able to finish the job in time may evoke emotional responses. Physical environmental factors may have direct effects on the energetical state of the organism.

Reactions. The reactivity to the work demands can be measured at three levels. *Performance and behavioral* measures can be obtained by registration (e.g., reaction times) and observation (e.g., with video). Performance refers to well-defined, often rather simple measures (speed and accuracy) used to characterize specific aspects of task-related behavior. Behavior denotes the rich complement of actions that constitute complex behavior (e.g., driving a car). Subjective measures, such as questionnaires and ratings, are used to obtain information on mood, effort

---

TABLE 2.1
The Position of Physiological Measures in Relation to Other Reactions to the Demands of the Work Environment That Are Processed at Different Levels and May Have Different Long-Term Outcomes

Demands	Processes	Reactions	Outcomes
Task	Cognitive	Behavioral	Productivity
Emotional	Affective	Subjective	Well-Being
Physical	Energetical	*Physiological*	Health

expenditure, workload, and feelings of fatigue. *Physiological reactivity* can be measured through a variety of physiological and biochemical variables, of which HR, respiration, blood pressure, and cortisol are the most popular.

Outcomes. Long-term effects on well-being and health may be evaluated by objective measures (e.g., absenteeism, visits to the doctor, morbidity, and mortality) and by questionnaires on well-being and health. Acute momentary reactions may be quite different from outcomes in the long run. For example, reactions to stress may consist of active coping accompanied by mental effort and intensive physiological reactions that may be beneficial, however, for well-being and health in the long run. In contrast, stress reactions consisting of passive coping (e.g., denial, withdrawal) may provide temporary relief and low physiological reactivity, but can induce chronic enhanced baseline values in the long run, which increase the risk of psychosomatic complaints and illness. Thus, the intensity of acute physiological reactions does not always predict long-term outcomes in terms of well-being, health, and productivity.

Most research areas have a clear preference for one level. Research on human factors concentrates on cognitive processing in a task environment where performance indices are the dominant dependent measures. Clinical and personality psychologists are interested in affective responses to emotional stimuli and the consequences for well-being, mostly based on the results of questionnaires. In medicine, physiological reactions to physical exertion are related to health indices. As described in more detail later, physiological reactions may be caused by the task-related processing, mental effort, emotions, or by physical changes in the internal (e.g., circadian rhythm) or external environment (e.g., noise).

Advantages and Disadvantages of Psychophysiological Methods

To show its feasibility the psychophysiological method is compared with other methods that are often applied in the field: performance indices, in particular the measurement of reaction times, questionnaires, and subjective ratings. The psychophysiological approach has the following advantages (see also Kramer, 1991):

- Most physiological measures are relatively unobtrusive. They only require the attachment of sensors to the body, which usually is quite acceptable to the operator.
- Most recording procedures do not interfere, to any discernible degree, with the performance of most tasks.
- The measurements can often be obtained continuously and do not require overt responses from the operator.
- Additional information is obtained that is multidimensional; when measures are obtained from different response systems, more information

becomes available on the different aspects of cognitive and energetical processes. For example, although changes in HR reflect workload effects in a diffuse way, components of the event-related brain potential (ERP) reflect specific aspects of workload.

- Comparisons between complex task environments on the basis of performance measures alone are often impossible. Given the trend to use humans as monitors of systems rather than as active controllers, this aspect becomes increasingly important. On the basis of physiological measures, operators of systems can be evaluated, with regard to demands, on their resources and effort.

The psychophysiological method also has the following disadvantages:

- Although the cost of recording systems has decreased dramatically, specialized equipment that is easy to use is needed to guarantee the availability of the results.
- The large variability of single responses, not only between persons, but also over time for the same person, poses substantial problems for an easy interpretation of the data. Indices of reactivity are obtained by referring the data to control conditions or baseline values. In most instances, the data can only be interpreted reliably after repeated measurements or after aggregation on a group level.
- Although standardized methods are available for sensors, procedures, and data analysis (see chap. 5, this volume), direct comparisons between studies are hampered by differences in the tasks and designs used. The interpretation of results can be facilitated with more standardization and when more normative data sets become available.
- Most measures are sensitive to interference from other response systems.
- Most measures are sensitive to more than one psychological process. Most measures are not only affected by workload and effort, but also by stress, emotions, physical environment, and physical exercise.
- In particular, the design of a study and the interpretation of the results still requires an extensive amount of expertise, which makes the method expensive and time consuming.
- The interpretation of findings still poses a problem because there is no generally agreed on theoretical framework (see section on states and processes). The ultimate goal is the evaluation of human performance or the understanding of the functioning of a particular operator. For the prediction and understanding of human behavior in response to work demands, it is necessary to have a strong conceptual link between physiological and psychological factors. Most psychophysiological theories, however, are generally under specified and only weakly related

to physiological mechanisms (Cacioppo & Tassinary, 1990; Hockey, Coles, & Gaillard, 1986).

- The validity of many physiological measures, particularly ERPs, only has been tested in the laboratory. They only refer to a particular process in laboratory tasks in a student population (Cacioppo & Tassinary, 1990). A P300, for example, only reflects task difficulty or capacity when similar task conditions are compared; it is not a direct measure of difficulty.

Although many psychophysiologists still prefer to restrict their research to the laboratory, a growing number of studies address applied problems and adhere to an integrated approach, relating physiological variables to performance and subjective ratings as a function of environmental conditions or personality attributes. Given the current technology, more studies use a multimeasure approach that focuses on solving the problems in the work environment by judiciously choosing central nervous system and autonomic nervous system measures depending on the theoretical questions or practical problems.

The psychophysiological approach may be used in the following different areas:

- The evaluation of cognitive processing, interference between tasks, and the effects of mental load in complex task environments.
- The monitoring of fitness of operators in vigilance situations.
- The assessment of environmental stressors such as noise, drugs, or sleep loss.
- The evaluation of stress reactions to time pressure.
- Emotional responses to psychosocial factors (e.g., social support) in the work environment.
- The influence of personality (e.g., temperament) and coping style on the effects just cited.

The information obtained in these areas may be used in a variety of applications, as follow:

- To optimize performance or (re)design the work environment.
- To improve task structure, human-machine interface, and task allocation.
- To prevent system damage and increase safety by preventing errors and maintaining an optimal state.
- Improvement of the psychosocial factors in the work environment.
- To optimize work-rest schedules and introduce short breaks at the appropriate times.
- Provide an index for the "costs" involved in executing tasks.
- Identification of factors in the work environment that increase the risk for stress effects in employees.

- Identification of workers with an increased risk of developing stress responses or becoming ill.
- Selection, training, and career planning of employees.

Because the techniques and expertise are now available, the time has come for psychophysiologists to engage in sophisticated studies oriented toward the solution of practical problems in the workplace.

Besides the problems cited here, there are other aspects that need attention:

1. Psychophysiological measures may be disturbed by other physiological variables (e.g., respiration, eye movements) and artifacts (e.g., physical activity) that cannot always be easily controlled.
2. The measures may be affected by changes in the internal (e.g., circadian rhythm) or external environment (e.g., temperature).
3. The same measure is often related to quite different concepts (e.g., HR has been proposed to reflect mental load, effort, resources, attention, emotion, etc.).
4. There are many dependent measures, whereas there are only a few theories; this is just the opposite with performance measures, where there are many theories and basically only two measures: speed and accuracy.

In general, performance measures have a more specific and direct relation to a presumed psychological concept than physiological measures. Increases in reaction time due to an increase in task difficulty are explained in terms of an increase in processing time. Also here aspecific effects (e.g., changes in motivation, time of day, time on task, etc.) may confound this relation, although they may be controlled or corrected.

STATES AND PROCESSES

The distinction between the intensive or energizing aspects of human behavior, as opposed to directional or content-specific processes, has a long history in psychology (see Hockey et al., 1986, for a review). According to Hebb (1955), arousal is a drive needed to energize behavior, but it is not a guide: It is an engine but not a steering wheel The function of arousal is to bring the brain into an optimal state for efficient information processing. The state of mind and body is regulated by a variety of energetical mechanisms. The term *energetical* is used generically to encompass all energizing mechanisms that regulate the organism and directly or indirectly influence psychological processing. This term is chosen because it does not have specific theoretical connotations and may prevent misunderstandings due to associations with theories in which commonly used labels, such as arousal, effort, fatigue, and activation that have specific meanings (Hockey et al., 1986).

Models in human factors are often based on the computer metaphor: Information coming from the work environment is transformed in sequential stages or modules into an output, resulting in an overt response (Sanders, 1983; Wickens, 1992). This metaphor may also illustrate the distinction between cognitive processes and energetical states: The software stands for the processing of cognitive information and the hardware for the state of the brain. The main reason for postulating that changes in state affect the efficiency of cognitive processing, is that information-processing models are not able to account for variations in human performance under demanding and threatening conditions. The efficiency of this processing is assumed to be dependent on the state of the operator. Therefore, investigators in engineering psychophysiology extend the models adopted from human factors research with "states," "resources," or "energetical mechanisms."

The psychophysiological approach may be used to evaluate the state of the organism and to examine the mechanisms that determine it. Energetical mechanisms may also be inferred indirectly from performance measures, through the effects of environmental stressors (e.g., sleep loss and long-term performance) or aspecific task variables (e.g., feedback, bonus). Changes in state may also be derived from subjective ratings (e.g., on mental effort, mood, and fatigue).

State Regulation and Performance Efficiency

Under normal circumstances, an individual's energetical state is in line with the activities he or she wants to undertake. Energetical mechanisms regulate the state of the brain and body in such a way that they are in an optimal condition to process information and execute the task. The execution, and even planning, of a task prompts energetical mechanisms to adapt our body to an appropriate state, which determines the efficiency and the capacity available to perform the task. The majority of bodily processes are regulated automatically and unconsciously. Because humans do not have to pay attention to these mechanisms, we have the opportunity to concentrate on the more interesting aspects of life. Only when our state is far from optimal, due to fatigue or strong emotions, do we realize how much cognitive processing can be influenced by our energetical state. Because these effects are mostly outside our control, we can only attempt to modulate them and adapt to the situation. When planning daily activities we may take into account possible fluctuations in our energetical state due to fatigue or time of day. For example, someone having problems getting started in the morning may make appointments only after coffee time.

Several ways in which state can be influenced by energetical mechanisms are as follows:

- Most energetical mechanisms that determine the state of the organism are autonomic; they are unconscious and not under voluntary control. The best known is the circadian rhythm that determines the sleep-wake state and fluctuates during the day. There are, however, many other regulatory loops of which we are hardly aware (e.g., blood pressure, blood sugar level, respiration, etc).

- The energetical state is also influenced by environmental factors that are either external (e.g., noise, light, vibration, etc.), or internal (e.g., sleep loss, drugs, alcohol, etc.). As with the autonomic processes, we are not able to influence the effects of these factors voluntarily, although we can attempt to modulate them, for example, by closing the curtains or drinking coffee.
- Performing a task indirectly influences the energetical state. When we start to perform a task, energetical mechanisms are triggered automatically to bring our mind to an optimal state. This adaptation process starts when we anticipate the performance of a particular task. Even mental tasks require energetical preparation comparable to warming up for physical exertion. No one enjoys giving an oral presentation on short notice (e.g., within 15 minutes after waking up), even when the talk has been well prepared.
- The only way the energetical state can be influenced voluntarily and consciously by cognitive processes is by mobilizing extra energy through mental effort. This "trying harder" action is voluntary and largely dependent on somone's motivation to attain a particular goal.
- Emotions, whether or not evoked by the task situation, have a strong influence on the energetical state. When the situation is interesting and challenging, positive emotions and mild anxiety have motivating and energizing effects. When the situation is threatening, boring, or irritating, negative emotions (e.g., anxiety, anger, worry) may debilitate the energetical state and reduce performance efficiency.

Most investigators describe their psychophysiological results in terms of fluctuations in the level of general activation. The relation between the energetical state and performance efficiency is assumed to be an inverted-U shaped curve (see also Hebb, 1955). A reduction in performance efficiency due to sleep loss or fatigue is assumed to be caused by a low level of general activation, whereas the negative influence of anxiety or stress is explained by an excessively high level of general activation. So far, the inverted-U shaped curve hypothesis has received scant empirical support, and a number of methodological problems have been raised against this type of research (e.g., Eysenck, 1982; Neiss, 1988). The most important is the lack of agreement among researchers as to how to determine objectively the different levels of arousal. It has also been questioned whether the inverted-U is a correlational or a causal hypothesis. Events or manipulations, either in the laboratory or in daily life, that enhance arousal may at the same time elicit strong emotions and distractions, which also reduce processing capacity and performance efficiency (see also Näätänen, 1973). A third problem is that arousal theories do not discriminate between different types of energy mobilization and hardly specify the effects on emotional and cognitive processes or the consequences this may have for the working behavior of the employee. As a result (one-dimensional) arousal theories are not able to explain why under some conditions efficient performance is possible even with high levels of arousal, whereas debilitating states that degrade performance may also occur at medium or low levels of arousal. It appears that strong negative emotions (e.g., worry, anxiety, and depression) reduce performance efficiency also at low levels of arousal.

Although most investigators agree that a multidimensional framework is necessary to describe the complex relationships between energetical mechanisms and psychological processes or between state and efficiency, only a few promising ideas have emerged about how this might be achieved (Hockey, 1997; Hockey et al., 1986). Instead of a one-dimensional model, an energetical state may be described as a multidimensional concept. Each particular state is the result of the activity of several energetical mechanisms and every task, whether physical or mental, has its specific state (in a multidimensional space) where performance is most efficient (Hancock, 1986; Hockey et al., 1986).

A state is the result of the adaptation of mind and body to the demands of the environment so that the processing required by the task is most efficient. This model of state regulation may be regarded as a multidimensional inverted-U curve (see also Hancock, 1986) in which (a) the state of the brain and body is determined by several energetical mechanisms; (b) each task has its own optimal level at which it can be performed most efficiently; (c) the closer the actual state is to the optimal state, the better the execution of the task; (d) deviations from the optimal state can be compensated for by mobilizing extra energy through mental effort; and (e) task situations with high attentional demands are more sensitive to changes in energetical state, because the area in which performance can be optimal is smaller.

Different types of states may be described in terms of the energetical mechanisms involved. This has been done at different levels of specificity.

• The general state of brain and body is still a powerful determinant of human performance. It may be used to explain the gross effects of environmental factors, in particular at the extremes of the arousal dimension. The general state is mainly determined by the autonomic nervous system and the effects of hormones. Attempts have been made to differentiate between different types of states. An energetical state should be characterized not only by intensity (high or low), but also by different patterns. A distinction is made between states dominated by the release of corticosteroids or catecholamines (Dienstbier, 1989; Frankenhaeuser, 1986; Ursin & Olff, 1993). On the basis of a frequency analysis of the HR variability, a distinction may be made between action and recovery states that are dominated by sympathetic or parasympathetic influences, respectively (see, e.g., Backs, 1998; Berntson, Cacioppo, & Quigley, 1993; Mulder, 1992).

• Different types of energy supply coming from different subcortical neuroanatomical structures were postulated by Pribram and McGuinness (1975; see also Boucsein, 1992; Sanders, 1983). When the brain receives information and reacts to the environment, these systems are stimulated to activate other parts of the brain in order to facilitate the processing of this information.

• The state of the neuronal structures where the information has to be processed, can be investigated by modern brain imaging techniques, such as positron emission tomography (PET) and functional magnetic resonance imaging (fMRI) (see Posner & Raichle, 1994).

Three Types of Energy Mobilization

This section describes the three ways by which energy may be mobilized that appear to be most important for engineering psychophysiology: task-induced activation, internally guided mental effort, and input-related emotional arousal (see also Gaillard & Wientjes, 1994).

Task-Induced Activation. When we know that a particular task has to be executed in the near future, we can prepare for certain activities on a cognitive and energetical level. Merely thinking about the task to be done affects the regulation of our state, where energetical mechanisms are activated to reach the optimal state needed to facilitate the task. The relation between state and efficiency is dependent on the demands of the task on the one hand and the availability of processing resources on the other. For simple, well-trained tasks that do not require many resources, a deviation from the optimal state will not result in a reduction of performance efficiency. However, in complex or novel task situations that require all our resources, even a small deviation from the optimal state may result in a performance decrement. Thus, the range in which efficient performance can be obtained is larger when there is abundant processing capacity than when the amount of resources needed for the task approaches the available capacity. We therefore prefer an easy task to a difficult one, especially when we are tired. In the evening for example, we may be too tired to write a technical paper, but we may have enough energy left to read a novel.

The Role of Mental Effort. When the actual state does not deviate too much from the optimal state, we are still able to perform the task, but at a slower pace or less accurately. To maintain performance, however, we may compensate for the deviation by mobilizing extra energy through mental effort. This "trying harder" response can only be maintained for a relatively short period because its physiological and psychological costs are high and result in mental fatigue.

Kahneman (1973) identified effort with the action of maintaining an activity in focal attention. This means that effort is needed to prevent lapses in atterntion resulting in performance deterioration. Mental effort plays an important role in the following, apparently quite different situations (see also Gaillard, 1993).

1. The energetical state is not optimal due to sleep loss, fatigue, or intensive emotions.
2. The task is attention-demanding due to incompatible, inconsistent, ambiguous task information.
3. Heavy demands on working memory and divided or sustained attention.
4. Skill acquisition or novel situations.

What these situations have in common is that participants may have problems with maintaining a particular task set, which may result in a decay in performance.

The Role of Emotion. In cognitive psychology, as well as in human factors and ergonomics, the role of affective processes has been largely neglected (e.g., Eysenck, 1982). As a consequence, theories in cognitive and engineering psychophysiology only distinguish between state and process, dichotomizing between energetical mechanisims (e.g., arousal, activation, or effort) on the one hand, and cognitive processes on the other. Affective processing in the work environment, however, plays an important role in the generation of efficient and reliable working behavior and can also be regarded as the most important mediator in the work-health relationship, as discussed later.

Positive emotions are needed to motivate people to do the task in the first place and to energize the organism. Negative emotions not only energize behavior, but may also reduce performance efficiency. This can take place in three ways:

1. Emotions may disrupt the regulation of the energetical state, which makes task performance less optimal.
2. Particular negative emotions (e.g., anxiety) have control precedence (i.e., continually demand attention) resulting in a reduction of the capacity available for the processing of task-relevant information.
3. Negative emotions may cause psychosomatic complaints that are also attention-demanding and reduce performance efficiency.

Two Approaches: Workload and Stress

In the working environment, psychophysiological techniques may be used to obtain indices reflecting the effects of work demands on the state of mind and body. Changes in state due to shift work, sleep loss, or work pressure may be used to explain decreases in performance efficiency. For example, deterioration in performance during long working hours is explained in terms of overload, fatigue, or stress. The three types of energy mobilization described earlier, induced by the task, mental effort, or emotions, are often used for the interpretation of psychophysiological results. Increases in physiological activity are associated with increases in task difficulty, mental load, or mental effort. When an increase in physiological activity is accompanied by a decrease in performance, it is often regarded as a stress reaction. The arousal level is too high for efficient performance. In applying psychophysiological techniques, two approaches may be distinguished—*warkload* and *stress*—that differ in theoretical and methodological background. Research on workload is concerned with the efficiency of performance in complex or mentally demanding tasks, mostly in a technical environment, whereas stress research concentrates on the work-health relationship and the psychosocial factors of the working environment.

Workload research uses cognitive-energetical models based on human factors and cognitive psychology (e.g., Hancock, 1986; Hancock & Meshkati, 1988; Hockey et al., 1986; Wickens, 1992), which describe how cognitive processes are affected by the energetical state, as indicated by psychophysiological measures. This kind of approach is especially relevant when employees have to work at the limits of their capacity. Mental load refers to the ratio between the available processing capacity of the operator and the capacity required by the task (e.g., Gopher & Donchin, 1986; Hancock, 1986; Kantowitz & Casper, 1988). The aim of this research is to reveal the factors that determine the workload of an employee and the risk of making errors. The effects of task demands on workload are examined in order to develop procedures to reduce the workload, redesign the work environment, and improve performance efficiency and safety.

Alternatively, stress theories are adopted from industrial, social, and personality psychology (Kahn, 1981; Karasek & Theorell, 1990; Lazarus & Folkman, 1984). Stress theories describe the relationship between person and environment. The evaluation of the situation (Lazarus & Folkman, 1984) or the perceived controllability of the situation (Karasek & Theorell, 1990) is central in these theories. Stress research examines the work-health relationship by investigating the influence of the work environment on well-being, psychosomatic complaints, and health risks.

At first sight, the two approaches appear to have the same focus, the examination of the balance between work demands and the available resources of the employee. The two areas differ, however, in their scope. In stress theories, demands and resources are conceptualized more broadly than in workload theories. Research on workload focuses on the capacity to process information in the task, and demands refer to the specific requirements of the task. In stress theories, demands refer not only to the task, but to the entire work environment, including its social and organizational aspects; resources refer not only to processing capacity, but also to personality traits, coping strategies, and social skills. Resources can also encompass the availability of means to perform the task (e.g., supplies, apparatus, information).

In both types of theories the operator is assumed to resolve the mismatch between demands and resources by either reducing the demands or increasing the resources. The two types of theories differ, however, in the type of factors involved. In workload theories the operator may change his or her work strategy by reducing the work pace, increasing the risk of making errors, or concentrating on the most relevant aspects of the job. The operator may also increase the available resources by mobilizing extra energy through mental effort. In stress theories, the adopted coping strategy is assumed to be the critical factor in the development of stress reactions. Emotion-oriented strategies are assumed to evoke more stress than active coping-oriented on the problems faced in the work environment. The adopted strategy may refer to all aspects of the environment, including the decision latitude the employee has to change the work environment in terms of supplies, apparatus, information, and manpower. The employee continuously monitors the environment for potential threats and danger on the one hand and interesting opportunities on the other. This appraisal is guided by the goals the individual has and is based on his or her norms and values. In addition to task information, the employee takes into account the psychosocial aspects of the situation (e.g., social approval or social support).

Although in both approaches, the discrepancy between demands and resources is assumed to evoke physiological reactions, the role psychophysiology plays is quite different. In workload research, physiological measures are used to examine changes in the energetical state induced by fluctuations in task difficulty and mental effort. In combination with subjective and performance measures, physiological measures are used to index mental effort and mental load and to provide information on the chronometry and strategy with which the task is executed. In stress research, the energetical state is dominated by negative emotions. The psychophysiological approach may be used to examine how affective processes and physiological mechanisms mediate between cognitive processes and bodily reactions, how the homeostasis of the energetical state may be disrupted, and how this results in psychosomatic complaints and increased health risks.

The two research lines described here constitute the baseline for the most important approaches in engineering psychophysiology: the investigation of mental workload with psychophysiological indices and the investigation of the work-health relationship in stress research. An outline of these two approaches is given in the next two sections.

PSYCHOPHYSIOLOGY OF WORKLOAD

As discussed in the previous section, mental workload has been defined in terms of an interaction between the demands of the task and the ability of an individual to fulfill these demands in a accurate and timely fashion (Gopher & Donchin, 1986; Kramer, 1991; Wickens, 1992). The study of mental workload has progressed along two distinct, but often interacting, routes. One route has concentrated on the understanding of the mechanisms that underlie variations in mental workload with relatively simple tasks. Examination of this issue has often taken place in well-controlled laboratory settings and has focused on the ability of participants to perform two or more tasks at the same time or to rapidly alternate between different tasks or skills. In the second case, the focus has been on the characterization of mental workload in either high-fidelity simulations or operational tasks and, in particular, in the context of the piloting of aircraft and car driving. Furthermore, in such cases there has often been an interest in either predicting variations in mental workload or characterizing changes in mental workload in realtime.

Psychophysiological measures have played an important role in both the examination of the mechanisms which underlie the performance manifestations of mental workload and in the measurement of mental workload in complex task environments. For example, psychophysiological measures have been instrumental in supporting changes in the theoretical conceptualization of mental workload from the notion of undifferentiated capacity or resources (Kahneman, 1973; Kantowitz & Casper, 1988) to the notion of multiple capacities or resources (Navon & Gopher, 1979; Polson & Freidman, 1988; Wickens, 1992). In this regard, some psychophysiological measures have proven to be uniquely sensitive to specific varieties of resource demands, whereas other measures appear to be sensitive to relatively general resources or processing operations. For example, the P3 component of the ERP appears to reflect perceptual-central processing while being relatively insensitive to response-related processing (Kramer & Spinks, 1991). Similarly, other ERP measures are uniquely sensitive to early attentional selection processes (i.e., the P100; Hillyard et al., 1996), the monitoring and detection of errors (i.e., the error related negativity; Gehring, Goss, Coles, Meyer, & Donchin, 1993; Scheffers, Coles, Bernstein, Gehring, & Donchin, 1996), the programming and execution of responses (i.e., lateralized readiness potential; Coles, Scheffers, & Fournier, 1995; Osman, Moore, & Ulrich, 1995), and the detection of semantically incongruent information (i.e., N400; Kutas & Van Petten, 1994). Other psychophysiological measures such as HR and HR variability (Mulder, 1992; Veltman & Gaillard, 1996) and electro-oculogram (EOG) measures (Brookings, Wilson, & Swain, 1996; Stern & Dunham, 1990) are sensitive to variations in mental workload, but not diagnostic with regard to specific varieties of workload that may be involved in the person-task interaction.

One important topic for future research is the interaction or interplay between energetical mechanisms and psychological resources that comprise the construct of mental workload. Given that psychophysiological measures appear to reflect both general and specific aspects of workload, they provide a potentially ideal methodology with which to examine the control and deployment of resources and mechanisms across varying levels and types of mental demands comprising the multidimensional construct of mental workload. Of course, success in accomplishing such an endeavor is contingent on the mapping of psychophysiological measures to psychological constructs and mechanisms, a continual process.

It is important to note that although the resource metaphor, in either an undifferentiated or multiple capacity sense, has served as the basis of the conceptualization of mental workload in the past, the concept of resources or capacity has been a hotly debated issue in recent years. For example, a number of researchers have proposed that interactions between parallel computational processes rather than graded sharing of capacity or resources provides the best account of multitask performance decrements (Allport, 1987; Navon & Miller, 1987). Other researchers, have argued that single-stage bottlenecks similar to that proposed by Broadbent (1958) can account for multitask deficits. In these models, a particular stage or process can only be allocated to a single task at a time. Therefore, when several tasks are to be performed at the same time one or more tasks must wait for this process, thereby increasing the amount of time required to perform multiple tasks relative to single-task baselines. Bottlenecks in central as well as peripheral (e.g., response execution) processes have been proposed in this class of models (Meyer & Kieras, 1997; Pashler, 1994).

This debate over whether resource- or process-based mechanisms underlie mental workload, has important implications for psychophysiological measurement. Given the present state of knowledge concerning the mechanisms that underlie variations in mental workload, it seems prudent to assume that both computational processes and energetical mechanisms are important. Indeed the strongest evidence for computational or bottleneck models has been obtained with relatively simple tasks and in nonstressful situations. Evidence in favor of resource or energetical contributions to mental workload has been obtained in complex tasks and with environmental stressors (e.g., sleep loss, fatigue, drug effects). Fortunately, as indicated here, different psychophysiological measures appear to be sensitive to energetical and computational processes. Therefore, it would appear that psychophysiology has the potential to provide insights into the relative contributions of energetical and computational processes to the mental workload experienced in different situations and with different tasks.

A neglected aspect of mental workload for which psychophysiological measures might also make a contribution concerns the strategies that individuals use to cope with variations in processing demands. Although there are, at present, far too few detailed studies of strategic changes in performance that accompany variations in mental workload in complex tasks, the studies that have been conducted suggest a number of changes in information processing that could be assessed with psychophysiological measures. Some of these changes include a narrowing of attention either to a particular area of space or to a class of events or objects, reorganization of the goals and priorities and a limited focus on the most task-relevant goals (e.g., such as passenger safety in air traffic control; Sperandio, 1978), a shift from higher level decision-making strategies to well-practiced and automatic skills (Rasmussen, 1983), and an increase in errors of omission and commission (Reason, 1990). Indeed, the error-related negativity (Scheffers et al., 1996) has already been employed to examine the monitoring and correction of errors in simple laboratory tasks. Other psychophysiological measures would appear to have the potential to assess other types of strategic changes that accompany variations in mental workload.

In addition to the use of psychophysiological measures in the examination of theoretical conceptualizations of mental workload in well-controlled laboratory settings, there has also been a good deal of research in the past decade that has focused on psychophysiological measures of workload in complex simulated tasks or operational environments. These studies have focused on two related, but distinct, issues with regard to mental workload in real-world tasks. One class of studies has focused on distinguishing between intermediate and low (or very low) levels of mental workload, whereas the other class of studies has focused on distinguishing between intermediate and high levels of workload. In the former case, the interest has been in assessing and predicting vigilance decrements in tasks such as sonar and radar monitoring and quality control inspection tasks. In the latter case, the main interest has been in assessing and predicting performance decrements that result from high-processing demands in aviation-related tasks such as piloting.

In the context of the assessment of vigilance decrements Makeig and colleagues (Makeig & Inlow, 1993; Makeig & Jung, 1996) have developed an EEG-based system that detects changes in the alertness of U.S. Navy sonar operators. The system that was based on the spectral decomposition of EEG was modestly successful in predicting the number of failures to respond over a 30-minute task. Furthermore, the EEG-based prediction algorithm that was developed in one session did a good job of predicting performance lapses in a second session. Other researchers have examined the use of other psychophysiological measures such as skin conductance (Yamamoto & Isshiki, 1992) and EOG activity (Morris & Miller, 1996) as indices of fatigue and vigilance. In general, although each of these psychophysiological techniques has shown promise in the evaluation and prediction of vigilance decrements (i.e., very low mental workload) each of the measures is limited in terms of its tempo ral resolution, reliability, and applicability to different environments. Therefore, there is clearly a need for additional research to determine how best to integrate such measures into a multimeasure alertness detection and prediction system.

Psychophysiological measures have also been used to assess changes in mental workload in demanding environments and, in particular, in piloting. Heart rate has the longest history as a measure of mental workload in simulated and actual flight (Roscoe, 1993). In fact, HR is quite useful in that it can provide a continuous record of fluctuations in mental workload without the need to introduce additional signals into the piloting task. However, the disadvantages of HR as a measure of mental workload include its lack of diagnosticity with regard to the variety of processing demands imposed on the individual as well as its sensitivity to factors other than mental workload (Veltman & Gaillard, 1996, 1998).

EEG as well as ERP measures have been studied in simulated and actual flight conditions (Fowler, 1994; Kramer, Sirevaag, & Braune, 1987; Sirevaag et al., 1993; Sterman & Mann, 1996; Wilson, Fullenkamp, & Davis, 1994) and have, for the most part, shown promise in assessing changes in mental workload under adverse conditions. However, although EEG, like HR, can provide a continuous measure of mental workload without the need for the imposition of task-irrelevant stimuli, it is not particularly diagnostic of the types of demands imposed on the individual by the task. This lack of diagnosticity by EEG may be countered, at least to some degree, by the development of techniques for the rapid analysis of dense arrays of EEG data (Gevins et al., 1995). ERPs, on the other hand, are quite diagnostic but suffer from low signal-to-noise ratios with single trial data. This problem may be overcome, at least in part, with more sophisticated pattern recognition and analysis procedures (Trejo & Shensa, 1993) and by the use of several ERP components, rather than just a single ERP component (i.e., P300) as has been the practice in the past. Indeed, the evaluation and prediction of mental workload in high workload environments would benefit from the incorporation of several different psychophysiological measures into assessment batteries.

WORK AND HEALTH

The ideas about the relationship between work and health have changed dramatically since the 1980s. In the classic view, the employee, in particular his or her "weak" personality, was held accountable for stress problems such as psychosomatic complaints, sleeping problems, and an unhealthy lifestyle. It has now been demonstrated (see Karasek & Theorell, 1990, for a review) that some factors in the work environment have negative effects on health and well-being. In a longitudinal study in which more than 7,000 Swedish employees were monitored during a 9-year period, it was found that the risk of cardiovascular diseases increased and life expectancy was estimated to be shortened by 7 years due to unhealthy psychosocial factors in the work environment (Johnson, Hall, & Theorell, 1989).

Little is known, however, about the underlying mechanisms that mediate work demands, stress reactions, and increased health risks. Classical epidemiological studies based on correlations between work characteristics and health indices provide only limited insight into these mechanisms. Most stress theories, such as the Michigan model (Kahn, 1981) and the demand-control model (Karasek & Theorell, 1990), minimally specify the mechanisms mediating work demands and health outcomes. The psychophysiological technique may play an important role in specifying the characteristics of the working environment that increase health risks and in revealing the mechanisms that mediate work demands and health risks. With the aid of psychophysiological measures, the links may be examined between distinctive response patterns to work demands, intermediate pathophysiological processes, the precursors of illness, and indices of ill health.

Given that stress and disease are interrelated via a complex interplay of psychological and physiological processes, associations between stress and disease cannot simply be described in terms of cause and effect It is assumed that the dysfunctional physiological activity under stress triggers pathophysiological processes, which in turn cause changes in regulatory mechanisms and lead to tissue damage, provoke the development of malignant processes, or suppress the immune system (see also Kamarck & Jennings, 1991; Krantz & Manuck, 1984; Steptoe, 1989; Ursin & Olff, 1993).

Arousal, Efficiency, and Health

Although most researchers agree that work demands influence physiological (re) activity and regulation in a complex manner, many theories still adhere to a one-dimensional concept of arousal, originally stemming from the flight-fight reaction described by Cannon (1932). States of high arousal are associated with emergency reactions, inefficient performance, and anxiety. In this tradition, negative views on physiological activity are predominant. The activity of the sympathetic nervous system has been coupled with anxiety, depression, and cardiovascular disease. Stress management has become almost identical to reduction of that arousal (Holmes & Rahe 1967; Selye, 1956). Also, more recent investigators such as Karasek and Levi (see Karasek & Theorell, 1990) still use a one-dimensional model of physiological arousal (the inverted-U shaped curve). This is particularly surprising because their stress model (described later) is multidimensional.

Furthermore, in human performance research changes in demands, mostly operationalized by increasing the difficulty of the task, are assumed to enhance the level of arousal, either directly or via mental effort. However, studies on the relation between physiological activity and performance efficiency have revealed ambiguous and contradictory results (e.g., Eysenck, 1982). Reliable results appear to be found only at the extremes: A reduction in performance is found at a low level due to sleep loss and fatigue and at a high level of arousal due to anxiety. Even then, performance deterioration and high arousal may be affected independently by the stressor; high levels of arousal are often accompanied by intense emotions that can be distracting and decrease performance.

The multidimensional view on energetical states, outlined earlier, is better able to accommodate a number of experimental observations that are difficult to explain in a one-dimensional concept. A one-dimensional theory cannot explain:

- Why some types of enhanced physiological activity (e.g., relaxed jogging) are assumed to be healthy, whereas other types are not.
- Why people are able to work very hard without experiencing negative effects.
- Why stress reactions may also occur at medium or low levels of arousal, for example when people perform a monotonous task under conditions of underload or work in physical and social isolation. People may even experience stress reactions when they work half time or are not working at all.
- Why there is no direct relation between the level of arousal and performance efficiency. For example, the number of working hours per week is not related to health outcomes. Employees in active jobs with high psychological demands that presumably, have a higher level of arousal, do not have increased health risks and psychological complaints, compared to employees in passive jobs. Employees in active jobs even appear to have more energy left in the evening to engage in active hobbies and social activities (see also Karasek & Theorell, 1990).

Thus, high levels of behavioral activation or physiological activity are not necessarily accompanied by a reduction in performance efficiency, well-being, and health. There appear to be different patterns of physiological reactivity to work demands that have different consequences for performance efficiency and health risks.

An integrative framework of how energetical states are regulated should be able to specify how the physiological reactions to work demands affect efficiency, well-being, and long-term health risks and how these effects are modulated by psychosocial factors (e.g., social support) and personality characteristics (e.g., temperament, coping). The construction of such a framework may not yet be possible because too little is known about the relation between psychological processes and energetical mechanisms. Instead of attempting to build an integrative, multidimensional theory, it may be more realistic to develop models of state regulation and a taxonomy of states. On the basis of such a taxonomy, it is possible to specify the negative effects that particular work demands may have for performance efficiency, well-being, and health. On the basis of this analysis, the work environment may be (re)designed and recommendations may be given to enhance the work efficiency of employees while reducing the probability that stress reactions occur.

Biobehavioral States

Physiological reactions to the work environment are determined by the type and amount of work on the one hand and by the characteristics of the employee (e.g., temperament and coping strategy) on the other. The current state is continuously modulated by several factors such as fluctuations in work demands, the work strategy of the employee, the progress made, and the feedback received. The work environment, however, also has enduring effects on the mind and body that appear to be specific for a particular type of work. If work demands consequently evoke a particular pattern of reactions one could speak of biobehavioral (BB) states (see also Gaillard & Wientjes, 1994; Leonova, 1994), psychobiological states (Neiss, 1988), or adjustment modes (Hockey, 1997). A BB state is characterized by a typical pattern that includes cognitive, emotional, and physiological reactions that are highly interactive and interdependent (see Table 2.1). The state notion differs from most arousal theories (see earlier section on state regulation and performance efficiency), in that arousal is no longer regarded as a causal agent; cognition and emotion may induce physiological reactions and the other way around. Thus, cognitions, emotions, and physiological reactions induced by work demands constitute a pattern of reactivity.

Although this is not the place to give an overview of studies that have dealt with BB-states, we offer a few examples. Frankenhaeuser (1986, 1989) identified states based on different neuroendocrine response patterns. These responses were collected in a series of studies in which the control that subjects had over their performance was manipulated. The "effort without distress" state involves effort and engagement and is often accompanied by positive emotions. This state is dominated by sympathetic activity and the release of catecholamines (e.g., epinephrine) by the adrenal-medullary system. The "distress" state is characterized by passive coping and feelings of helplessness. This state is dominated by the release of cortisol by the adrenal-cortical system. In the third state, "effort with distress," the work demands require effort, but also evoke negative feelings (e.g., anxiety). The state is associated with sympathetic dominance and increased excretion of both catecholamines and cortisol. Dienstbier (1989) made a similar distinction between a "tough" arousal pattern related to efficient performance and a pattern related to defensiveness. Physiological toughness is characterized as an arousal pattern encompassing strong catecholamine responses to challenging and stressful situations, but also a fast recovery and low resting values. A "tough" arousal pattern is associated with efficient performance, positive temperament, emotional stability, and immune system enhancement. This pattern is contrasted with a defensive pattern dominated by high cortisol responses, protracted recovery, and increased baselines in rest conditions. Also Karasek's demand-control model results in reactivity dimensions (activity and strain; Karasek & Theorell, 1990). According to this model, stress reactions will occur when the work environment encompasses high psychological demands in combination with a low level of control. The symptoms vary from reduced work satisfaction and well-being to life-threatening cardiovascular diseases. In contrast, active jobs that are highly demanding but have control possibilities do not produce stress reactions or psychosomatic complaints and do not increase health risks.

Although these results originate from quite different disciplines and research areas, they share the following aspects:

1. Reactivity patterns are specified at different levels: behavioral, physiological, and emotional.

2. Reactivity patterns are defined in terms of enduring states using two dimensions, activity, ranging from a passive to an active state, and affectivity, ranging from a negative to a positive mood state; combining the two dimensions results in four states that could be labeled *enthusiastic* (active, positive), *relaxed* (passive, positive), *strain* (active, negative), and *depressive* (passive, negative).

3. High levels of physiological activity are not necessarily accompanied by negative emotions, psychosomatic complaints, and increased health risks. Thus, a work environment that is exacting, but stimulating and challenging, may evoke efficient performance, active coping, enhanced activation, and positive emotions. In contrast, working below level of competence, may evoke noncreative behaviors and passive coping, which result in reduced activation accompanied by apathy and depression. Similarly, a hectic environment in which employees have limited control over their behaviors and where they experience little social support will evoke stress reactions involving anxiety, psychosomatic complaints, and dysfunctional physiological reactions.

To illustrate the concept of BB states we relate the three types of energy mobilization, described earlier, to the demands of the work environment and to the mood and coping style these demands may generate. We distinguish five states in the working environment: regular performance, high mental load, underload, adverse environment, and stress (see also Gaillard, 1993).

Regular Performance. Under normal working conditions, the energy mobilization is tuned by the demands of the task. This task-induced activation is sufficient for a proper execution of the task. Because the task is data limited, devoting more energy to it via mental effort does not improve performance. Given a balanced work-rest schedule, the employee does not experience psychosomatic complaints or feelings of fatigue and health risks remain unaffected.

Mental Load. Under high levels of mental load, extra energy is mobilized by mental effort, mostly in situations that are attention-demanding. Mental effort is used to improve or maintain performance efficiency. Under mental load, the situation is experienced as a challenge, which results in positive emotions and feelings of accomplishment. Mental effort is a normal and healthy aspect of an active coping strategy to meet work demands. This type of mobilization is largely under the control of the employee, which promotes well-being and reduces the risk of psychosomatic complaints. Also the recovery is rather fast, and feelings of fatigue may be accompanied by satisfaction when the task has been completed successfully.

Underload. When employees have to work under their competence level, or when the task and the work environment are not stimulating or motivating, the activation induced by the task may be too low for efficient performance. In combination with fatigue, activation may drop below an acceptable level, as may be evident from physiological indices (see section on psychophysiology of workload). Although the employee has very little to do, he or she may have negative emotions. Extra energy may be mobilized via mental effort, which is only possible when the employee is well motivated by incentives outside the task (e.g., salary, social control, etc.).

Adverse Environment. The task has to be executed under the influence of environmental stressors, such as heat, noise, vibration, or working at night, which may disrupt the energetical state of the employee physically. These are direct effects on the organism and are not mediated via cognitions or emotions as is the case with stress reactions.

Stress. Under stress, energy is mobilized by negative emotions over which the employee has limited control. In contrast to mental load, the situation is not perceived as challenging, but as threatening, which results in psychosomatic complaints, dysfunctional physiological reactions, and inefficient coping strategies. The enhanced physiological activity is not instrumental to the execution of the task, it may even be distracting and, therefore, reduce efficiency. Because the employee has limited control over this type of energy mobilization, it may easily result in maladaptive activity patterns, such as overreactivity, protracted recovery, or sustained activation, which are associated with performance deterioration, helplessness, depression, burn-out, and increased health risks.

A taxonomy of states may shed light on the relation between work demands and performance efficiency and may explain why some types energy mobilization reduce well-being and endanger health. Psychophysiological techniques may be used to identify the BB states that prevail in the work environment. Physiological measures should not only be used as an indicator of heightened arousal, but as corroboration for the existence of a BB state. In particular, a distinction should be made between adaptive and maladaptive patterns. Adaptive patterns enhance performance efficiency and promote well-being, whereas maladaptive patterns are associated with the disruption of the homeostasis, inefficient functioning, and negative health effects. Research efforts should concentrate on the emotional and physiological mechanisms that mediate cognitive processes and bodily reactions and the discovery of the underlying mechanisms of overreactivity, recovery, and accumulation of stress effects over time and between stressors.

CONCLUDING REMARKS

Psychophysiological techniques provide an interesting approach toward the assessment of the functioning of employees in their working conditions. With the aid of ambulatory methods (see chap. 5, this volume), the reactions of employees may be observed and recorded, either during their daily activities or in a simulated environment. Given the small and user-friendly apparatus now available, psychophysiological techniques are nonintrusive and likely to be accepted by the employees being investigated. Psychophysiological techniques have ecological validity because they interfere minimally with the ongoing activities of the employee. In addition, measures may be obtained before and after a working day or in pauses between work periods. This enables the possibility of assessing the recovery from workload, which is a major determinant of stress reactions. As illustrated in the framework presented in Table 2.1, the approach should be integrative. Physiological measures should be related to other dependent measures, such as subjective ratings, performance indices and observations, and, if possible, long-term outcomes, such as absenteeism, employee incapacity, turnover, and so on. For the interpretation of the results in terms of psychological processes, it is essential that the effects of work demands on psychophysiological measures be compared with the effects on other measures. The sort of measures that should be involved depend on the aims of the study and the type of approach (see also the section on three types of energy mobilization). In this way psychophysiological techniques provide useful tools to solve a variety of problems that may be encountered in the work environment.

REFERENCES

Allport, A. (1987). Selection for action: Some behavioral and neurophysiological considerations of attention and action. In H.Heuer & A.Sanders (Eds.), *Perspectives on perception* (pp. 395–419). Hillsdale, NJ: Lawrence Erlbaum Associates.

Andreassi, J.L. (1995). *Psychophysiology: Human behavior and physiological response* (3rd ed.). Hillsdale, NJ: Lawrence Erlbaum Associates.

Backs, R.W. (1998). A comparison of factor analytic methods of obtaining cardiovascular autonomic components for the assessment of mental workload. *Ergonomics, 41,* 733–745.

Berntson, G.G., Cacioppo, J.T., Quigley, K.S. (1993). Cardiac psychophysiology and autonomic space in humans: empirical perspectives and conceptual implications. *Psychological Bulletin, 114,* 296–322.

Boucsein, W. (1992). *Electrodermal activity.* New York: Plenum Press.

Boucsein, W., Luczak, H., Stern, J.A., & Yagi, A. (1996). *Psychophysiology in Ergonomics PIE.* Lengerich, Pabst Science Publishers.

Broadbent, D.E. (1958). *Perception and communication.* London: Pergamon Press.

Brookings, J.B., Wilson, G.F., & Swain, C.R. (1996). Psychophysiological responses to changes in workload during simulated air traffic control. *Biological Psychology, 42,* 361–377.

Caccioppo, J.T., & Tassinary, L.G. (1990). Inferring psychological significance from physiological signals. *American Psychologist, 45,* 16–28.

Caldwell, J.A., Wilson, G.F., Cetinguc, M., Gaillard, A.W.K., Gundell, A., Lagarde, D., Makeig, S., Myhre, G., & Wright, N.A. (1994). *Psychophysiological assessment methods.* Neuilly sur Seine, AGARD Advisory Report 324.

Cannon, W.B. (1932). *The wisdom of the body.* New York: Norton.

Coles, M.G.H., Scheffers, M., & Fournier, L. (1995). Where did you go wrong? Errors, partial errors, and the nature of human information processing. *Acta Psychologica, 90,* 129–144

Dienstbier, R.A. (1989). Arousal and psychophysiological toughness: Implications for mental and physical health. *Psychological Review, 96,* 84–100.

Eysenck, M.W. (1982). *Attention and arousal* Berlin: Springer Verlag.

Fowler, B. (1994). P300 as a measure of workload during a simulated aircraft landing task. *Human Factors, 36,* 670–683.

Frankenhaeuser, M. (1986). A psychobiological framework for research on human stress and coping. In M.H.Appley & R.Trumball (Eds.), *Dynamics of stress* (pp. 101–116). New York: Plenum.

Frankenhaeuser, M. (1989). A biopsychosocial approach to work life issues. *International Journal of Health Services, 19,* 747–758.

Gaillard, A.W.K. (1993). Comparing the concepts of mental load and stress. *Ergonomics, 9,* 991–1005.

Gaillard, A.W.K., & Wientjes, C.J.E. (1994). Mental load and workstress as two types of energy mobilization. *Work & Stress, 8,* 141–152.

Gaillard, A.W.K., Boucsein, W., & Stern, J. (Eds.). (1996). Psychophysiology of workload. *Biological Psychology, 42,* 245–352.

Gehring, W., Goss, B., Coles, M.G.H., Meyer, D., & Donchin, E. (1993). A neural system for error detection and compensation. *Psychological Science, 4,* 385–390.

Gevins, A., Leong, H., Du, R., Smith, M., Le, J., DuRousseau, D., Zhang, J., & Libove, J. (1995). Towards measurement of brain function in operational environments. *Biological Psychology, 40,* 169–186.

Gopher, D., & Donchin, E. (1986). Workload-An examination of the concept. In K.R.Boff, L.Kaufman, & J.P.Thomas (Eds.), *Handbook of perception and human performance* (pp. 41:1–41:49). New York: Wiley.

Hancock, P.A. (1986). Stress and adaptability. In G.R.J.Hockey, A.W.K.Gaillard, & M.G. H.Coles (Eds.), *Energetics and human information processing* (pp. 243–251). Dordrecht, The Netherlands: Nijhoff.

Hancock, P.A., & Meshkati, N. (Eds.). (1988). *Human mental workload.* Amsterdam: North-Holland.

Hebb, D.O. (1955). Drives and the C.N.S. (conceptual nervous system). *Psychological Review, 62,* 243–254.

Hillyard, S.A., AnlloVento, L., Clark, V., Heinze, H.J., Luck, S., & Mangun, G.R. (1996). Neuroimaging approaches to the study of visual attention: A tutorial. In A.F.Kramer, M.

G.H.Coles, & G.D.Logan (Eds.), *Converging operations in the study of visual selective attention* (pp. 107–138). Washington, DC: APA Press.

Hockey, G.R.J. (1997). Compensatory control in the regulation of human performance under stress and high workload: A cognitive-energetical framework. *Biological Psychology, 45,* 73–93.

Hockey, G.R.J., Coles, M.G.H., & Gaillard, A.W.K. (1986). Energetical issues in research on human information processing. In G.R.J.Hockey, A.W.K.Gaillard, &. M.G.H.Coles (Eds.), *Energetics and human information processing* (pp. 3–21). Dordrecht, The Netherlands: Nijhoff.

Holmes, T.H., & Rahe, R.H. (1967). The social readjustment rating scale. *Journal of Psychosomatic Research, 11,* 213–218.

Johnson, J.V., Hall, E.M., & Theorell, T. (1989). Combined effects of job strain and social isolation on cardiovascular disease morbidity and mortality in a random sample of the Swedish working population. *Scandinavian Journal of Work and Environmental Health, 15,* 271–279.

Kahn, R.L. (1981). *Work and health.* New York: Wiley.

Kahneman, D. (1973). *Attention and effort.* Englewood Cliffs, NJ: Prentice-Hall.

Kamarck, T., & Jennings, J.R. (1991). Biobehavioral factors in sudden cardiac death. *Psychological Bulletin, 109,* 42–75.

Kantowitz, B.H., & Casper, P.A. (1988). Human mental workload in aviation. In E.L. Wiener & D.C.Nagel (Eds.), *Human factors in aviation* (pp. 157–187). San Diego: Aca-demic Press.

Karasek, R.A., & Theorell, T. (1990). *Healthy work.* New York: Basic Books.

Kramer, A.F. (1991). Physiological metrics of mental workload: A review of recent progress. In D.L.Damos (Eds.), *Multiple task performance* (pp. 279–328). London: Taylor & Francis.

Kramer, A.F., Sirevaag, E., & Braune, R. (1987). A psychophysiological assessment of operator workload during simulated flight missions. *Human Factors, 29,* 145–160.

Kramer, A.F., & Spinks, J. (1991). Capacity views of information processing. In R.Jennings & M.Coles (Eds.), *Psychophysiology of human information processing: An integration of central and autonomic nervous system approaches* (pp. 179–250). New York: Wiley.

Krantz, D.S., & Manuck, S.B. (1984). Acute psychophysiologic reactivity and risk of cardiovascular disease: A review and methodologic critique. *Psychological Bulletin, 96,* 435–464.

Kutas, M., & Van Petten, C. (1994). Psycholinguistics electrified: Event-related brain potential investigations. In M.Gernsbacher (Ed.), *Handbook of psycholinguistics* (pp. 83–144). San Diego: Academic.

Lazarus, R.S., & Folkman, S. (1984). *Stress, appraisal, and coping.* New York: Springer.

Leonova, A.B. (1994). Industrial and organizational psychology in Russia: The concept of human functional states and applied stress research. In C.L.Cooper & I.T.Robertson (Eds.), *International review of industrial and organizational psychology* (pp. 183–212). Chichester, England: Wiley.

Makeig, S., & Inlow, M. (1993). Lapses in alertness: coherence of fluctuations in performance and EEG spectrum. *Electroencephalography and Clinical Neurophysiology, 86,* 23–35.

Makeig, S., & Jung, T.P. (1996). Tonic, phasic and transient EEG correlates of auditory awareness in drowsiness. *Cognitive Brain Research, 4,* 15–25.

Meyer, D.E., & Kieras, D.E. (1997). A computational theory of executive control processes and multipletask performance: Part I. Basic mechanisms. *Psychological Review, 104,* 3–65.

Morris, T.L., & Miller, J.C. (1996). Electrooculographic and performance indices of fatigue during simulated flight. *Biological Psychology, 42,* 343–360.

Mulder, L.J.M. (1992). Measurement and analysis methods of heart rate and respiration. *Biological Psychology, 34,* 205–236.

Näätänen, R. (1973). The inverted-U relationship between activation and performance: A critical review. In S.Kornblum (Ed.), *Attention and performance* (Vol. 4. pp. 155–174). London: Academic Press.

Navon, D., & Gopher, E. (1979). On the economy of the human processing system. *Psychological Review, 86,* 214–225.

Neiss, R. (1988). Reconceptualizing arousal: Psychobiological states in motor performance. *Psychological Bulletin, 103,* 345–366.

Osman, A., Moore, C., & Ulrich, R. (1995). Bisecting RT with lateralized readiness potentials: Precue effects after LRP onset. *Acta Psychologica, 90,* 111–127.

Pashler, H. (1994). Dualtask interference in simple tasks: Data and theory. *Psychological Bulletin, 116,* 220–244.

Polson, M., & Freidman, A. (1988). Task sharing within and between hemispheres: A multiple resource approach. *Human Factors, 30,* 633–643.

Posner, M.I., & Raichle, M.E. (1994). *Images of mind.* San Francisco: Freeman.

Pribram, K.H., & McGuinness, D. (1975). Arousal, activation and effort in the control of attention. *Psychological Review, 82,* 116–149.

Rasmussen, J. (1983). Skills, rules and knowledge: Signals, signs and symbols and other distinctions in human performance models. *III Transactions on Systems, Man and Cybernetics, 13,* 257–266.

Reason, J. (1990). *Human error.* New York: Cambridge University Press.

Roscoe, A.H. (1993). Heart rate as a psychological measure for inflight workload assessment. *Ergonomics, 36,* 1055–1062.

Sanders, A.F. (1983). Towards a model of stress and human performance. *Acta Psychologica, 53,* 61–97.

Scheffers, M., Coles, M.G.K, Bernstein, P., Gehring, W., & Donchin, E. (1996). Eventrelated brain potentials and error related processing: An analysis of incorrect responses to go and no-go stimuli. *Psychophysiology, 33,* 42–53.

Selye, H. (1956). *The stress of life.* New York: McGraw-Hill.

Sirevaag, E., Kramer, A., Wickens, C., Reisweber, M., Strayer, D., & Grenell, J. (1993). Assessment of pilot performance and workload in rotary wing helicopters. *Ergonomics, 9,* 1121–1140.

Sperandio, J.C. (1978). The regulation of working methods as a function of workload among air traffic controllers. *Ergonomics, 21,* 193–202.

Steptoe, A. (1989), Psychophysiological interventions in behavioral medicine. In G.Turpin (Ed.), *Handbook of clinical psychophysiology* (pp. 215–239). New York: Wiley.

Sterman, M.B., & Mann, C. (1996). Concepts and applications of EEG analysis in aviation performance evaluation. *Biological Psychology, 40,* 115–130.

Stern, J.A., & Dunham, D.N. (1990). The ocular system. In J.T.Cacioppo, & L.G.Tassinary (Eds.), *Principles of psychophysiology: Physical, social and inferential aspects* (pp. 513–553). Cambridge: Cambridge University Press.

Trejo, L.J., & Shensa, M.J. (1993). Linear and neural network models for predicting human signal detection performance from eventrelated potentials: A comparison of wavelet transforms with other feature extraction methods. *Proceedings of the International Simulation Conference.* San Diego: Society for Computer Simulation.

Ursin, H., & Olff, M. (1993). The stress response. In C.Stanford, P.Salmon, & J.Gray (Eds.), *Stress: An integrated response* (pp. 3–22). New York: Academic Press.

Veltman, J.A., & Gaillard, A.W.K. (1996). Physiological indices of workload in a simulated flight task. *Biological Psychology, 42,* 323–342.

Veltman, J.A. & Gaillard, A.W.K. (1998). Physiological workload reactions to increasing levels of task difficulty. *Ergonomics, 41,* 656–669.

Wickens, C.D. (1992). *Engineering psychology and human performance.* New York: Harper Collins.

Wilson, G.F., Fullenkamp, P., & Davis, I. (1994). Evoked potential, cardiac, blink and respiration measures of pilot workload in air to ground missions. *Aviation, Space and Environ-mental Medicine, 65,* 100–105.

Yamamoto, Y., & Isshiki, H. (1992). Instrument for controlling drowsiness using galvanic skin reflex. *Medical and Biological Engineering, 30,* 562–564.

Chapter 3

The Design and Analysis of Experiments in Engineering Psychophysiology

Julian F.Thayer[1]
University of Missouri-Columbia
National Institute of Aging
Bruce H.Friedman
Virginia Polytechnic Institute
and State University

Biobehavioral research is among the most challenging of scientific endeavors. The study of interactions between living systems and their environments has tested the limits of research methodologies and theoretical models. Research designs in engineering psychophysiology, as in most biobehavioral research, have been characterized by the use of single dependent variables (DV) or multiple DVs treated singly that are averaged over subjects and discrete experimental epochs on a single measurement occasion. Although these designs have provided a great deal of useful information, their origins in agronomy suggest that they may hold limited generalizability to the study of humans as complex living systems (Ford & Ford, 1987). Importantly, in the context of engineering psychophysiology one often wants to make inferences about transactions between individual human beings and their environments. The typical research design oversimplifies the complexity of these relationships and thus does not unambiguously allow for such inferences. Rather, these designs tend to obscure underlying processes by shrouding rich individual data with group data aggregation procedures (Glass & Mackey, 1988; O'Connor, 1990).

---

[1] The contributions of the co-authors of this chapter were equal; order of authorship was arbitrary.

It is a truism in the biobehavioral sciences that no single measure or aspect of responding can adequately represent a complex latent construct (Nesselroade & Ford, 1987; Schwartz, 1986). Rather, such constructs must be represented by an entire pattern of manifestations (Cattell, 1966). Yet, psychophysiological responses have traditionally been operationalized with individual DVs in relative isolation from each other. Rendered in this manner, the classical psychophysiological paradigm is dependent on individual measures that reflect mean values collected over discrete experimental epochs, across many organisms, on a single experimental occasion (Friedman, 1998).

In contrast, systems approaches stress the variety of interactive mechanisms that can lead to similar physiological and behavioral states (Globus & Arpaia, 1992; Mandell & Selz, 1992). In studying the behavior of such systems, description and elucidation of the nature of underlying regulatory processes should precede any premature hypothesis testing with group means in individual variables (Barton, 1994). In view of the prevalence and importance of rhythmicity in biological regulatory mechanisms, inclusion of time-varying or temporal aspects of responding is crucial to accurately portray such activity (Glass & Mackey, 1988; Goldbeter & Decroly, 1983; Hrushesky, 1994).

Unfortunately, the analysis of variance (ANOVA) design, a staple of psychophysiological research, muddles temporal information by parsing physiological events into discrete epochs, thereby disrupting the continuity of responding (O'Connor, 1992). Furthermore, precise mapping of physiological activity from these distinct periods onto experimental manipulations is fraught with hazards such as delayed physiological responses and compensatory homeostatic processes (Levenson, 1988). Alternatively, all recorded activity might be considered as relevant; functional relationships among ongoing physiological processes could then be extracted across observations (O'Connor, 1990).

In response to these various concerns, an alternative framework for engineering psychophysiology is offered here: a multivariate, systems perspective that emphasizes the study of individuals. A distinctive feature of this approach is its focus on intra-individual variability in the behavioral and physiological processes of an organism. Moreover, within-groups variance (typically treated as error in traditional experimental psychology) is also investigated because it contains a wealth of relevant information (Cronbach, 1957). Hence, it is appropriate first to address these basic methodological issues in research design and analysis in engineering psychophysiology.

RESEARCH: WHAT ARE WE REALLY TRYING TO DO?

In this section, basic issues in research design and analysis are outlined, beginning with a discussion of the research hypothesis and its relationship to the partitioning of variance. Attention is then shifted to the elements of a statistical test, followed by an examination of the selection perspective on research design and the distinction between *intra*-individual and *inter*-individual variability. In conclusion, a rationale for usage of multivariate, replicated, repeated-measures, single-subject designs in engineering psychophysiology is offered. It is imperative to keep in mind that the most important aspect of any statistical test is the informed judgment of the researcher. As such, a clear understanding of these basic issues goes a long way toward the appropriate design and analysis of experiments in engineering psychophysiology.

Variance Partitioning and Hypothesis Testing

It is important at the outset to clearly define ones' basic goals in the scientific study of living systems. Indeed, these objectives are often difficult to discern from research methodology textbooks. Simply put, this enterprise is primarily concerned with the detection of systematic relationships amidst the morass of variability in biobehavioral responses. This task calls for the partitioning of observed variability into systematic and random components, which in turn will yield patterns of associations such that events can be described, predicted, controlled, and ultimately understood. In the simplest case, a research question or hypothesis is tested by an investigation of the existence, direction, and magnitude of a relationship between an independent or predictor variable (IV) and a DV or criterion variable.

It is a basic and often implicit assumption of most scientific research that the whole is equal to the sum of its parts. It then follows that the partitioning of observed variability will lead to an explicit set of relationships between IVs and DVs. This premise is the basis of all common statistical procedures and is the foundation of the general linear model; that is, the presumption of linearity allows for unambiguous partitioning of variability. As such, small causes lead to small effects and large causes to large effects. This concept of proportionality is a cornerstone of the general linear model (West, 1990). However linearity, and thus this ability to neatly partition variance, often does not hold in nature, and so arose the impetus for contemporary work on nonlinear dynamics, the science of complexity, or "chaos theory" (Gleick, 1987). In spite of the notoriety that has been attained by this field, most research methodologies are based on the presumption of linearity. We attempt here to provide a glimpse into the world of chaos by outlining a framework for research that would allow, in conjunction with the appropriate statistical methods, the examination of nonlinear dynamic relationships.

ANOVA: Variance Partitioning in a Linear World

When trying to assess the relative importance of a particular association, one typically asks if this systematic relationship is large relative to random fluctuations.

In a simple t-test or one-way between-subjects ANOVA this question becomes a test of the ratio of two variances: one that represents systematic variability (between subject variance) and the other representing unsystematic variability (within subject variance). In the one-way between-subjects ANOVA case, this ratio is expressed as:

$$F = \text{between-subject variance/within-subject variance}$$

When this ratio is much greater than one, with the appropriate degrees of freedom *(df)*, there is evidence for a statistically significant association between IV and DV However, in numerous situations there are many possible systematic relationships, but only some of interest. Thus, it is preferable to think of a statistical test as the ratio of the variance of interest to the variance of non-interest:

$$F = \text{variance of interest/variance of non-interest}$$

Another important consideration is the effect size. Because one is not only attempting to estimate the probability, but also the direction and magnitude of relationships, a direct index of the latter is very useful. Unfortunately, the tradition of null-hypothesis testing has tended to divert the focus of research away from the dimension of magnitude. In fact, any significance test represents the confluence of four mathematical components:

The size of the study refers to the number of subjects under investigation and is often represented in the significance test by the *df* associated with the denominator of the ratio of variances.

The effect size refers to the magnitude of the relationship between the IV and DV This quantity is the amount of variability in the DV that is due to variation in each IV

The alpha level or Type I error rate: By convention, the level of alpha is usually set at .05, but this figure is arbitrary. The Type I error rate is the probability of falsely rejecting the null hypothesis; that is, the chance of concluding that a systematic relationship exists in the population when it, in fact, does not. This possibility tends to predominate the consciousness of investigators and many post-hoc techniques (e.g., the Bonferoni inequality, the Newman-Keuls test, etc.) have been developed to control it.

The power and Type II error rate (beta): Power and its related Type II error rate are probably the most neglected aspects of a statistical test. Power refers to the probability that a relationship in the population will be detected when one in fact exists. Therefore, power might reflect what R.A.Fisher called the "sensitivity" of an experiment (Fisher, 1942). Importantly, the power of an obtained statistical test reflects the probability that such a result can be replicated (Good-man, 1992). The effects of low statistical power on the reproducibility of research findings has been well documented (Goodman, 1992; Harris, 1997). The Type II error (beta) occurs when one fails to reject a false null hypothesis.

This error is inversely related to the power of a test:

$$power = 1 - beta$$

One factor that leads to low power or inflated Type II error risk is an inadequate sample size. As a convention, Type I error rates of 0.05 and implicit Type II error rates of 0.20 (power of 0.80) are adopted. However, in practice in many areas of investigation, one rarely has adequate power (i.e., power greater or equal to 0.80) and, therefore, the Type II error rate is much higher (Cohen, 1992).

The following equation reveals the relationship between the statistical test value on the one hand and the effect size and size of study on the other:

$$statistical\ test = effect\ size \times size\ of\ the\ study$$

("the equation of the experiment"; Rosenthal & Rosnow, 1984) or more concretely for a two group test:

$$F = (mean\ square\ of\ interest / mean\ square\ of\ non\text{-}interest) \times n$$

with the appropriate F distribution based on the numerator and denominator *dfs* and the mean squares representing the variability per degree of freedom. Thus the numerator is an estimate of the variability of interest, the denominator is an estimate of the variability of non-interest, and n is the sample size per group.

What is clear from the "equation of the experiment" is that "statistical significance" is a function not only of the magnitude of the relationship (the effect size), but also the size of the study. Thus, one way to achieve a statistically significant result is to increase the size of the study. A consequence of this fact is that any non-zero association between IVs and DVs can be statistically significant if one runs enough subjects (Meehl, 1978)! In fact, the probability that the so-called "null" hypothesis, if taken literally, is truly false is essentially zero. Perhaps the best information derived from a significant statistical test result is that many subjects were run, which one knows without having do to any mathematical calculations at all.

Moreover, all other things being equal, the probability of a Type I error is inversely related to the probability of a Type II error. Thus, as one decreases the risk of a Type I error, for example by decreasing the alpha level considered acceptable, the risk of a Type II error increases. The appropriate balancing of Type I and Type II errors is very much content-area specific (Rosenthal & Rosnow, 1984). However, using the conventional values for Type I and Type II errors (0.05 and 0.20, respectively), a Type I error is deemed to be four times more egregious than a Type II error. In areas such as engineering psychophysiology, where the cost per datum tends to be relatively high, this level of balance in which we are more apt to accept a Type II error and thus conclude that no relationship exists when in fact one does may not be cost effective.

Although an extended discussion of the relative merits of statistical hypothesis testing is beyond the scope of this chapter (see Harris, 1997, for a recent debate), suffice it to say that knowledge of the nature of these tests is crucial for the design of the experiment. At the very least, this awareness can help to determine the number of subjects needed to have a fair chance of detecting an effect of a certain size while balancing the risk of Type I and Type II errors. Moreover, reasoned consideration of the effect size can greatly enhance the ability to determine the practical significance, and not just the statistical significance, of the results of an experiment.

How Do We Choose What to Measure?

Of all the aspects of observable phenomena, how does one choose which to examine? That is, what makes an observation a datum? This issue has been the topic of much deliberation and has generally been covered under the rubric of measurement theory, scaling, dimensional analysis, or, more generically, *data theory* (Jacoby, 1991). This work has mainly been concerned with topics related to the DV. On the IV side, Cattell (1988) presented a system of relationships based on three dimensions (persons, occasions, variables) termed the *data box.* Common to both data theory and the data box is the notion that the researcher, explicitly or more often implicitly, selects from a broad range of possible dimensions or modes of interest. This critical decision involves selection from a universe of possible scores those that will be the subject of investigation. That this choice occurs often without full knowledge of the various selection effects threatens not only the validity of inferences draw from such experiments, but also ultimately the quality of research that serves as the collective database of the field.

Nesselroade and Jones (1991) have provided a cogent exposition on the nature of these selection effects. They noted that a single datum can be characterized as a "'draw' of one piece of information from a universe of information" (p. 21). Minimally, this hypothesized universe represents the scores for every conceivable person and variable on all possible measurement occasions. Thus, the three dimensions of *persons, variables,* and *occasions* represent a minimum universe from which data are selected. Because each possible score is not likely to be available due to constraints such as time, money, population base rates, and the like, most of the data will remain unrealized. Therefore, in attempting to make generalizations from one's results, the necessity of selecting a representative subset of scores to comprise the data cannot be overestimated.

Indeed, the effects of selection can have a profound influence on the inferences drawn from any particular study (Nesselroade, 1988). A familiar example of such consequences is the attenuation of a correlation coefficient produced by a restricted range of scores on one or both of the variables involved (Nesselroade & Jones, 1991). An observed correlation coefficient represents at least two influences: the "true" magnitude of the relationship and the effect of selection of scores from the universe for observation. If the actual correlation between two variables is already zero, the selection of a narrow range of scores will not reduce this association further. However, if the true correlation is 0.80, for example, this association will likely be underestimated by the observed correlation based upon a selected subset of scores. This difference between the true and observed relationship comprises the selection effect.

A thornier situation exists in biobehavioral research, where one is confronted with myriad possible dimensions from which to select. Cattell (1988) mentioned 10 such dimensions and innumerable others are possible. Thus, this decision is a complex *multi-modal selection operation* (Nesselroade & Jones, 1991). For example, if an investigator attends exclusively to the persons mode, which is often the case, even appropriate sampling methods do not protect against selection effects on the variables or occasions modes.

Known Systematic Sources of Variability

These three sources of variance (persons, variables, and occasions) are present in nearly all experimental designs. Their relationship with a set of scores should be explicitly investigated and the systematic variance associated with each accounted for before valid inferences can be drawn (Cattell, 1988). The most familiar source is that of persons: the effects of inadequate selection on this dimension are widely known. However, the exclusive emphasis on this mode has tended to distract investigators from selection effects on the variables and occasions dimensions, inadvertently providing a false sense of security to experimenters who have accounted for bias on person selection while neglecting these other equally important dimensions.

Multivariate techniques are required to assess multiple sources of variance. Yet, there has generally been a paucity of multivariate studies in experimental psychology (Harris, 1992). Multiple indicators of constructs have become salient due to increasing interest in structural equation modeling in biobehavioral research where multiple indicators are necessary to estimate latent variables and their associated error variance. In addition, the occasions dimension holds particular special significance for psychophysiology because most studies involve repeated measurements to some degree (Vasey & Thayer, 1987).

Importantly, the dynamics of a system cannot be investigated unless the organism is observed repeatedly over time. *Organismic theory* views the human being as a unified entity whose component parts function according to laws that direct the whole organism (Goldstein, 1939). These principles guide the aspects of the environment to which the organism attends and reacts. Goldstein asserted that in-depth investigations of single individuals over a wide range of observation conditions are necessary to apprehend the operation of these superordinate functions in naturalistic environments.

Indeed, a full description of human responding requires an observation on some measure or measures in an certain individual at a particular time and place (occasion; Nesselroade & Ford, 1987). Thus, three known sources of variability must be taken into account in the design and analysis of psychological studies. For example, in field research in engineering psychophysiology, whether a measurement is taken at work versus at home can have important implications for the meaning of that score. Admittedly, all experiments represent a compromise among a number of competing forces; no single experiment can address all questions equally well. Programmatic research is marked by the systematic investigation of multiple influences on a particular phenomenon.

Cognizance of the multiple sources of variability in a particular study can lead to several consequences. First, experiments should be designed so as to highlight the particular variability of interest. For example, if one is interested in individual differences, persons should be observed over many different occasions to estimate the variance in scores that is relatively unchanging. Moreover, one may want to use "raw" rather than "change" scores because partial *ipsitisation* (from subtraction of base values; see Cattell, 1988) tends to minimize the variance due to persons. Second, focus should be on indices of the magnitude of the association and less reliance placed on tests of statistical significance. Third, a program of research studies should be examined in aggregate; that is, the results of several studies should be combined to produce estimates of the magnitude of associations. Meta-analytic procedures for this purpose can produce more stable estimates of the importance of various factors and aid in the accumulation of knowledge that lies at the heart of any scientific enterprise (Guzzo, Jackson, & Katzell, 1987). Furthermore, a significant p-value found in any one study yields essentially no useful information regarding the probability of replicating that finding (Goodman, 1992; Guttman, 1985).

Inter-Individual Versus Intra-Individual Variability

An important and often overlooked distinction that has led to confusion in the literature is that between inter-individual and intra-individual variability. That is, relationships that exist among individuals may not be the same as those relationships that exist within individuals. From a research design perspective, it is important to be clear which of these associations is being investigated. A common mistake is to formulate a hypothesis concerning an intra-individual association, say, on the effects of different treatments on an individual's blood pressure, but to conduct an experiment in which each person receives a different treatment or manipulation.

As another relevant example, few studies have truly tested the James model of emotion (1884), which suggests that the physiological responses and the subjective experience of a given emotion are highly related (an intra-individual hypothesis). Instead, most studies have involved an emotional response elicited from a group of individuals to several manipulations, such as viewing pictures of facial expressions (see Ellsworth, 1994, for a review). Subjects may have various physiological responses measured such as the facial electromyogram (EMG) from the brow and cheek, and then are asked to rate their subjective emotional state after viewing each picture. The data are then analyzed by correlating physiological and self-reported responses aggregated across all individuals. In this case, inter- and intra-individual differences are confounded. A more appropriate test of the Jamesian hypothesis would be to correlate the physiological and subjective responses within individuals and then combine the results of these within-person correlations. Using this within-person approach with a multivariate measure of association (*redundancy analysis*; Lambert, Wildt, & Durand, 1988), responses aggregated across subjects revealed little relationship (shared variance on the order of magnitude of 10%) between EMG and subjective responses (Thayer, 1988). When the same data were analyzed using within-subject measures of association, the shared variance increased to approximately 80%!

Clearly, different data aggregation procedures lead to vastly different inferences about the Jamesian hypothesis. A related study reported that the physiological measures that best discriminated the group emotion profiles were not those that best discriminated among any individual's emotion profiles (Thayer & Faith, 1994). These findings suggest that the effects of confounding inter-individual variability and intra-individual variability can have enormous consequences for the generalizations and conclusions reached in any particular study.

Multivariate, Replicated, Repeated-Measures Single-Subject Designs

In any one study, it is generally not feasible to represent all possible modes of data classification in a completely satisfactory manner. Therefore, it behooves investigators to make informed choices on these dimensions rather than let chance and expediency dictate research design. Preparation for conducting large group studies may involve prior intensive study of individuals with multiple measures on numerous occasions in order to discern information on sampling of variables and occasions (Nesselroade & Jones, 1991). Indeed, the history of psychological research is replete with prominent examples in which principles of broad applicability emerged from the intensive study of individuals (see Barlow & Hersen, 1984, for a review). Importantly, it is only through intensive studies of individuals that behavior patterns can be examined as they unfold over time, a critical feature in the study of nonlinear dynamics. Finally, the individual has long been recognized as the ultimate entity in psychology, for it is there that processes occur and applications are made (Rosenzweig, 1958). This point is especially salient when applied to engineering psychophysiology.

In sum, multivariate, replicated, repeated-measures, single-subject designs are highly compatible with the aims of engineering psychophysiology. Central to these aims is the desire to develop applications that are relevant to specific individuals (persons) as they carry out complex, multivariate (variables) behaviors that evolve over time (occasions). Repeated-measures designs and the collection of multiple interrelated measures are common in psychophysiology, and so the considered application of this research paradigm can enhance the quality of research that is already being conducted. Thus, no radical change in experimental procedures is required; only a more reasoned data extraction from the rich corpus of already available information is necessary.

Furthermore, recent statistical advances have expanded the repertoire of tools with which to analyze data from these designs. For example, hierarchical linear models (J.E.Schwartz, Warren, & Pickering, 1994), random regression models (Jacob et al., in press), or pooled cross-sectional time series (Dielman, 1983) allow for the partitioning of inter-individual and intra-individual variability from a number of different sources. Complemented by *set analytic* techniques that allow for the examination of multiple dependent variables (Cohen, 1982), these methods offer many data analytic strategies for multivariate, replicated, repeated-measures, single-Subject designs. Several of these techniques are illustrated in the next section.

ILLUSTRATIVE EXAMPLES

The following three examples depict the range of possible analytic approaches to multivariate, replicated, repeated-measures, single-subject designs. Other techniques such as dynamic factor analysis (Wood & Brown, 1994) allow for the examination of multiple dependent variables simultaneously, as well as the estimation of lagged effects. As these techniques become more accessible, researchers will need to grow more aware of the impact of design decisions on the inferences that can be drawn from a particular study. It is hoped the preceding sections will aid in increasing appreciation of these issues and lead to more informed decisions.

Example 1: Structural Equation Models and Variance Partitioning

In attempting to partition the variance in a data set, a number of factors must be considered. First, the design must allow for variability along the dimension of interest. Next, multiple indicators must be obtained. Finally, a measurement model must be available that specifies the unambiguous estimation of the variance associated with each source. A model has been constructed that allows for the simultaneous estimation of variability due to three common sources in behavioral data (Steyer, Ferring, & Schmitt, 1992): (a) the person (inter-individual variability), (b) the situation and/or person-situation interaction (intra-individual variability), and (c) method variance and/or measurement error. In this model, the total variance [Var (Y_{ik})] of observed variable Y measured in person i at occasion k is decomposed into three parts: (a) variance due to persons, and thus consistent; (b) variance due to the situation or per-son-situation interaction, and thus specific, and (c) reliable variance, the inverse of which is variance due to error or a particular method. The model is estimated with structural equations and indices of consistency, specificity, and reliability are obtained.

Cortisol has been used to investigate a number of topics in engineering psychophysiology such as the effects of mental effort and controllability (Peters et al., 1998), sleep deprivation (Leproult, Copinschi, Buxton, & Van Cauter, 1998), and acceleration stress (Tarui & Nakamura, 1987). The above measurement model has recently been applied to several studies of salivary cortisol (Kirschbaum et al., 1990). In the first study, two simultaneous saliva samples were collected on two occasions in the early afternoon, 6 weeks apart. Results indicated that cortisol showed high levels of specificity (75%) and thus reflected primarily situation or person-situation variance. The consistency (person variance) was estimated to be approximately 21%, whereas the measurement unreliability was only approximately 4%. In the second study, two simultaneous measures of salivary cortisol were collected in a group of young mothers in the morning, afternoon, and evening for 3 consecutive days after the delivery of a healthy baby. Person variance accounted for approximately 60% of the variability in morning cortisol, whereas person variance accounted for about half as much (30%) of the variability in afternoon and evening cortisol measurements.

Taken together, these studies illustrate how design and analysis complement each other in the partitioning of variability into its constituent sources. The two simultaneous measures of saliva allowed for estimation of method or measurement error variance; multiple occasions of measurement allowed for estimation of variability due to consistent person factors and situation/person-situation interaction variance. This design demonstrates the need for variability along a dimension in order to estimate the influence of that dimension on an observation.

Example 2: Data Aggregation Across Tasks

One topic that has been investigated extensively in engineering psychology is the human-computer interface (see chap. 14, this volume). The health effects of working with video display terminals (VDTs) is of particular interest to engineering psychophysiology. Headache, eye strain, muscle tension, and tachycardia are a few of the negative health consequences associated with VDT use. To better understand these findings we have examined the link between the eyes and the heart. A pooled cross-sectional time-series design has been used to investigate the relationship between oculomotor and cardiovascular variables during a series of stressors (Tyrrell, Thayer, Friedman, Leibowitz, & Francis, 1995). Three subjects participated in four experimental sessions during which they were exposed to a diverse set of stressors. Multiple oculomotor and cardiovascular measures were taken during each stressor, the recovery from each stressor, and one baseline per session. Hierarchical regression analysis was used to estimate the inter- and intra-individual variability associated with the prediction of oculomotor variables from cardiovascular variables.

There were 81 (3 participants×4 sessions×7 conditions–3 observations with missing data) simultaneous observations of these measures. Because the research focus was on intra-individual associations among the variables, the mean of the predictor variable was entered into the regression equation on the first set, which served as an estimate of the inter-individual variability. Next, the cardiovascular measures were entered together in a second step, which gave an estimate of intra-individual variability.

Results indicated that all of the variability in one oculomotor variable *(dark vergence)* was due to inter-individual factors, and in the second oculomotor variable *(dark focus)* most of the variability (56%) was due to inter-individual factors. However, a statistically reliable additional 8.7% of the variability in dark focus was due to intra-individual factors. These findings suggest that only inter-individual differences in cardiovascular activity can predict dark vergence, but both inter- and intra-individual variability in cardiovascular measures can predict variability in dark focus. Hence, this multivariate, replicated, repeated-measures, single-subject design allowed for estimation of both inter- and intra-individual influences on the variability of a particular DV These results suggest that the eyes and the heart are linked and may have an influence on autonomic nervous function related to the negative health effects of VDT use.

A similar methodology was employed to extract patterns of autonomic cardiovascular control across diverse conditions (Friedman et al., 1993; Friedman & Thayer, 1998a). In these studies, data were first aggregated and standardized within subjects, across tasks selected to elicit heterogeneous patterns of autonomic activity. Then, the data were aggregated within preselected anxiety groups (one group characterized by panic attacks, the other by blood-phobic syncope). At the intra-individual level, consistent autonomic patterns within individuals were found across diverse tasks. Inter-individually, the anxiety groups could be distinguished by distinct modes of cardiovascular control. Thus, the complementarity of intra- and inter-individual approaches was demonstrated.

Example 3: Multilevel Models

Recent advances in statistical software have brought the use of hierarchical linear models and stochastic regression models into easy reach. PROC MIXED in the SAS software package (SAS Institute, Inc.) and BMDP 5V in the Biomedical software package (SPSS, Inc.) can be used to estimate these types of models. There are numerous advantages to these approaches compared to previous methods. First, inter-individual and intra-individual variability can be simultaneously estimated. Second, the use of random coefficients allows these estimates to be generalized beyond the particular data sample. Third, differing numbers of observations per participant can be accommodated. Finally, missing data can be easily handled in these models.

Ambulatory monitoring of physiological responses has had a major influence on engineering psychophysiology (see chap. 5, this volume). Researchers are no longer confined to the laboratory and subjects can now be monitored during actual work situations. Several versions of these multilevel models have been applied to study the effects of various factors such as mood, location, and postural effects on ambulatory heart rate and blood pressures (Jacob et al., in press; Schwartz et al, 1994). Schwartz et al. (1994) present a simple illustration of the model; for a single within-person factor such as location (work vs. home), the model is:

$$Y_{ij(k)} = (\mu + \alpha_i) + (\beta_k + \delta_{ik}) + \varepsilon_{ij(k)}$$

where:

- The first term on the right hand side $(\mu + \alpha_i)$ is the inter-individual variance.
- The second term on the right hand side $(\beta_k + \delta_{ik})$ is the intra-individual variance.
- $Y_{ij(k)}$ is the jth blood pressure reading for person i, taken in the kth location.
- μ is the weighted grand mean of an individual's average awake blood pressures weighted by the number of readings per person.
- α_i is the deviation of person i's average awake blood pressure (from the grand mean μ).
- β_k is the average intra-individual (main) effect of being in location k (the weighted average of the β's equals zero).
- δ_{ik} is the deviation of the effect for person i of being in location k from β_k, the person-by-location interaction effect (the weighted average of the δ's equals zero for each person).
- $\varepsilon_{ij(k)}$ is the deviation of person i's jth observation from its predicted value based upon the preceding parameters (the mean of these deviations for all observations of person i taken in the kth location equals zero).

Full details for estimation of the model are given in Schwartz et al. (1994, p. 214). Application of this model toward multiple-parameter estimation is illustrated by Schwartz et al. (1994); a similar model and estimation procedure has been applied to longitudinal regression (Jacob et al., in press). These models allow for the estimation of multiple influences on a DV in a nonarbitrary metric. For example, Schwartz et al. (1994) found that the average intra-individual effect of being at work versus at home on systolic blood pressure was 2.4 mm Hg. Thus, the implications of this result are easy to comprehend, whereas traditional ANOVA-type models may state results in standard deviation units or other derived indices, the practical significance of which is often difficult to gauge.

SYSTEMS CONCEPTIONS
OF PSYCHOPHYSIOLOGICAL RESPONDING

We close this chapter with a brief introduction to the implications of work in dynamical systems theory for experimental design and analysis. This section is meant to portray a systems perspective that may be a fruitful worldview from which to approach research. The multivariate, replicated, repeated-measures, single-subject design can be used to provide data for examination within this dynamical systems perspective.

Mechanistic and Reductionistic Models

A crucial distinction between the dominant psychophysiological paradigm and a systems approach parallels the difference between mechanistic and holistic models of organismic function (Goldstein, 1939). Mechanistic representations portray psychophysiological responses as discrete events, consistent with a reductionist view that breaks nature down into its smallest parts and considers them separately (Pepper, 1942). These elements can be reconstructed according to principles that apply at the lower level of analysis. Mechano-reductionist approaches are consistent with the aforementioned linear model of nature and assume that changes in input will yield proportional changes in output (Briggs & Peat, 1989). In the context of psychophysiological responses, this view presumes a similar proportional relationship between the IV (e.g., the presentation rate of information on a computer display) and the DV (e.g., change in heart rate).

Conceptualization and operationalization of this paradigm is relatively straightforward, but its applicability is limited by the pitfalls inherent in one-to-one mapping of IV manipulations onto DV responses (Levenson, 1988). These obstacles pose a serious challenge to customary notions of linear "dose-response relationships," and IV-DV associations. Such concerns have substantial implications for research designs that rely on average changes from a resting "baseline" to a discrete "experimental" condition as indicators of "reactivity" (e.g., Manuck, 1994). Furthermore, many physiological responses reflect the simultaneous input of multiple intrinsic rhythms (Saul, 1990) and so confound the unequivocal linkage of IV and DV.

Consideration of these factors is essential to an ecologically valid comprehension of psychophysiological response dynamics relevant to experimental manipulations or, more generally, to organism-environment transactions. These notions underscore the importance of viewing complex behavior as an integrated system reaction, rather than as the additive output of component processes. Indeed, reducing a complex system to its constituent elements does not invariably allow for reconstructing that system (Anderson, 1972).

Temporality and Repeated Measures Designs

Repeated observation is required to fully capture adaptive as well as rhythmic processes in organisms. Studies that use repeated measurements yield many benefits, in view of the importance of these time-dependent processes. However, traditional ANOVA approaches to repeated-measures data are fraught with problems that can lead to an increase in the Type I error rate, losses in conceptual and statistical power, and overall inefficiency (Vasey & Thayer, 1987). Importantly, traditional ANOVA is based on the assumption of the independence of observations. However, repeated observations of ongoing activity are frequently correlated with each other often in complex ways. This pattern of dependencies among observations is a source of information that is lost when using traditional ANOVA. The multivariate analysis of variance (MANOVA) approach takes these dependencies into account, but the multiple observations of the same DV are analogous to the multiple DVs discussed previously.

The MANOVA approach to repeated-measures analysis has become standard in many psychophysiological journals (e.g., *Psychophysiology* and *Journal of Psychophysiology*), largely because of its conspicuous advantages in this type of research. First, biobehavioral changes over time begin to reveal complex response patterns to manipulations that are missed in single-occasion measurements. Thus, compensatory and other self-regulatory responses can be observed. In addition, by using the organism as its own control, inter-organism variability that increases error variance and reduces statistical power can be minimized. Finally, the increase in statistical power associated with repeated-measures designs can be utilized more fully by the M ANOVA approach, thereby increasing sensitivity to subtle effects with far fewer subjects required.

Systems Theory and the Study of Individuals

Systems theory seeks principles that are widely applicable across diverse complex systems (Miller, 1978; Schwartz, 1982). The ideal venue for modeling such organismic systems may be the single-subject paradigm (Denenberg, 1982; Friedman, 1998; Goldstein, 1939; Nesselroade & Ford, 1987). A basic advantage of such designs is their sensitivity to temporal patterns in biobehavioral processes. Multivariate, multioccasion, single-subject paradigms have the resolving power necessary to portray patterns of stability and change that characterize organism-environment transactions (Nesselroade, 1991). Replication of these patterns across individuals can bridge specific and general applicability. This spiraling process is therefore congruent with the quest for principles that are relevant at multiple levels of analysis.

It is fundamental to systems theory that basic processes operating at the level of the individual will also be manifested at both lower and higher levels of analysis (Schwartz, 1982). In nonlinear dynamics terminology, these similarities are referred to as *fractals* and occur frequently in nature (Barton, 1994; Bassingthwaighte, 1988; Goldberger, 1992; Losa, Nonnenmacher, & Weibel, 1994; West, 1990). In general, systems approaches have revealed the utility of seeking correspondences across multiple layers of biopsychological inquiry (Friedman & Thayer, 1998a, 1998b; Kandel, 1983; Mandell, Stewart, & Russo, 1981). For example, perception of the relationship between individual and population disease-prevention strategies has been underscored in epidemiological research (Rose, 1992). Clearly, the study of individuals is an integral component in the scientific quest for general laws of behavior: They are complementary parts of a whole (Rosenzweig, 1958).

Another aspect of this area that is of great concern to engineering psychophysiology is the study of individual differences. Biological organisms display wide-ranging individual differences in physiology (Fahrenberg, 1986; Sargent & Weinman, 1966; Woodhead, Blackett, & Hollaender, 1985). The importance of individual differences for engineering psychophysiology can be illustrated by the effects of differences in visual acuity on cardiovascular responses to a computer display (Tyrrell, Pearson, & Thayer, in press).

A thorough exploration of psychophysiological responding necessitates the extensive study of individuals over time, a highly problematic enterprise in large-N designs. Beyond pragmatic concerns, these designs constrain individual response patterns into group molds. To take advantage of the emerging field of dynamical systems, experiments must be designed in such a way as to mine the rich inter- and intra-individual variability inherent in living systems.

SUMMARY AND CONCLUSION

In this chapter, we have attempted to expose assumptions that are often deceptively implicit in the design and analysis of experiments. Engineering psychophysiology, with its focus on person-environment interactions, has a pressing need to explicate those factors that contribute to inter-individual differences and distinguish them from sources of intra-individual variability. The search for associations among IVs and DVs can be expressed as the partitioning of system variability into factors that contribute to this observed variation. Furthermore, although the assumption of linearity has been useful in promoting well-controlled studies of psychophysiological variables, it also represents a limiting influence in the burgeoning study of complexity and dynamical systems. However, designs that can be used to partition data into linear estimates of variance can also be used to investigate the dynamics of person—environment transactions. We hope that this chapter will help researchers in engineering psychophysiology take advantage of contemporary analytic techniques for the study of dynamical systems, as well as serve to aid in the understanding and appropriate use of extant data analytic tools.

REFERENCES

Anderson, P.W. (1972). More is different. *Science, 177,* 393–396.

Barlow, D.H., & Hersen, M. (1984). *Single case experimental designs: Strategies for studying behavioral change.* New York: Pergamon.

Barton, S. (1994). Chaos, self-organization, and psychology. *American Psychologist, 49,* 5–13.

Bassingthwaighte, J.B. (1988). Physiological heterogeneity: Fractals link determinism and randomness in structures and functions. *News in Physiological* Science, 3, 5–9.

Briggs, J., & Peat, F.D. (1989). *Turbulent mirror.* New York: Harper & Row.

Cattell, R.B. (1966). Multivariate behavioral research and the integrative challenge. *Multivariate Behavioral Research, 1,* 4–23.

Cattell, R.B. (1988). The data box: Its ordering of total resources in terms of possible relational systems. In J.R.Nesselroade & R.B.Cattell (Eds.), *Handbook of multivariate experimental psychology* (2nd ed., pp. 69–130). Chicago: Rand McNally.

Cohen, J. (1982). Set correlation as a general multivariate data-analytic method. *Multivariate Behavioral Research, 17,* 301–341.

Cohen, J. (1992). A power primer. *Psychological Bulletin, 112,* 155–159.

Cronbach, L.J. (1957). The two disciplines of scientific psychology. *American Psychologist, 12,* 671–684.

Denenberg, V.H. (1982). Comparative psychology and single-subject research. In D.W. Fiske (Series Ed.) & A.E.Kazdin & A.H.Tuma (Vol. Eds.), *New directions for methodology of social and behavioral sciences: Single-case research designs* (Vol. 13, pp. 19–31). San Francisco: Jossey-Bass.

Dielman, T.E. (1983). Pooled cross-sectional and time series data: A survey of current statistical methodology. *The American Statistician, 37,* 111–122.

Ellsworth, P. (1994). William James and emotion: Is a century of fame worth a century of misunderstanding? *Psychological Review, 101,* 222–229.

Fahrenberg, J. (1986). Psychophysiological individuality: A pattern analytic approach to personality research and psychosomatic medicine. *Advances in Behavior Research and Therapy, 8,* 43–100.

Fisher, R.A. (1942). *The design of experiments* (3rd ed.). Edinburgh: Oliver & Boyd.

Ford, M.E., & Ford, D.H. (Eds.). (1981). *Humans as self-Constructing living systems: Putting the framework to work.* Hillsdale, NJ: Lawrence Erlbaum Associates.

Friedman, B.H. (1998). *An idiodynamic portrayal of cardiovascular reactivity.* Manuscript submitted for publication.

Friedman, B.H., & Thayer, J.F. (1998a). Anxiety and autonomic flexibility: A cardiovascular approach. *Biological Psychology, 47,* 243–263.

Friedman, B.H., & Thayer, J.F. (1998b). Autonomic balance revisited: Panic anxiety and heart rate variability. *Journal of Psychosomatic Research, 44,* 133–151.

Friedman, B.H., Thayer, J.F., & Borkovec, T.D., Tyrrell, R.A., Johnsen, B.H., & Colombo, R. (1993). Autonomic characteristics of nonclinical panic and blood phobia. *Biological Psychiatry, 34,* 298–310.

Glass, L. & Mackey, M.C. (1988). *From clocks to chaos.* Princeton, NJ: Princeton University Press.

Gleick, J. (1987). *Chaos: Making a new science.* New York: Penguin Books.

Globus, G.G., & Arpaia, J.P. (1992). Psychiatry and the new dynamics. *Biological Psychiatry, 35,* 352–364.

Goldberger, A.L. (1992). Applications of chaos to physiology and medicine. In J.H.Kim & J. Stringer (Eds.), *Applied chaos* (pp. 321–331). New York: Wiley.

Goldbeter, A., & Decroly, O. (1983). Temporal self-organization in biological systems: Periodic behavior vs. chaos. *American Journal of Physiology, 245,* R478–R483.

Goldstein, K. (1939). *The organism.* New York: American Book Co.

Goodman, S.N. (1992). A comment on replication, P-values, and evidence. *Statistics in Medicine, 11,* 875–879.

Guttman, L. (1985). The illogic of statistical inference for cumulative science. *Applied Stochastic Models and Analysis, 1,* 3–10.

Guzzo, R.A., Jackson, S.E., & Katzell, R.A. (1987). Meta-analysis analysis. *Research in Organizational Behavior, 9,* 407–442.

Harris, R.J. (1992). Multivariate statistics: When will experimental psychology catch up? In S.Koch & D.Leary (Eds.), *A century of psychology as a science.* Washington, DC: American Psychological Association.

Harris, R.J. (Ed.). (1997). Special section: Ban the significance test? *Psychological Science, 8,* 1–20.

Hrushesky, W.J.M. (1994, July/August). Timing is everything. *The Sciences, 34,* 32–37.

Jacob, R.G., Thayer, J.F., Manuck, S., Muldoon, M., Tamres, L., Williams, D., Ding, Y., & Gatsonis, C. (in press). Ambulatory blood pressure responses and the circumplex model of mood: A four-day study. *Psychosomatic Medicine.*

Jacoby, W.G. (1991). *Data theory and dimensional analysis.* Newbury Park, CA: Sage.

James, W. (1884). What is an emotion? *Mind, 9,* 188–205.

Kandel, E.R. (1983). From metapsychology to molecular biology: Explorations into the nature of anxiety. *American Journal of Psychiatry, 140,* 1277–1293.

Kirschbaum, C., Steyer, R., Eid, M., Patalla, U., Schwenkmezger, P., & Hellhammer, D.H. (1990). Cortisol and behavior: Vol. 2. Application of a latent state-trait model to salivary cortisol. *Psychoneuroendocrinology, 15,* 297–307.

Lambert, Z.V., Wildt, A.R., & Durand, R.M. (1988). Redundancy analysis: An alternative to canonical correlation and multivariate multiple regression in exploring interset associations. *Psychological Bulletin, 104,* 282–289.

Leproult, R., Copinschi, G., Buxton, O., & Van Cauter, E.V. (1998). Sleep loss results in an elevation of cortisol levels the next evening. *Sleep, 20,* 865–870.

Levenson, R.W. (1988). Emotion and the autonomic nervous system: A prospectus for research on autonomic specificity. In H.L.Wagner (Ed.), *Social psychophysiology and emotion: Theory and clinical applications* (pp. 17–42). Chichester, England: Wiley.

Losa, F.A., Nonnenmacher, T.F., & Weibel, E.R. (Eds.). (1994). *Fractals in biology and medicine.* Cambridge, MA: Birkhauser-Verlag.

Mandell, A.J., & Selz, K.A. (1992). Dynamical systems in psychiatry: Now what? *Biological Psychiatry, 31,* 299–301.

Mandell, A.J., Stewart, K.D., & Russo, P.V. (1981). The Sunday syndrome: From kinetics to altered consciousness. *Federation Proceedings, 40,* 2693–2698.

Manuck, S. (1994). Cardiovascular reactivity in cardiovascular disease: "Once more unto the breach." *International Journal of Behavioral Medicine, 1,* 4–31.

Meehl, P.E. (1978). Theoretical risks and tabular asterisks: Sir Karl, Sir Ronald, and the slow progress of soft psychology. *Journal of Consulting and Clinical Psychology, 46,* 806–834.

Miller, J.G. (1978). *Living systems.* New York: McGraw-Hill.

Nesselroade, J.R. (1988). Sampling and generalizability: Adult development and aging research issues examined within the general methodological framework of selection. In K. W.Schaie, R.T.Campbell, W.Meredith, & S.C.Rawlings (Eds.), *Methodological issues in aging research* (pp. 13–42). New York: Springer.

Nesselroade, J.R. (1991). Interindividual differences in intrainidividual change. In L.M. Collins & J.L.Best (Eds.), *Best methods for the analysis of change* (pp. 92–105). Washington, DC: American Psychological Association.

Nesselroade, J.R., & Ford, D.H. (1987). Methodological considerations in modeling living systems. In M.E.Ford & D.H.Ford (Eds.), *Humans as self-constructing living systems: Putting the framework to work* (pp. 47–79). Hillsdale, NJ: Lawrence Erlbaum Associates.

Nesselroade, J.R., & Jones, C.J. (1991). Multi-modal selection effects in the study of adult development: A perspective on multivariate, replicated, single-subject, repeated measures designs. *Experimental Aging Research, 17,* 21–27.

O'Connor, K. (1990). Towards a process paradigm in psychophysiology. *International Journal of Psychophysiology, 9,* 209–223.

O'Connor, K. (1992). Design and analysis in individual difference research. In A.Gale & M. W.Eysenck (Eds.), *Handbook of individual differences: Biological perspectives* (pp. 45–78). Chichester, England: Wiley.

Pepper, W.C. (1942). *World hypotheses.* Berkeley: University of California Press.

Peters, M.L., Godaert, G.L., Ballieux, R.E., van Vliet, M., Willemsen, J.J., Sweep, F.C., & Heijnen, C.J. (1998). Cardiovascular and endocrine responses to experimental stress: Effects of mental effort and controllability. *Psychoneuroendocrinology, 23,* 1–17.

Rose, G. (1992). Strategies of prevention: The individual and the population. In M.Marmot & P.Elliot (Eds.), *Coronary heart disease epidemiology* (pp. 311–324). New York: Oxford University Press.

Rosenthal, R., & Rosnow, R.L. (1984). *Essentials of behavioral research: Methods and data analysis.* New York: McGraw-Hill.

Rosenzweig, S. (1958). The role of the individual and of idiodynamics in psychology: A dialogue. *Journal of Individual Psychology, 14,* 3–20.

Sargent, F., & Weinman, K.P. (1966). Physiological individuality. *Annals of the New York Academy of Sciences, 134,* 696–719.

Saul, J.P. (1990). Beat-to-beat variations of heart rate reflect modulation of cardiac autonomic outflow. *News in Physiological Science, 5,* 32–37.

Schwartz, G.E. (1982). Cardiovascular psychophysiology: A systems perspective. In J.T.

Cacioppo & R.E.Petty (Eds.), *Perspectives in cardiovascular psychophysiology* (pp. 347–372). New York: Guilford.

Schwartz, G.E. (1986). Emotion and psychophysiological organization: A systems approach. In M.G.H.Coles, E.Donchin, & S.W.Porges (Eds.), *Psychophysiology: Systems, processes, and applications* (pp. 354–377). New York: Guilford.

Schwartz, J.E., Warren, K., & Pickering, T.G. (1994). Mood, location, and physical position as predictors of ambulatory blood pressure and heart rate: Application of a multi-level random effects model. *Annals of Behavioral Medicine, 16,* 210–220.

Steyer, R., Ferring, D., & Schmitt, M.J. (1992). States and traits in psychological assessment. *European Journal of Personality Assessment, 8,* 79–98.

Tarui, H., & Nakamura, A. (1987). Salivary cortisol: A good indicator of acceleration stress. *Aviation, Space, and Environmental Medicine, 58,* 573–575.

Thayer, J.F. (1988, March). *Concordance of self-report and physiological responses to emotion.* Paper presented at Emotions and Attention, University of Florida, Gainsville, FL.

Thayer, J.F., & Faith, M. (1994). Idiographic nonlinear pattern classification of autonomic and self-report measures of emotion [Abstract]. *Psychosomatic Medicine, 56,* 178.

Tyrrell, R.A., Pearson, M.A., & Thayer, J.F. (in press). Behavioral links between the oculomotor and cardiovascular systems. In O.Franzen & H.Richter (Eds.), *Accommodation/vergence mechanisms in the visual system.* Stockholm, Sweden: Birkhauser Ferlag.

Tyrrell, R.A., Thayer, J.F., Friedman, B.H., Leibowitz, H.W., & Francis, E.L. (1995). A behavioral link between the oculomotor and cardiovascular systems. *Integrative Physiological and Biological Science, 30,* 46–67.

Vasey, M.W., & Thayer, J.F. (1987). The continuing problem of false positives in repeated measures ANOVA in psychophysiology: A multivariate solution. *Psychophysiology, 24,* 479–486.

West, B.J. (1990). *Fractal physiology and chaos in medicine.* Teaneck, NJ: World Scientific.

Wood, P., & Brown, D. (1994). The study of intraindividual differences by means of dynamic factor models: Rationale, implementation, and interpretation. *Psychological Bulletin, 116,* 166–186.

Woodhead, A.D., Blackett, A.D., & Hollaender, A. (Eds.). (1985). *Molecular biology of aging.* New York: Plenum.

Chapter 4

Signal Processing and Analysis in Application

Holger Luczak
Matthias Göbel
Aachen University of Technology, Germany

ENGINEERING PSYCHOPHYSIOLOGY AND THE ROLE OF ENGINEERS

Originally, measures of psychophysiological interest were intended to clarify elementary human performance characteristics and behavior structures (e.g., Donders, 1869/1968; von Helmholtz, Gullstrand, Kries, & Nagel, 1909–1911). With the availability of more sophisticated technical equipment, focus was expanded to analyze endogenous processes by externally accessible measures, obtaining more detailed knowledge about the principles of human actions and reactions. A large ensemble of measures and methods have now been developed, both for continued elementary studies on the methodological side and for application-oriented objective workload measurements.

Considering the widespread literature on psychophysiological contributions, two interesting dilemmas may be stated:

1. On the methodological side, discussions of appropriate methods and interpretations of psychophysiological measures have continued for several decades. The large variety of work and working conditions to be studied requires extensive research, and the sometimes apparently divergent or even contradictory results need clarification by further research.

2. On the application side, a large number of basic studies performed in the laboratory stand against a very small number of design-oriented and field studies

that use psychophysiological analysis. A "scientifically grounded" design of work processes and workplaces would benefit significantly from objective and detailed information about human actions as well as human stress-strain processes (in addition to simulation models and data tables). However, engineers and designers often decide against the use of psychophysiological methods in engineering for several reasons: Their analysis appears rather complex, their interpretation seems nontheoretical, and statistically insignificant results are not uncommon.

Both of these situations are caused by the complex nature of signal recording and its indirect decomposition according to physiological and behavioral functions and by the large variability caused by individual and situational factors. During laboratory studies, simplified and strictly controlled experimental conditions help in obtaining reliable results. Systematic variation of task conditions allows the individual study of single effects, whether driven by a theoretical framework or not. In contrast, engineering and design usually focus on real work tasks of considerable complexity where nonuniform conditions occur almost as a matter of principle. In this case, a conceptual framework is required to bring together the numerous variables of the task itself (i.e., task components), the performance and environmental conditions, an individual's constitution and disposition, and psychophysiological signal recording and processing (see Fig.4.1).

Engineers might contribute to psychophysiology through the analytical and systematic approach of their discipline. From an engineer's point of view, the study of human reactions as well as any kind of signal processing, should be supported by quantitative modeling or mathematical equations. Such a statement can be easily accepted for signal processing (if it is performed), but it appears to be somewhat illusory or even inappropriate for human reactions. However, from a fundamental point of view, this is not true: Any kind of human behavior is based on some type of rule. Considering physiological reactions, basic behavior characteristics originate from the organic system and its neuronal control. Hence,

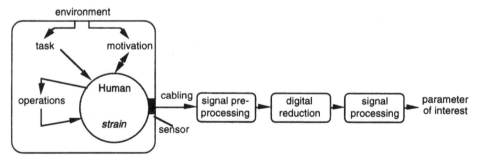

FIG. 4.1. Simplified structure of emerging psychophysiological measures.

such reactions correspond to the functional structure of the organic system and should be interpreted accordingly.

The necessity of a systematic operational procedure for psychophysiological analysis becomes obvious when considering the large number of variables that contribute to a measurement result, from the origin of the signal to a final statistical evaluation. Along the way, transformations, distortions, crossover effects, and other influences occur during signal recording and signal processing in combination with the external influences of task conditions and individual behavior.

This chapter is intended to provide a background for the practical application of psychophysiological measures based on the analytical methods of the engineering discipline, starting from signal recording by sensors up to the analysis and interpretation approaches for complex tasks and long-term studies. This approach may help bring together the results of different studies with possibly divergent outcomes and is a key to mastering the complexity of design-oriented approaches and field studies.

SIGNAL RECORDING AND PROCESSING

Psychophysiological reactions are measured by sensors placed on the body surface or by sensors recording body movements. The external recording of psychophysiological measures causes noise, distortions, and other artifacts that may obscure the signal information. Furthermore, signal amplification and processing can induce additional errors. Hence, the attributes of the recorded signals largely determine any further kind of processing and interpretation. Therefore, a set of criteria is presented to evaluate the suitability and quality of the processed signals.

Precision and Distortion of Raw Signals

A precise data set is required to reproduce even small-intensity variations in the physiological system being monitored. Available equipment permits signal amplification, digital sampling and data storage without significant loss of information, but noise and artifacts may still occur, primarily on the sensor side. Consider the signals from surface electrodes, for example, where noise already occurs in the body (e.g., on the way from the heart muscles to the skin and, significantly, at the contact areas between the skin and the electrode). Hence, electrode material and electrolytes effect noise level (Zipp, 1982, 1988). Electrically, the electrode represents a signal source with high impedance (up to 0.5 MOhm; Rau, 1973; Silny & Rau, 1977) and low output amplitude (down to several micro-volts; Rohmert & Luczak, 1973). Although the sensitivity and input impedance of the amplifiers makes it possible to process signals without significant loss of information, the cables and connectors can still represent weak points in terms of signal quality. Unshielded cables are very sensitive to electromagnetic emissions. Even in a cable 10 cm in length, radiations from AC power and electronic devices can induce significant potentials. Shielded cables, on the other hand, cause a capacitive load (usually 5 to 100 pF/m) that may lead to an attenuation of high-frequency signal components. To minimize distortions caused by cables and connectors, as well as electromagnetic emissions, active sensors with a first amplification device close to or directly within the sensor have been developed (e.g., Göbel, 1996a; Hagemann, Luhede, & Luczak, 1985). Figure 4.2 shows an active EMG surface electrode and the possibility to detect singular activation spikes.

Distortions inherent to the sensor reduce the available processing resolution, similar to an independent signal source fed into the sensor. However, noise and technical artifacts have distinguishing characteristics. Noise is of a stochastic nature with a total amplitude that increases almost proportionally to signal bandwidth. For this reason, appropriate low-pass filtering that only reduces the processed signal bandwidth to the minimum required range can provide a useful method for noise minimization, as is detailed later. Artifacts, on the other hand, consist of mostly systematic components, such as crossover components and motion artifacts on the sensor side. In the case of different frequencies of artifacts and signal information, specific low- and high-pass filtering or notch filters can provide attenuation of artifacts. Figure 4.3 shows the effect of a highly selective Fourier notch filter to eliminate distortions of 50 Hz AC power supply.

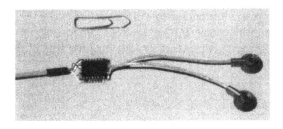

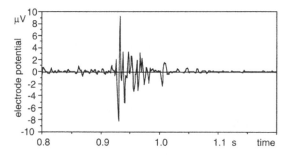

FIG. 4.2. "Active electrodes" enable measurement of very small potentials (example from an EMG recording of m. flexor carpi ulnaris).

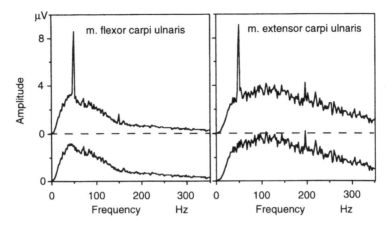

FIG. 4.3. Signal spectra of two EMGs before and after 50-Hz-notch filtering using a digital Fourier filter.

Alternatively, specific correlation filters may help to distinguish signals of different characteristics within the same frequency range, but this method requires careful adjustment to work properly (Cohen, 1986).

Data Sampling

Complementary to amplitude resolution, temporal resolution must also be considered. This is relevant for both the raw sensor signal and the processed parameters. At first, a suitable sensor signal bandwidth is required so that variations of signal amplitude can be processed correctly. For example, the accuracy of heart rate measurements based on electrocardiogram (ECG) results depends directly on the temporal resolution of the ECG signal for a precise identification of the Q-R peak. A low sampling rate of 50 Hz in this case would, at best, lead to a detection of a heart beat between 10 ms before and 10 ms after the real beat. Consequently, a real heart rate of a constant 60 bpm would appear as a stochastically varied frequency between 58.8 and 61.2 bpm (depending on the random synchronization of heart beat and data sampling). Although the mean heart rate would be 60 bpm, the processing of heart rate variability or other beat-to-beat interval analysis parameters would be seriously disturbed. The size of this type of error decreases linearly to an increase in sampling frequency; therefore, fast sampling rates should be applied in the case of high accuracy demands.

Similarly, estimating eye-movement fixations requires high-speed data sampling if spatially close fixations must be distinguished. Otherwise, short saccades would be artificially lengthened and, thus, computed saccade speed would be reduced by signal incompleteness (see Fig. 4.4). In general, the temporal resolution of the sensor signal directly determines parameter accuracy if any time criterion is used for parameterization.

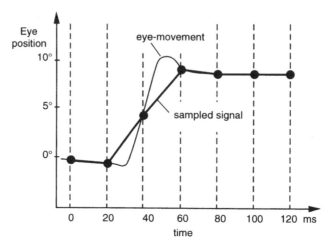

FIG. 4.4. Effect of sampling fast eye movements by low sampling frequency (50 Hz, dotted vertical lines: sampling times).

If digital data sampling is employed, any signal components of a frequency higher than 50% of the sampling rate must be eliminated. Otherwise, high frequency components would be transformed into the sampled signal bandwidth and thus cause significant distortions (Shannon & Weaver, 1949). Using the example of Fig. 4.4 for explanation, low frequency components occur in the sampled signal during the saccade phase that are of a higher frequency in the original eye movement (effect of "mirrored" signal components within the frequency domain). All subsequent processing would indicate a false (i.e., alias) eye movement frequency or speed, respectively.

Any limitation of resolution for data parameters must be considered a noise component that prevents signal variations below the noise level from being distinguished. Again using Fig. 4.4 for explanation, a number of such saccades shall be considered that follow a similar function. Obviously, sampling would happen at different points relative to the function for each saccade. Depending on the "synchronization" between saccade occurrence and data sampling, the data samples would indicate different eye movement speeds in spite of an identical movement function.

Distortions During Signal Processing

Low-pass filtering reduces noise in raw signals and may also reduce noise in data parameters, but at the cost of temporal resolution. Continuing the example of heart rate, the estimation of beat-to-beat intervals contains all distortions of the Q-R peak discrimination. If heart rate is to be estimated for a longer period of time (e.g., 1 min), averaging all beat-to-beat intervals within this period (similar to a low-pass-filter of the first order) reduces the effects of stochastic and periodic distortions relative to the period of time averaged. Obviously, no signal changes within this period may be extracted anymore. Using overlapped averaging intervals seems to produce an apparent noise reduction without loss of temporal resolution. However, it should be noted that even though the amount of data may be increased by signal processing, the amount of information within data may not be increased as a matter of principle. Consequently, no elementary improvement of signal quality may be achieved this way.

A similar effect occurs during data processing. Any signal processing of raw data for deriving psychophysiological parameters is based on the recomputation of multiple input data samples for each output data sample (e.g., the time line of eye movements for fixation and saccade discrimination). The more input samples that contribute to output data, the less changes in one input sample (e.g., due to noise) will affect output data. This is of particular importance if singular peaks (e.g., maximum and minimum values) are extracted from a large number of samples. Because only one or two samples of input data determine the output measure in this case, such a method is highly sensitive to any distortion of data. A significant smoothing of output data can be achieved by the consideration of multiple input samples, for example, if the 90th percentile is used instead of the maximum peak, which represents a single output sample.

Another source of distortions may occur during signal processing and parameterization due to interference generated by nonlinear data processing. An example of EMG processing is presented to explain such complex phenomena (see Göbel, 1996a, for details).

EMG signals originate from cellular depolarization pulses within human motor units. Some 10 to 100 motor units in each muscle produce an interference pattern that arrives at the surface electrodes. Although specific electrode and signal processing designs can permit analysis of motor units in a detailed fashion (Reucher, 1988), EMG is mostly employed for the analysis of total muscle activity. The number of active motor units and their firing frequency corresponds physiologically to muscle strength and to the total number of electrical spikes. Therefore, muscle strength is measured by integrating the incoming electrical potentials of the surface electrodes. Using a common procedure, the bipolar signal is rectified after amplification and then integrated (see Fig. 4.5).

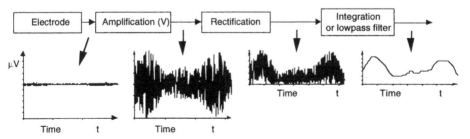

FIG. 4.5. Standard signal processing for integrated EMG (according to Laurig, 1970).

Integration (or low-pass filtering with similar effects) is required to smooth firing spikes that do not seem to contribute to information due to their quasi-Stochastic nature. Increased integration time eliminates such noise, but at the cost of temporal resolution, as described previously. For higher temporal resolution (and correspondingly shorter integration times), a significant amount of noise due to the remaining spikes seems to be inevitable. Additionally, a high-pass filter is included for the raw signal to attenuate motion artifacts of electrodes and cables (Zipp, 1988). Questions arise about how to set the filter and integration time, as well as about how to distinguish information and noise to optimize information content. In fact, a detailed signal analysis of even a simple processing procedure (see Fig. 4.5) demonstrates that only a minor part of noise occurs in the final signal due to the spikes of the EMG pattern.

The raw EMG signal represents an amplitude-modulated signal with a stochastic carrier. The rectification procedure then extracts the amplitude function equivalent to the demodulation of amplitude-modulated radio signals. Using a sinusoidal carrier function of higher frequency, the carrier signal frequency is doubled during rectification, and the frequency range of the rectified carrier components differs from the frequency range of the intensity function (see Fig. 4.6). Simple low-pass filtering then permits extraction of the amplitude function only. In contrast to technical applications of amplitude-modulated signals, the carrier signal of the EMG is of a stochastic nature. The same demodulation process may be used for such types of signals as well, but in this case additional interference components of a lower frequency than the carrier components occurs during rectification. Hence, relevant information and interference components may not be separated completely by the subsequent low-pass filter. Such interference components are of a stochastic nature similar to the carrier signal itself, appearing as a source of noise (Luczak & Göbel, 1996). The characteristics of such "artificial" noise may be computed (see Equation 1, Fig. 4.7) and depend on filter functions for the raw signal as well as for the rectified signal. It can be shown that 50% to 90% of the energy of stochastic components within the output signal results from the rectification process and are not part of the original intensity function of EMG (see Fig. 4.8).

$$s_{SD}(f_{Out}) = \frac{\sqrt{\pi+1}}{\sqrt{3}} \bullet \sqrt{\sum_{h=1}^{\infty} \left(\left(\frac{1}{(2h-1)\bullet(2h+1)}\right)^2 \bullet \frac{C+D}{2\bullet\int\limits_{-\infty}^{\infty} s_{Cr}^2(f)\bullet F_{In}^2(f)df}\right)} \tag{1}$$

$$C = \int\limits_{-\infty}^{\infty}\left[s_{Cr}^2(f)\bullet F_{In}^2(f)\bullet s_{Cr}^2\left(\frac{f_{Out}}{h}-f\right)\bullet F_{In}^2\left(\frac{f_{Out}}{h}-f\right)\right]df$$

$$D = \int\limits_{-\infty}^{\infty}\left[s_{Cr}{}^{2}(f) \bullet F_{In}{}^{2}(f) \bullet s_{Cr}{}^{2}\left(\frac{f_{Out}}{h} + f \right) \bullet F_{In}{}^{2}\left(\frac{f_{Out}}{h} + f \right) \right] df$$

s_{SD} *(fOut)* Amplitude of output noise as a function of frequency
s_{Cr} *(f)* Spectral function of motor unit spike pattern
F_{In} *(f)* Filter function of raw signal (including tissue and electrode)

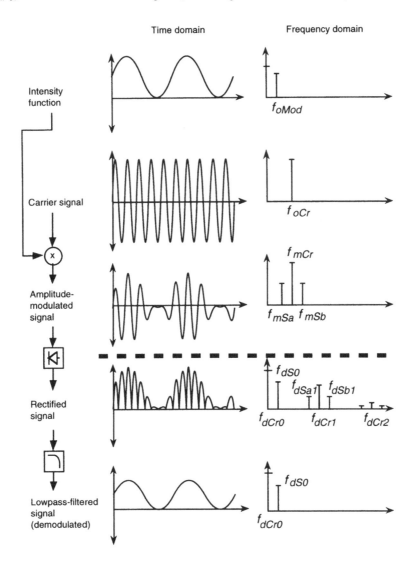

FIG. 4.6. Amplitude modulation and demodulation of sinusoidal signals.

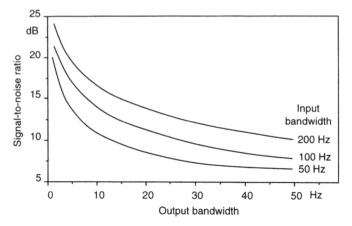

FIG. 4.7. Signal-to-noise ratio as a function of input and output bandwidth, computed by Equation 1 (accuracy $< \pm 1$ dB compared to measurement of simulated raw EMG signals processed by the integration method).

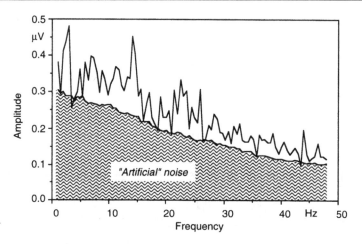

FIG. 4.8. Frequency spectrum of the integrated EMG (m. deltoideus during a position control task) with computed spectrum of interference components caused by the rectification process.

Recognizing the origins of this type of noise, at a minimum permits the optimization of the signal processing procedure according to the measurement purposes at hand, and it also permits estimation of the quality of the information derived. Figure 4.9 shows the effect of adaptive filtering (according to Equation 1) for an artificially generated raw EMG with an intensity function consisting of two sinusoidal components of 1 Hz and 10 Hz.

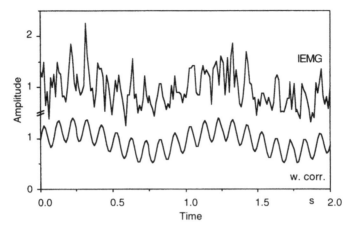

FIG. 4.9. Output signal of a simulated raw EMG (with sinusoidal intensity function) using the integration method and after an adaptive Fourier filtering according to Equation 1.

In conclusion, the detailed application of signal theory complicates analysis, but it is of significant relevance for data quality.

QUANTITATIVE MODELING
OF PSYCHOPHYSIOLOGICAL RESPONSES

The indirect nature of derived psychophysiological parameters prevents a straightforward interpretation concerning the functional aspects of the human organic system. Adequate functional modeling may bridge this gap for a more aggregated interpretation (e.g., in terms of workload). Two typical examples of engineering psychophysiology, the estimation of mental workload by heart rate variability analysis and the analysis of sensory feedback in movement control, are used to explain how adequate functional modeling helps transform variations of physiological measures into psychophysiological scaling.

The Case of Arrhythmia/HRV Depression
by Mental Workload

Psychophysiological responses are frequently discussed on the basis of "pure phenomenology." This means that a relation between input and output variables is described by a statistical approach without deeper insight into the underlying physiological structures and neural functions (Luczak & Laurig, 1973). This evaluation applies to most studies on arrhythmia or heart rate variability (HRV) depression under mental workload. A way to move from a purely experimental and correlational approach to a more causal view of the phenomenon can be

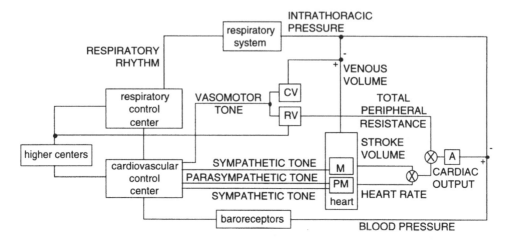

FIG. 4.10. Qualitative model by Mulder (1980).

provided by qualitative models. This is done by splitting the "black box" (between input and output) into system elements and relations in the form of a qualitative system description. For arrhythmia depression, this approach seems to be useful in order to generate hypotheses about the cause of the phenomenon, such as the baroreflex-sensitivity hypothesis as formulated by Mulder (1980, see Fig. 4.10). In conjunction with results from spectral analysis about frequency bands in the HRV spectrum and their assignment to different control loops, Mulder hypothesized that arrhythmia depression is a consequence of an "emotional responsive pattern." During the defense reaction, the sensitivity of the baroreflex in blood pressure control may be reduced and may be regarded as a reason for arrhythmia or HRV depression during mental workload.

The Mulder hypothesis is oversimplified and vague compared to quantitative HRV or arrhythmia modeling (Luczak, 1975; Luczak & Phillip, 1980; Luczak & Raschke, 1975). Taking one of the latest variants of this model as published by Ohsuga, Terashita, Shimono, and Toda (1995) and its use for the explanation of HRV or arrhythmia phenomena, a variety of experimental effects can be explained (see Fig. 4.11).

The responses to mental workload in this model depend, except for the reduction of baroreflex-sensitivity, on the interference of respiration and blood pressure control on feedback gains, sympatic and vagal tone, and so on. "More than hundred combinations of their values were tested and the results showed various patterns of changes in HR, MBP and their RF and MF's [mid frequency component of HRV] that explained the divergence of experimental observations" (Ohsuga et al., 1995, p. 774).

By using such a quantitative model, the "nature of tasks" (Luczak & Rohmert, 1997) can be brought together in terms of a different physiological cost

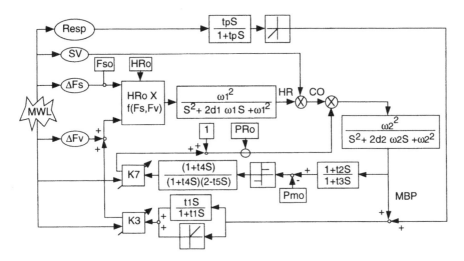

MWL   = Mental workload
Resp   = Respiration volume/ RF = respiratory frequency
SV   = Estimate of stroke volume
HR   = Instantaneous heart rate
MBP   = Mean arterial blood pressure
K3/K7 = Feedback gains
ΔFs   = Symphatic tone
ΔFv   = Vagal tone

FIG. 4.11. Block diagram of the revised model from Luczak (Ohsuga et al., 1995).

structure, the HRV or arrhythmia response, and a differentiation of physiological implications and reactions in the relevant systems.

Analysis of Sensory Feedback in Movement Control

An analogous variant of quantitative modeling based on less precise structural information is the analysis of physiological feedback loops in human motor control by noninvasive measures.

Any kind of motor action requires sensory feedback information for successful execution. Many debates have occurred regarding the question of whether movements are generated with or without continuous feedback (Abernethy & Sparrow, 1992; Laszlo, 1992). However, there is agreement on the fact that feedback is required in the case of precise movements. Obviously, this is true for visual control, but proprioceptive and haptic senses are also involved. The meaning of such multimodal feedback may be illustrated by the example of steering control during fast driving: Control performance is worse if power assistance isolates

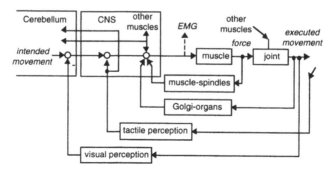

FIG. 4.12. Simplified structure of the role of sensory feedback in motor control.

force feedback to the driver compared to the case of a less intensive assistance with perceptible force feedback. However, measuring the feedback sources used for movement control is feasible in a direct way only for eye movements, and, even for this modality, the estimation of how visual perception is employed as a source of feedback remains vague. External performance measures in general do not provide information about the different senses involved in movement control.

The structure of movement control is a cascaded feedback control circuit (multiple feedback loops within another) shown in a simplified form in Fig. 4.12. In fact, this system is much more complex. All perceptions, except for the visual modality, have a multivariate origin due to the numerous sensory elements within the human body. A suitable ensemble of muscles needs to be orchestrated and controlled, even if only a simple finger movement to a target point is performed. Furthermore, transmission characteristics and feedback signal processing are not necessarily of a linear nature. For instance, negative feedback may be considered only if deviations exceed a tolerable limit. Altogether, this prevents a complete analytical description. For functional modeling, only the cascaded structure of feedback loops (see Fig. 4.12), and physiologically caused processing and transmission delays can be stated.

Information about the motor processes can be measured noninvasively only by the output information of the motor system using motion analysis or EMG. Thus, the activation patterns of each muscle have to be regarded as the most detailed sources of information available. From this motor output, signal components have to be extracted that do not rely on the movement generation itself, but on movement corrections caused by sensory feedback.

Due to reaction delays as a structural attribute of human information transmission and processing, any attempt to correct deviations becomes effective with a delay and any subsequent corrections are delayed again. Consequently, deviation correction causes at least slight oscillations if negative feedback is

processed delayed. The frequency of oscillations corresponds directly to the delay
within each information feedback loop (see Equation 1, Fig. 4.13).

$$F_{osc} = \frac{1}{2 \bullet t_{del}}$$ (2)

F_{OSC} Oscillation frequency
t_{del} Transmission delay

Increased amplification of the feedback loop improves precision of control
but causes increased amplitude of oscillations. If stability limits are exceeded,
permanent oscillation will result. The occurrence of such oscillatory components
must be considered a matter of fact due to the severely restricted control quality
of feedback systems with internal delay. However, such oscillations do not affect
external movements significantly because the low-pass characteristic of mass
inertia leads to a strong attenuation of oscillations. Feedback control remains
effective even if operated close to stability limits.

The oscillation components are integral parts of the motor output signal and,
thus, may be used for analysis. Oscillation frequency is mainly determined by
the transmission delay. Oscillation amplitude depends on the loop amplification
(the influence of feedback), which again determines control precision. Other
feedback parameters (e.g., non-linearities, elements with integral or differential
characteristics, etc.) do not significantly affect oscillations. In the case of cascaded
feedback, no interactions between the different control loops occur in terms
of oscillations.

The oscillation frequencies can be estimated, according to Equation 2, based
on the reaction delays for the main feedback loops in human motor control (see
Table 4.1). According to this functional model, the higher frequency components
of the integrated EMG should contain information about the effects of feedback.
The components of the integrated EMG that relate directly to the external
movement must be of a lower frequency than the oscillations if feedback control
is to be effective (usually below 2 Hz). The part of total signal energy in each of

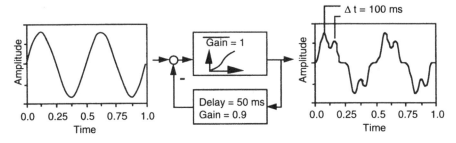

FIG. 4.13. Response of a simulated negative feedback control system with internal delay.

TABLE 4.1

Reaction Time and Oscillation Frequency of Sensory Feedback Loops

Feedback Mechanism	Reaction Time	Oscillation Frequency
Visual perception	150–230 ms	2.17–3.3 Hz
Haptic perception	80–150 ms	3.3–6.25 Hz
Polysynaptic reflexes	45–80 ms	6.25–11 Hz
Monosynaptic reflexes	15–45 ms	11–33 Hz

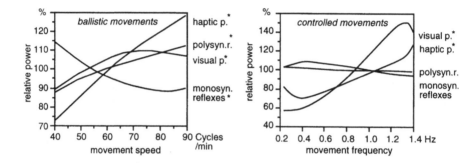

FIG. 4.14. Feedback intensity as a function of movement speed for the sensory feedback loops according to Table 1 (mean values of 10 subjects and 8 hand and arm muscles; *: 30 to 80% of total variance explained).

the four frequency ranges mentioned in Table 4.1 can be computed as measures of the degree of feedback utilization for the different sensory modalities.

An experimental investigation based on this method for analysis (Göbel, 1996a, 1996b) demonstrates significant effects of work pace during ballistic movements (repeated peg-hole fitting; see Fig. 4.14, left side) as well as during continuous tracking (see Fig. 4.13, right side).

A study of learning effects during the execution of a manual assembly task (20 hours of work on 3 subsequent days) points out the development in the usage of sensory feedback as shown in Fig. 4.15. During the first hours of work (when subjects had little practice), the execution speed was significantly slowed by monosynaptic reflex activity. This is also true for polysynaptic reflexes in a later phase of work. Although reflex activity no longer affected work performance when subjects were experienced, visual perception became relevant after approximately 12 hours of practice. Haptic feedback affected working pace, again in a delayed fashion, with the effect occurring after having worked for 16 hours.

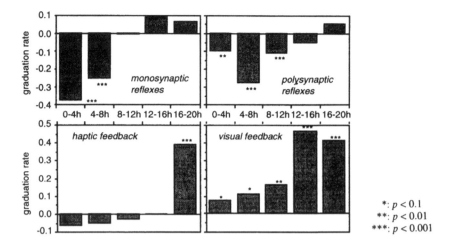

FIG. 4.15. Correlation between feedback and execution speed during the learning phase of a manual assembly task (mean values of 10 subjects and 8 hand and arm muscles).

A question still remaining is the interpretation of feedback in terms of workload. Although the results shown in Fig. 4.14 and Fig. 4.15 indicate a strong influence of feedback is associated with high degrees of workload, the reverse relationship should not be assumed to be evident.

PSYCHOPHYSIOLOGICAL ANALYSIS OF COMPLEX TASKS

In addition to suitable signal recording and processing as discussed earlier, the measured psychophysiological parameters need to be associated with the external conditions (e.g., the types of tasks or task elements performed, working time, etc.).

During laboratory studies, effects of interest are usually studied in an isolated way. This means that experimental tasks are often simplified to the points of interest, and performed with controlled and, if necessary, systematically varied conditions. Thus, a relation between external conditions and psychophysiological measures can be evaluated directly. In contrast, real work tasks and field studies involve considerable complexity. A complex task is at first characterized by numerous task elements performed sequentially and, possibly, simultaneously. In most cases, subjects are, at least to a certain degree, autonomous to process a task in an individual manner and with individually varied pace. Obviously, the manner of task execution may have an effect on psychophysiological reactions and, consequently, psychophysiological measures. Even more, situation-dependent action planning may be part of the task itself, resulting in additional effects on psychophysiological reactions. Examples for this are urban bus driving or drawing with a CAD-system (Göbel, Springer, & Scherff, 1998; Springer, Langner, Luczak, & Beitz, 1991).

This section is devoted to two important issues concerning the psychophysiological analysis of complex tasks: (a) the decomposition of complex tasks in order to specify how workload is related to different task factors and (b) a conceptual framework to explain interactions between task, work strain, and performance.

Assignment of Psychophysiological Measures to the Components of Complex Tasks

Workload analysis of complex tasks usually starts with the question of how the various stressor variables (e.g., task elements, working conditions, etc.) contribute to work strain. An isolated execution and analysis of all stressor variables is usually neither reasonable (due to interactions with other stressor variables and the working context) nor feasible (due to the large number of settings required). Thus, the psychophysiological measure[1] obtained during the execution of a complex task needs to be decomposed according to the different stressor variables (Luczak, 1983; Luczak et al., 1986). A source of variation (as the basis for decomposition) can be obtained by the time line of the psychophysiological measure that reflects, at least partially, the different composition of stressor variables during different periods of time.

The approach proposed is based on a compilation of all factors in the form of a polynomial equation (Equation 3). Using the time samples of the psychophysiological measure and of the stressor variables as sources of variation, the coefficients were computed by a polynomial regression:

$$pm(t)=c0+c1 \cdot str1(t)+c2 \cdot str2(t)\ldots+cn \cdot strn(t) \qquad (3)$$

pm Psychophysiological measure
$str_1 \ldots str_n$ Stressor variables

For application, some restrictions of this approach and possible refinements need to be mentioned.

Data Recording. The stressor variables fed into Equation 3 should not only consider the variables of interest, but, as far as possible, all relevant stressor variables (meaning all factors that contribute to work strain). A first set of stressor variables is related to the actions performed or task elements. Data can be acquired by videoanalysis, direct observation, keystroke or event protocols,

---

[1]   In the following subsection, the singular form "psychophysiological measure" is used for its plural form, too.

specific action measures (e.g., motion analysis and eye movement analysis[2]), and so on. Recordings of environmental conditions or specific work conditions should be included if conditions vary during the observation period and if this variation is assumed to affect work strain (e.g., traffic density and vehicle speed for the analysis of driving tasks). Variables might be scaled continuously (e.g., movement speed) or dichotonously (true or false). It should be noted that not all stressor variables may be observed objectively, but require the use of questionnaire or interview methods (e.g., subjective calculation and decision processes and self-induced time-stress).

The degree of detail needed to describe the stressor variables depends on the aspects of interest and the number of data samples and measures available. It might range from a rough description of different task sequences (e.g., functionality check, starting procedure, position control, and landing phase during aircraft flight) up to a sophisticated action and situation analysis (e.g., observation of different displays, decision phases, and movements to controls).

Figure 4.16 shows the results of such a decomposition for urban bus driving according to the tasks performed (Göbel et al., 1998, additional data processing as explained later is included). This example further illustrates the consequences of a missing environmental variable: The extraordinarily high mental work strain while activating the wiper is obviously not caused by the process of pushing the button, but by the mental effort required to look through wet windows while at the same time activating the wiper.

Temporal Fluctuations. In many cases, adaptation and fatigue of the subjects or drift of the measurement devices cause additional variability of the psychophysiological measures. In the case of a linear variation with time, an additional factor in Equation 3 ("time") can express this effect. In the more likely case of a nonlinear function to time, a polynomial or exponential regression to time should be computed first This function expresses the coefficient c_0 in Equation 3 as a function of time. The situational decomposition is then computed on the basis of the difference between the original function and the regression function (getting $c_0=0$). It should be noted that all factors included in Equation 3 must be approximately balanced over time to ensure their differentiation. If the fluctuations are of a periodic character (with low frequency), high-pass filtering instead of the polynomial regression to time might lead to a more accurate compensation.

Temporal Resolution. Processing tasks with high action frequency may induce an additional source of error. The variation of a psychophysiological measure that is caused by a stressor variable represents a subject's reaction. Thus, a delay between the occurrence or variation of a stressor variable and psychophysiological

---

[2] In this context, used to track down the attention focus on the basis of visual information acquisition.

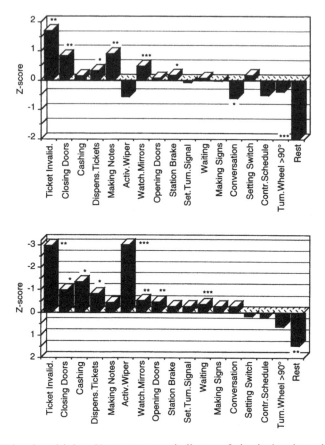

FIG. 4.16. Urban bus driving. Heart rate as an indicator of physical and psychoemotional work strain (top graph), and heart rate variability as an indicator of psycho-mental work strain (bottom graph), both fractionated to the executed tasks (measures were individually z-transformed before averaging, baseline was set to the values of driving; 8 subjects each 3 h of driving; *: $p < 0.05$, **: $p < 0.01$; ***: p < 0.001 compared to driving).

reaction must be assumed. Without any compensation, the task decomposition is distorted the greater the duration of an action approaching the reaction delay. The quantification of a possible delay (for compensation) may only be estimated by plausibility. In the bus driver example (Fig. 4.16), the computation of Equation 3 was tested for different delays of the ECG measures. The variance of the ECG parameters between different tasks was assumed to reach a maximum when the tested delay corresponds best to the physiological delay. Otherwise systematic variances would be eliminated due to the approximate stochastic duration and order of the tasks. The best differentiation between different tasks was obtained for very short delays of 0 to 2 seconds. Possibly, the adaptation of the drivers to their job allowed them to anticipate stressors and thus to minimize reaction delays in most cases.

Precision and Linearity. The correspondence tested by Equation 3 is based on a linear relation between all variables and does not consider any interactions. The introduction of additional terms of higher order and mixed terms of different factors could express nonlinearities and interactions if a sufficient amount of highly accurate data was available. In most cases, the restricted reliability of the psychophysiological measures (e.g., caused by internally evoked reactions) limits the possibilities to extract such additional information.

An example for a slightly varied application of this method is shown in Fig. 4.17 (Göbel, Suzuki, & Luczak, 1997). The psychophysiological measures derived during a simple choice-reaction task and a mental arithmetic task were extracted for the reaction periods only and were later classified according to the occurrence of errors.

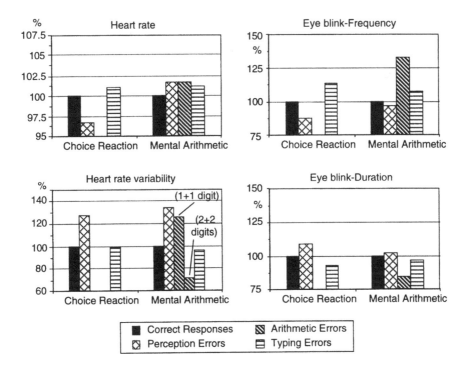

FIG. 4.17. Psychophysiological measures derived during a simple choice reaction task and a mental arithmetic task, cut for the reaction phase only and classified according to the occurrence of errors (methodological study with 4 subjects, no statistical validation).

Another method to explain peaks of work strain during complex tasks that cannot be characterized be external observation (e.g., design tasks) is described in another chapter (see chap. 15, this volume).

Interactions Between Task, Work Strain and Performance

The autonomy of executing a complex task in an individual manner obviously complicates the interpretation of psychophysiological measures. On the basis of the stress-strain concept, a deeper insight into the backgrounds of action control will be outlined to explain the variations in and interactions between task execution, work strain, and performance.

The general idea of the stress-strain concept, introduced by Rohmert, Laurig, and Luczak in the early 1970s (Luczak, 1975; Rohmert, 1973; Rohmert, 1987; Rohmert & Luczak, 1973), expresses the meaning of nonlinear and individually different reactions (strain) as a consequence of task demands and task conditions (stress). Originally, this concept focused on physical work with a fixed task procedure. Although Rohmert himself admits the consideration of flexible action/operation strategies is a weak point of this concept (Rohmert, 1982, 1984), the extended version of the stress-strain concept (Luczak, 1975; see Fig. 4.18) accounts for the inner relations between the working task and the strain that can be obtained by psychophysiological measures.

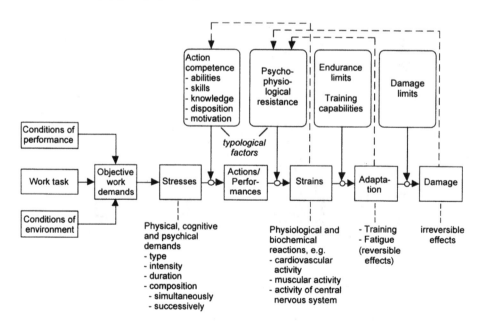

FIG. 4.18. Extended stress and strain concept (according to Luczak, 1975).

As already mentioned, the basic stress-strain concept explains the correspondence between an external demand to a subject (causing stress) and the individual psychophysiological reactions while fulfilling the demands (resulting in strain).[3] A nonlinear relationship between stress and strain is caused by:

1. A lower threshold of strain due to a minimum effort to perform a task (e.g., attention, walking),
2. Individually different intensities of physiological reactions to stressors (psychophysiological resistance in Fig. 4.18),
3. A bottleneck in the physiological processes, leading to a leveling off in output performance (while strain still increases or approaches an upper limit), and
4. The complex interactions between multiple stresses[4] and multiple types of strain reactions according to the different physiological resources involved (Luczak, 1982).

Regardless of any analytical constraints to measure strain (which imply additional nonlinearities when derived from psychophysiological measures due to physical and technological attributes; Luczak, 1987), the meaning of stress still needs to be clarified for the interpretation of strain.

Stress directly corresponds to task demands and task conditions only for very simple tasks without any autonomy for operation. Otherwise, individuals may modulate stress by selecting different operation alternatives and orders or by varying their effort for task performance. This implies variability of external behavior as well as adaptation of endogenous processes. During complex tasks, different types of tools that evoke different stresses might be used to optimize strain and performance.

The selection of appropriate operation strategies (the regulation of stress in the context of the stress-strain concept) is determined by two probably conflicting aims: (a) to maximize performance and (b) to minimize strain. Subjects will try to optimize total efficiency (performance vs. strain) depending on the primary goals, the actual conditions of performance and the actual gradient of psychophysiological resistance (see Fig. 4.18 for the latter term). Thus, a strong repercussion of strain and performance on action selection must be assumed. For real work tasks, performance and strain are considered to be of a multifactorial nature. Performance is typically defined by quantity (e.g., work pace) and quality (e.g., accuracy, number of mistakes) parameters. In many cases, additional interactions exist between the factors describing performance (e.g., if assembly parts did not fit, the perception of the mistake leads to a correction at the cost

---

[3] A demand in the context of the stress-strain concept includes actions to be performed as well as environmental conditions that persist.
[4] Multiple stresses might be evoked even by a single task.

of time). Strain is caused by the exploitation of multiple physical and cognitive resources that might become balanced by changing operation strategy and operation order, respectively.

For the selection of operation strategies, the following conclusions can be drawn:

1. If increased momentary performance does not contribute to increased to tal performance (e.g., if process parameters limit speed and quality criteria are fulfilled), operation strategies will be aimed at strain minimization.

2. Work pace may be varied only within a small range if waste of resources is to be avoided. A lower limit of strain is caused by the minimum effort needed to perform (e.g., attention), even if performance capacity is not utilized (a mostly time-related strain component). When approaching upper performance limits, the relation between effort and performance becomes more ineffective. In both cases, subjects will feel uncomfortable largely due to the obviously ineffective operation procedures. Hence, a tendency toward faster or slower task performance is to be expected. If the speed of task performance is limited by external conditions, it might be an effective strategy to switch between different tasks during complex work processes or to execute task sequences intermediately, but each with appropriate speed. For psychological strain, similar effects might be observed to avoid monotony.

3. Endurance limitations are to be expected within the limited range of strain variability. Depending on the knowledge or estimation of total work duration, changes in operation behavior might be caused by the aim to keep performance resources available for the total working period (Schmidtke, 1965).

4. If performance requirements continuously overload performance capabilities, subjects will operate at the upper performance limit (e.g., during continuous tracking). The upper performance limit is once again determined by maximum strain. Consequently, changes of output performance directly reflect strain limitations.

5. Strain is a consequence of stress, but does not necessarily occur simultaneously with stress. Stress, at first, affects an exploitation of resource capacities with subsequent initiation of compensatory strain reactions. Thus, strain follows stress with a time lag as discussed earlier. The exponentially delayed increase of heart rate after starting heavy physical work and the respective decreased function of heart rate after work termination illustrate this effect. The repercussion of strain on activity and the subsequent effect on strain represents a negative feedback loop with an internal delay due to the time lag of strain and possible hysteresis of compensatory reactions. This relationship explains the somewhat periodic variations of performance and strain (see next section for experimental findings).

6. Changes in behavior aimed to optimize performance and strain must at first be worked out and estimated for improved efficiency (Luczak & Rohmert, 1976).

This requires testing and evaluating different behavior strategies at intervals. Somewhat irregular changes in behavior might be caused by such kinds of trials.

However, high performance demands will also conflict with efforts to optimize strain. In this case, work motivation plays a particular role for the balance of performance against strain. The effects mentioned earlier contribute to a phenomenon of approximately constant load (invariant strain) whereas performance and operation strategies vary with task demands and working time. For example, no significant change of extremity movement speed was observed during the learning phase of a manual assembly task, whereas performance in terms of number of units per hour increased by 30% during a total of 22 h of work (Göbel, 1996a). In another study (Boucsein, Figge, Göbel, Luczak, & Schaefer, 1998), different process delay indicators were examined during work with a multitasking computer system. In the case of a simple task structure, the psychophysiological measures and output performance varied significantly for different types of process delay indicators. During the performance of a more complex task (with more autonomy for action arrangement), a significant variation in performance output was found for different types of process delay indicators, but psychophysiological strain measures did not vary significantly in this case. Obviously, the subjects used increased autonomy to control action coordination in a way that kept strain at an approximately constant (possibly optimum) level on the one hand and benefited from a higher performance and, consequently, a quicker work termination on the other hand.

In the preceding part of this section, the autonomy to arrange work actions was discussed for its effects on the balance of strain and performance. Furthermore, this autonomy represents an additional stressor for action planning and decision making. Consequently, increased action autonomy allows subjects to minimize strain (relative to performance) more effectively, but causes increased strain for action planning (Fig. 4.19). Due to the complexity of this topic, only broad conclusions may be drawn from a general point of view.

Any strain analysis must be interpreted together with performance if task demands or conditions of performance allow an individual opportunity for operation management. Stress analysis in the form of recordings of the actions performed might help to outline performance optimization with respect to the working individual. To a certain extent, it is likely that the strain level is kept approximately constant in order to keep total efficiency high. Thus, strain measures would mainly reflect the ability for process optimization with respect to the conditions of performance.

It might be easier to outline relationships between task conditions and strain measures if action autonomy is restricted. Conversely, it is likely that the generalization of the measures for the case of less restricted tasks will fail. For example, fixation of the work pace in order to stabilize performance is difficult because preferences vary individually. Depending on the training level and

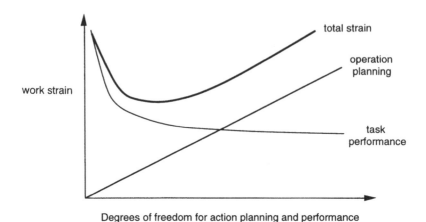

Degrees of freedom for action planning and performance

FIG. 4.19. Schematic relation between degrees of freedom for action planning and performance and the composition of work strain.

the individual constitution, a particular work pace may overload one subject while failing to challenge another subject. Consequently, strain reactions would vary considerably between individuals. If working pace could be controlled individually, this would be much less the case for work strain and performance.

To find the most effective working procedure, it is likely that subjects will optimize their overall performance within the externally permitted margins of behavior. Although it is difficult to prove this hypothesis objectively, it is implicitly plausible according to Darwin's law of natural selection (Darwin, 1859/1969). Thus, maximum freedom to perform a task should be permitted to discover the most effective operation strategies. After work procedures have reached a stable state, their structure would represent a relative perfect operation strategy. However, individually different behavior patterns are to be expected. Furthermore, changes of the performance conditions (e.g., switching from a constant amount of work to a constant work duration) might distort the measures. This is why psychophysiological measures can assist evaluation, but do not outline appropriate design solutions directly.

Another very important cause of irregular behavior is induced by conflicting aims. For example, during work in control rooms, ambiguous situations occur from time to time. The basis of explicit decision processes in the form of past experiences, estimation of success, and so on is not represented directly by physiological reactions that depend, rather, on effort, activation, and attention (Kahneman, 1973; Pribram & McGuinness, 1975). Thus, psychophysiological measures turn out to be valid means to analyze skill and rule-based tasks, whereas knowledge-based tasks are associated with some uncertainty or incompleteness for interpretation (for a definition, see Rasmussen, 1976; Rasmussen & Jensen, 1974).

SPECIFIC ISSUES OF TIME-DOMAIN ANALYSIS
AND INDIVIDUAL CHARACTERISTICS:
LEARNING, FATIGUE, AND RHYTHMICS

Finally, engineering psychophysiology applied to the study of work tasks often needs to consider temporal effects which occur during long-term task performance, for example learning, adaptation, and fatigue. Thus, specific types of data processing are required to analyze repeated measurements as a function of time. This section is devoted to the extraction of appropriate parameters that characterize temporal variabilities in a larger number of time-based samples.

The simplest method, splitting a whole data set into different intervals, is accompanied by the necessity of examining predefined sequences. Using this method, neither time-related information nor functional interpretations may be derived. Using a more phenomenologically oriented approach, functional relations can be derived by calculating regression functions against time. A polynomial regression provides a close approximation to the original data samples if an equation of appropriate order is computed. In the case of a sufficient amount of available data samples, a regression of high order (e.g., 5 to 20 coefficients) would, from a mathematical point of view, explain more variance than a regression of low order. But from an analytical point of view, the question remains how to express numerous coefficients in terms of learning, fatigue, adaptation, and so on.

Considering two main sources of long-term variability, adaptation and fatigue, two second-order coefficients appear appropriate for monotonous regression functions in the beginning and during later phases of task execution. Strictly speaking, exponential functions should be expected for learning and fatigue effects, but nonlinearity of the measures may indicate other characteristics. Such a second-order function explains long-term tendencies of parameter variation as it is shown by the trend function in the upper diagram of Fig. 4.20.

In the case of regression of exponential functions, the parameters of the learning function (e.g., outset value, expected long-term value) and of the fatigue reaction (e.g., beginning of fatigue, speed of fatigue increase) can be derived directly from the equation coefficients.

As mentioned before, this type of regression may not explain the maximum possible variance. The remaining variance consists of two parts with a periodic and a stochastic nature. The periodic part can be explained by the reflection of strain to operation strategy in the form of a negative feedback loop similar to the effects of sensory feedback discussed previously.

In the manual assembly task experiment already mentioned (Göbel, 1996a), 20 parameters describing performance, movement, muscle activity and movement control were processed for such period components by a Fourier function. The explained part of variance due to periodic changes (28±3%) was generally higher than the part of variance explained by the long-term trend (19±4%). Similarly,

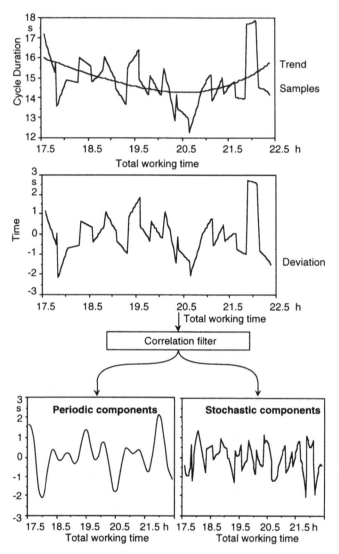

FIG. 4.20. Decomposition of long-term trend, periodic, and stochastic components using the example of cycle duration of a manual assembly task.

the frequency of periodic changes remained mostly stable for all parameters. Duration of one periodic cycle was 69.9 min during the first 4 h working period and increased to 74.2 min during the fifth working period (standard deviation between all parameters was 4.5 min). Although no further correlations of the periodic components of the different parameters could be established in this case, a significant interaction of resource exploitation and task performance with a remarkable time lag is to be assumed.

CONCLUDING REMARKS

Obviously, the examples and methods described here cannot provide a complete summary of all the possible contributions of the engineering discipline to application. Rather, this chapter merely sampled situations to illustrate the long process from the recording of the psychophysiological signals to the interpretation of psychophysiological measures according to external conditions. On the one hand, engineering methods might help to outline information of relevance and to ensure high quality data processing. But on the other hand, the application of engineering methods is somewhat complicated. Finally, even detailed modeling will not explain the range of human physiological reactions completely.

Considering the number of studies using psychophysiological measures and the potential of the application, a more concrete conceptual framework is required to integrate past studies, differentiate more precisely between insufficient measurement qualities and nonsignificant, but true, results, and obtain a more exhaustive explanation of human behavior. Only a multidisciplinary effort promises progress in this direction.

REFERENCES

Abernethy, B., & Sparrow, W.A. (1992). The rise and fall of dominant paradigms in motor behavior research. In J.J.Summers (Ed.), *Approaches to the study of motor control and learning.* Amsterdam: Elsevier.

Boucsein, W., Figge, B.R., Göbel, M., Luczak, H., & Schaefer, F. (1998). Beanspruchungskompensation beim Multi-Tasking während der Bearbeitung einer CAD-Simulation [Compensation of strain by means of multitasking during a simulated CAD task]. *Zeitschrift für Arbeitswissenschaft, 52*(4), 227–230.

Cohen, A. (1986). *Biomedical signal processing.* (Vols. 1 & 2). Boca Raton, FL: CRC-Press.

Darwin, C. (1969). *On the origin of species by means of natural: Or the preservation of favored races in the struggle for life.* London: Bruxelles. (Original work published 1859)

Donders, F.C. (1969). On the speed of mental processes. (W.G.Koster, Trans.). *Acta Psychologica, 30,* 412–431. (Original work published 1868)

Göbel, M. (1996a). *Elektromyografische Methoden zur Beurteilung sensumotorischer Tätigkeiten* [Electromyographic analysis of precise motoric tasks]. Dokumentation Arbeitswissenschaft, Bd. 40. Cologne: Otto Schmidt.

Göbel, M. (1996b). Electromyographic evaluation of sensory feedback for movement control. *Proceedings of the XIth International Ergonomics and Safety Canference.* Zurich, Switzerland.

Göbel, M. (1996c). Noise and distortion in integrated EMG—Methods of ascertaining and compensation. *Proceedings of the 1st International Conference on Psychophysiology in Ergonomics PIE 96.* Wuppertal.

Göbel, M., Springer, J., & Scherff, J. (1998). Stress and strain of short haul bus driv-ers—Psychophysiology as a design oriented method for analysis. Ergonomics, *41*(5), 563–580.

Göbel, M., Suzuki, S., & Luczak, H. (1997). Psychophysiological correlates to human errors.

In P.Seppälä, T.Luopajärvi, C.-H.Nygard, & M.Mattila (Eds.), *Proceedings of the 13th triennial Congress of the International Ergonomics Association* (Vol. 4, p. 341). Tampere, Finland: Finnish Institute of Occupational Health.

Hagemann, B., Luhede, G., & Luczak, H. (1985). Improved "active" electrodes for recording bioelectric signals in work physiology. *European Journal of Applied Physiology, 54,* 95–98.

Kahneman, D. (1973). *Attention and effort.* Englewood Cliffs, NJ: Prentice Hall.

Laszlo, J.I. (1992). Motor control and learning: how far do the experimental tasks restrict our theoretical insight? In J.J.Summers (Ed.), *Approaches to the study of motor control and learning.* Amsterdam: Elsevier.

Laurig, W. (1970). *Elektromyografie als arbeitswissenschaftliche Untersuchungsmethode zur Beurteilung von statischer Muskelarbeit* [Electromyography as an ergonomic method to evaluate static muscle load]. Berlin, Cologne, Frankfurt: Beuth-Verlag.

Luczak, H. (1975). *Untersuchungen informatorischer Belastung und Beanspruchung des Menschen* [Analysis of mental workload]. Angewandte Informatik-Fortschritt-Berichte der VDI-Zeitschriften Reihe 10, Nr. 2. Düsseldorf: VDI-Verlag.

Luczak, H. (1982). Grundlagen ergonomischer Belastungssuperposition [Principles of ergonomic workload superposition]. In W.Rohmert (Ed.), *Ergonomie der kombinierten Belastungen.* Cologne: Verlag Otto Schmidt.

Luczak, H. (1983). *Schiff der Zukunft/Arbeitswissenschaftliche Begleitforschung* [Ship of the fu-ture—Ergonomic reasearch]. In Hansa Herford, Germany: Zentralorgan für Schiffahrt, Schiffbau, Hafen.

Luczak, H. (1987). Psychophysiologische Methoden zur Erfassung psychophysischer Beanspruchungszustände [Psychophysiological methods for psychophysical workstrain]. In U.Kleinbeck & J.Rutenfranz (Eds.), *Enzyklopädie der Psychologie, Serie Wirtschafts-Organisations-und Arbeitspsychologie,* Band 1 (pp. 185–259). Göttingen, To ronto, Zürich: C.J.Hogrefe.

Luczak, H., Baer, K., Hagemann, B., Luhede, G., Klug, H., Schütte, M., Schwier, W., & Wildt, U. (1986). *Belastungs-und Beanspruchungsuntersuchungen zum Schiff der Zukunft* [Workload and workstrain measures for the ship of the future]. Bremerhaven: Wirtschaftsverlag.

Luczak, H., & Göbel, M. (1996). Psychophysische Aspekte sensumotorischer Tätigkeiten [Psychophysiological aspects of fine-metric tasks]. In K.Landau, H.Luczak, & W.Laurig (Eds.), *Ergonomie der Sensumotorik,* Festschrift anläßlich der Emeritierung von Herrn Prof. Dr. I.W.Rohmert (pp. 34–56). Munich, Vienna: Hanser Verlag.

Luczak, H., & Laurig, W. (1973). An analysis of heart rate variability. *Ergonamics, 16*(1), 85–97.

Luczak, H., & Phillip, U. (1980). Decomposition of heart rate variability under the ergonomic aspects of stressor analysis. In R.I.Kitney & O.Rompelmann (Eds.), *The study of heart rate variability* (pp. 123–177). Oxford: Clarendon Press.

Luczak, H., & Raschke, F. (1975). Regelungstheoretisches Kreislaufmodell zur Interpretation arbeitsphysiologischer und rhythmologischer Einflüsse auf die Momentanherzschlagfrequenz: Arrhythmie [Theory of a control model to explain physiological and rhythmological effects on heart rate frequency: Arrhythmia]. *Biological Cybernetics, 18*(1), 1–13.

Luczak, H., & Rohmert, W. (1976). Anpassungsreaktionen von Arbeitspersonen bei ergonomischen Feldstudien informatorischer Arbeitsinhalte [Ergonomic field studies about the adaption of workers performing cognitive tasks]. *European Journal of Applied Physiology and Occupational Physiology, 35*(1), 33–47.

Luczak, H., & Rohmert, W. (1997). Belastungs-Beanspruchungs-Konzepte [Stress and strain concepts]. In H.Luczak & W.Volpert (Eds.), *Handbuch Arbeitswissenschaft* (pp. 326–332). Stuttgart: Schaeffer Poeschel Verlag.

Mulder, G. (1980). *The heart of mental effort.* Groningen: University of Groningen.

Mulder, G., & Mulder, L.J. M. (1981). Information Processing and cardiovascular control. *Psychophysiology, 18*(4), 392–401.

Ohsuga, M., Terashita, H., Shimono, F., & Toda, M. (1995). Assessment of mental workload based on a model of autonomic regulations on the cardiovascular system. In Y.Anzai, K. Ogawa, & H.Mori (Eds.), *Symbiosis of human and artifact* (pp. 771–776). Amsterdam: Elsevier.

Pribram, K.H., & McGuinness, D. (1975). Arousal, activation and effort in the control of attention. *Psychological Review, 82,* 116–149.

Rasmussen, J. (1976). Outlines of a hybrid model of the process operator. In B.Sheridan & G. Johannsen (Eds.), *Monitoring behavior and supervisory control* New York: Plenum.

Rasmussen, J., & Jensen, A. (1974). Mental procedures in real-life tasks: A case study of electronic troubleshooting. *Ergonomics, 17,* 293.

Rau, G. (1973). Der Einfluß der Elektroden- und Hautimpedanz bei Messungen mit Oberflächenelektroden (EMG) [Effects of electrode and skin impendence using surface electrodes for EMG]. *Biomedizinische Technik, 18,* 23–27.

Reucher, H. (1988). *Neue Ansätze zur Erfassung elektromyografischer Signale mit räumlich filternden nichtinvasiven Me?anordnungen* [New approaches to analyze electromyographic signal by spatially filtering electrode configurations]. Fortschrittsberichte VDI, Reihe 17, Nr. 49. Düsseldorf: VDI Verlag.

Rohmert, W. (1973). Psychophysische Belastung und Beanspruchung von Fluglotsen [Psychophysiological workload assessment of air traffic controllers]. Schriftenreihe *Arbeitswissenschaft und Praxis,* Band 30. Berlin, Cologne: Beuth.

Rohmert, W. (1982). Ergonomie der kombinierten Belastungen. Vorbemerkungen zu einem arbeitswissenschaftlichen Kolloquium [Ergonomics of combined streeors. Remarks to a human factors colloquim]. In W.Rohmert (Ed.), *Ergonomie der kombinierten Belastungen.* Cologne: Verlag Otto Schmidt.

Rohmert, W. (1984). Das Belastungs-Beanspruchungs-Konzept [The stress-strain concept]. *Zeitschrift für Arbeitswissenschaft, 38*(4), 193–200.

Rohmert, W. (1987). Physiological and psychological work load measurement and analysis. In G.Salvendy (Ed.), *Handbook of human factors.* New York: Wiley.

Rohmert, W., & Luczak, H. (1973). Zur ergonomischen Beurteilung informatorischer Arbeit [Ergonomic assessment of mental work]. *Internat. Zeitschrift für angewandte Physiologie, 31,* 209–229.

Schmidtke, H. (1965). *Die Ermüdung: Symptome-Theorien-Messversuche* [Fatigue: Symptoms-Theory]. Bern, Stuttgart: Verlag Hans Huber.

Shannon, C.E., & Weaver, W. (1949). *The mathematical theory of communication.* Urbana: University of Illinois Press.

Silny, J., & Rau, G. (1977). Messung des komplexen Hautwiderstandes [Measurement of complex skin impedance]. *Biomedizinische Technik, 22,* 409–410.

Springer, J. (1992). *Systematik zur ergonomischen Gestaltung von CAD-Software* [A systematic approach for ergonomical design of CAD-software]. Fortschrittberichte VDI, Reihe 20, Nr. 60. Düsseldorf: VDI Verlag.

Springer, J., Langner, T., Luczak, H., & Beitz, W. (1991). Experimental comparison of CAD systems by stressor variables. *International Journal of Human-Computer Interaction, 3*(4), 375–405.

von Helmholtz, H., Gullstrand, A., von Kries, J., & Nagel, W. (1909–1911). *Handbuch der physiologischen Optik* [Handbook of physiology optics]. (Vols. l-3). Hamburg and Leipzig: Voss.

Zipp, P. (1982). Effect of electrode geometry on the selectivity of myoelectric recordings with surface electrodes. *European Journal of Applied Physiology, 50,* 35–40.

Zipp, P., (1988). *Optimierung der Oberflächenableitung bioelektrischer Signale* [Optimization of surface electrodes and equipment for bioelectric signals] Fortschritt-Berichte VDI, Reihe 17, Nr. 45. Düsseldorf: VDI Verlag.

Chapter 5

Recording Methods
in Applied Environments

Jochen Fahrenberg
Forschungsgruppe Psychophysiologie,
University of Freiburg, Germany

Cornelis J.E. Wientjes
TNO Human Factors Research Institute,
The Netherlands

Recording methods in applied environments make increasing use of new techniques for the acquisition of physiological and psychological data. Portable multichannel recorder/analyzer systems and handheld computers provide new means of ambulatory assessment and foster a more ecologically valid approach to many issues in the applied fields. In this chapter, such developments are reviewed with particular emphasis on ambulatory techniques.

In applied and environmental physiology and psychophysiology, many investigations have been conducted using standard laboratory equipment moved to and installed at the workplace. However, bulky electronic apparatus and connecting cables impose clear restrictions on the subject's behavior. The development of lightweight portable recorders that allow for monitoring of the electrocardiogram (ECG) or blood pressure without restrictions on mobility have furthered the application of ambulatory assessment methodology beyond the medical field.

A comprehensive overview of physiological measurement at the workplace, including issues like choice and placement of electrodes and sensors, recorder technology, signal processing and parameter abstraction, would require more space than is available here. There are, however, a number of suitable books, manuals and guidelines on psychophysiological methods (Cacioppo & Tassinary,

1990; Hugdahl, 1995; Martin & Venables, 1980; Rösler, in press; Schandry, 1988; and more specific contributions concerning certain response systems, e.g., Berntson et al., 1997; Boucsein, 1992; Jorna, 1992; Mulder, 1992; Pickering, 1991; Schneiderman, Weiss, & Kaufmann, 1989).

Methodological guidelines have also been provided by Committees of the Society for Psychophysiological Research concerning, for example, heart rate (HR) and HR variability (Berntson et al., 1997) and blood pressure (Shapiro et al., 1996). Ambulatory monitoring techniques, biotelemetry and radio tracking were reviewed by Amlaner and Macdonald (1980), Littler (1980), and Webster (1988). An overview on methodology and applications is included in the book *Ambulatory Assessment* (Fahrenberg & Myrtek, 1996). Reviews on methods, especially psychophysiological methods in human engineering and work physiology, were performed by Hancock and Meshkati (1988), Luczak (1987), Manzey (1998), and Miles and Broughton (1990), and a number of contributions to methodological issues are available (Boucsein, 1991; Hinton & Burton, 1997; Richter & Hacker, 1998).

Contributions concerning psychological data, that is, the acquisition of behavioral measures, behavior ratings, and self-report data in the natural environment have been made in the literature on behavior assessment (cf. de Vries, 1992; Nelson & Hayes, 1986; Suen & Ary, 1989). The scope, however, is mostly clinical and new techniques, such as electronic diaries and physical activity monitoring, have been rarely employed in this context. Therefore, the development of computer-assisted methods suitable for ambulatory assessment studies are reviewed here in greater depth.

The recording of physiological and psychological variables is but one aspect of assessment in applied environments. Basic issues in methodology include the features of field studies as opposed to those of laboratory experiments, the development of adequate assessment strategies, issues in multimethod assessment, subject compliance, and method-dependent reactivity. In addition to addressing these issues, several examples of research have been selected in order to delineate a new orientation in ambulatory assessment.

APPLIED ENVIRONMENTS

Applied environments are settings in which behavior (including in a very broad sense physiological measures and self-report data) is assessed in order to answer practical questions in contrast to basic research in the laboratory. Such settings focus on the workplace and leisure time activities in particular, whereas clinical and forensic applications and consumer or media research is not included. Workplace settings are those in which all manner of activities involved in the pursuit of an occupation take place. Such settings might include a surgeon's operating theater or an artist's stage, man-machine operating systems, or the field of special services (e.g., pilots, parachute jumpers, astronauts, aquanauts, firemen, military services, etc.). Activities in workplace settings might also include driving and commuting or involve work under extreme conditions, such as high altitude or high/low humidity. Other applied environments might include personnel assessment (e.g., personnel selection and/or assessment centers), commercial outdoor and teambuilding programs (evaluation of training programs), public transport (e.g., workload studies among employees and studies of the fear of victimization among employees and users), environmental design (e.g., evaluation of the public's response to shopping centers, buildings, public safety measures such as video cameras), simulated and virtual environments (e.g., simulator or virtual environment validation by way of comparison of subjective or physiological responses in simulated environments with responses to the real task environment), and scenario simulations (e.g., playing of disaster scenarios by civil or military authorities).

The demand for ecological validity in behavioral research has encouraged field studies and the development of appropriate methodologies that are feasible in applied environments.

BASIC ISSUES

The focus of attention here is restricted primarily to workplace settings. Four basic issues are under consideration:

1. Workplace settings are nonlaboratory, that is, behavior is not elicited in an experimental design and in highly standardized tasks. Instead, naturalistic observation of behavior is preferable. The behavior under consideration is already part of the individual's repertoire and occurs in a familiar environment. However, the obvious advantages of ecological validity are usually accompanied by shortcomings in internal validity due to the complexity of effects, problematic control of independent and dependent variables, covariates, and extraneous variables (artifacts, confounds). In some areas, there is a noticeable trend to supplement field studies and quasi-experimental designs with field experiments that attain higher levels of control so that causal relationships may be detected more precisely.

2. Real-life assessments may involve more social interaction and issues of compliance and reactivity (i.e., method-dependent artifacts) and other sources of unwanted variance. Behavior assessment at the workplace is not entirely naturalistic because the participants usually know that recordings are made. Furthermore, they may also be aware of the specific hypothesis under investigation or assume motives and intentions (Christie & Todd, 1975; Rosenthal & Rosnow, 1969). Such attitudes and attribution processes may affect the compliance and reactivity or may even lead to reactance (i.e., defensive and uncooperative behavior). On the other hand, participants may be especially interested in such recordings.

3. Workplace assessments usually require special attention to task demands, performance, and load. The *demand measures* form the counterpart of the *performance measures;* they specify the task or demand requirements of the environment (input), whereas the performance/outcome measures define the way the operational output is assessed. *Load measures,* which are often the prime focus of ambulatory studies, can only properly be interpreted in the context of task requirements, on the one hand, and performance, on the other.

In many studies, task demands are only assessed with subjective techniques. This approach is not adequate because there may often be, for various reasons, a considerable difference between the objective workload and the subjective evaluation by the employees. Hence, it is in most cases desirable to obtain objective data on task demands. The operationalization of demand as well as performance measures deserves very careful consideration. It is often necessary to consult experts and to carry out an operational task analysis in order to obtain sufficient insight into the relevant specifications.

Performance measures may often be obtained in a quantitative manner by measuring relevant task performance or output parameters (chosen on the basis of a task analysis), such as the number of actions taken or items produced per time unit. Zeier (1994) and Wientjes, ter Maat, and Gaillard (1994) have used these demand measures in studies of workload among air traffic controllers. However, more relevant measures may be based on a theory on organizational stressors rather than on a task analysis. Such measures are often much harder to obtain, although they may also include relatively simple quantitative measures. Obtaining these measures may sometimes even require the development of new assessment methods, such as those cited earlier.

The design of load measures (subjective, behavioral, and physiological responses) requires an especially careful approach. It should be acknowledged that a single valid indicator of load (overload, strain, monotony) does not exist. Behavioral measures may supplement the conventional self-reports and psychophysiological recordings. Measuring the performance on a secondary task may, in some settings, be a suitable method of assessing the workload on the primary tasks. For example, Veltman and Gaillard (1996), in their pilot workload studies, included an additional auditory task (a continuous memory task) as a secondary task (Wientjes, Veltman, & Gaillard, 1996).

4. Assessment of individual differences (or average response profiles for certain tasks) in applied environments generally seeks an empirically based answer to practical questions. Such objectives include monitoring and self-monitoring of performance, evaluation of mental and physical load and overload, selection and classification of personnel, detection of health risks, and evaluation of risk behavior. Especially when costly psychophysiological methods are involved, the empirical validity and, more importantly, the incremental validity (as compared to data already available) has to be discussed on theoretical grounds within the framework of cost-benefit analyses.

The earlier sections provide the basic background to the methodology of ambulatory measurement approaches and emphasize key methodological issues. So far, such discussion is very brief and rather abstract. Therefore, we provide a more detailed and concrete account on a selected number of important issues by referring to recent contributions to research work that illustrate a number of concepts and assessment strategies. A more detailed discussion of basic methodological issues was given by Fahrenberg (1996a), Stemmler (1992, 1996, in press), Stemmler and Fahrenberg (1989), Patry (1982), and Pawlik (1988).

PHYSIOLOGICAL RECORDINGS

Instrumentation

Physiological recordings in applied environments and ambulatory monitoring in medicine have a long history. The equipment used was either stationary, barely portable devices, or biotelemetry systems that were usually restricted to near-distance transmission. Portable analog recorders suitable for freely moving subjects came into use in the 1960s. Since then, multichannel recorders of modular design (e.g., the Medilog-System by Oxford Instruments) have become available. Such equipment has been used in a great number of investigations for the recording of, for example, HR, ECG, respiration, temperature, electrodermal activity, electromyogram (EMG), and physical activity (for reviews, see Fahrenberg & Myrtek, in press; Hankins & Wilson, 1998; Littler, 1980; Miles & Broughton, 1990; Turpin, 1990; the Journal of Ambulatory *Monitoring* [until 1995]; and the *Journal of Medical Engineering & Technology*).

The development of multichannel digital recorder/analyzer systems began in the 1980s and has progressed directly so that every market survey is soon out of date. Such recorders have proven to be very useful because they can be employed not only in ambulatory studies, but also in laboratory studies (data transferred to the computer via a serial port).

A general purpose digital recorder/analyzer for long-term recordings has been designed by G.Mutz (see Jain, Martens, Mutz, Weiß, & Stephan, 1996). In addition to general purpose analog input channels and marker channels, specific dedicated channels for recording electrodermal activity for example, may be available. Specific features include software programmable amplifier gain and high-pass and low-pass filtering plus valuable preprocessing options and real-time analysis of input.

Digital storage is usually on RAM, a Flash Card, or a PCMIA mini hard disk (e.g., a 131 MB disk allows an eight-channel recording with a sampling rate of 100 Hz for 36 hours; larger disks with 260 MB, 340 MB, or more are available). The postprocessing is dealt with by programs developed by the user. However, standard software for parameterization, statistical analysis, and graphic presentation of results will be increasingly important in this field.

Recent progress in ambulatory assessment that has utilized the real-time analysis of physiological recordings for feedback and control has proven most interesting. The subject/patient is prompted by a signal to respond to a set of questions about the setting, their present behavior, and subjective state, when a distinct increase in HR (accounting for effects due to physical activity) occurs (Myrteket al, 1988). The available equipment makes many innovative assessment strategies possible; for example, blood pressure measurement could be triggered in accordance with predefined critical values of the ECG (e.g., HR change, arrhythmia's, ST-depression). Such interactive strategies are readily accessible but have hardly been applied in psychophysiology.

An overview on general purpose recorder systems, which is not exhaustive, is depicted in Table 5.1. Besides devices suitable for ambulatory recordings and use

TABLE 5.1

Multichannel Digital Recorder/Analyzer for Physiological Measurement (Modular Design)

Name	Manufacturer
AMS	Free University, Amsterdam, The Netherlands
BIOLOG	UFI, Morro Bay, CA, United States
BIOPORT	Zak GmbH, Simbach, Germany
CARDIOTENS	Meditech, Budapest, Hungary
Daily Activity Monitor	McRoberts BV, The Hague, The Netherlands
EMBLA	Flaga, Reykjavik, Iceland
Flexilog	Oakfield Instruments, Witney, United Kingdom
HALLEY	Esaote Biomedica, Firenze, Iceland
Mini Logger	Sunriver, OR, United States
Myodata	Le Mazet St Voy, France
Medilog	Oxford Instruments, Witney, United Kingdom
PAR-PORT	Par-Elektronik GmbH, Berlin, Germany
PHYSIO-LOGGER	med-natic GmbH, München, Germany
PHYSIO Modul	Rimkus, Riemerling, Germany
VITAPORT, VARIOPORT	Vitaport GmbH, Erftstadt, Germany (Temec Instruments BV. Gemert, The Netherlands)

Note. Technical specifications were omitted here due to the multiplicity of parameters. The reader is referred to the manufacturers' information outlets (e.g., Internet homepages). Ambulant ECG (Holter) recorder, ambulatory blood pressure monitors, and sleep apnea recorders were not included.

in 24-hour or long-term monitoring, a wide range of portable (mobile) equipment, designed for in-field measurement, does exist. For example, continuous finger arterial pressure (Portapres 2), measurement of tidal volume and other spirometric variables (Respitrace), O_2/CO_2 gas exchange and energy expenditure, capnometry and pulse oximetry, polysomnography, cardiorespiratory polygraphic systems, 24-hour pH testing, or fetal monitoring. Furthermore, monitoring systems have been designed for use by patients, for example, as blood glucose monitoring systems, respiration and asthma monitoring systems, and EMG monitors. Again, technology is developing so quickly that the reader is referred to *Journal of Medical Engineering & Technology* and similar journals or report materials like the *International Hospital Equipment*. Neither the device technology in digital recorder/analyzer systems nor the progress in ambulatory assessment of endocrine responses are within the scope of this introduction (cf. Kirschbaum, Read, & Hellhammer, 1992).

Physiological Measures

A wide selection of physiological variables have been measured in applied environments, using mostly noninvasive methods. However, some investigators have made ambulatory recordings of arterial (inter-brachial) blood pressure or gastrointestinal functions. Even more specific sensor techniques and registrations have been used and conducted in medical research.

However, the great majority of investigators have dealt with a few selected measures. Such physiological measures and derived indices are compiled in Table 5.2. Neither the specific preparations, sensor placement, amplifier settings, filtering, preprocessing and biosignal analysis nor the biometric and physiological issues necessary to extract appropriate parameters can be reviewed here (see chap. 4, this volume).

Physiological measurement under field conditions faces various difficulties that are either not encountered or may be more easily controlled in the laboratory (Fahrenberg, Foerster, & Müller, 1996; Wilson, 1992). Such difficulties may be related to the equipment, signal processing, and general aspects of ambulatory monitoring, as follows:

Equipment and Recording

- Selection of appropriate electrode/sensor type and electrolyte, electrode/sensor placement, leads and connections (wiring).
- Adjusting methods for prolonged recordings, thereby minimizing artifacts and method-dependent reactivity.
- Synchronization of recordings if more than one device is used.
- Coping with battery problems.

TABLE 5.2
Physiological Measures Predominantly Used in Applied Environments

Biosignal/Variable	Abbreviation
Heart rate	HR
Heart rate variability (e.g., 0.10 Hz band)	HRV (bands)
Respiratory sinus arrhythmia	RSA
Blood pressure (systolic, diastolic, mean)	BP
Pulse wave velocity	PWV
Electrocardiogram (e.g., ST depression, T-wave amplitude)	ECG (ST, TWA)
Impedance cardiography (stroke volume index, systolic time intervals)	ICG (SV-Index, PEP, LVET)
Respiration (rate, volume, minute ventilation)	RR, VT, MV
O_2/CO_2 exchange	
Transcutaneous arterial O_2 partial pressure (saturation measurement)	ptcO2
Transcutaneous arterial CO_2 partial pressure	ptcCO2
Temperature (finger, core)	
Electrodermal activity (skin conductance)	EDA (SC)
Posture	
Motion (actometer, accelerometer)	
Speech activity (throat micro)	
Electromyogram (neck tension, eyeblink rate, etc.)	EMG
Electrooculogram (fixation, saccades, etc.)	EOG
Electroencephalogram (evoked potentials)	EEG (EP)

Note. For recording techniques, the reader is referred to the introductions from Cacioppo and Tassinary (1990), and Martin and Venables (1980). More specific literature is available for each measure.

Signal Processing

- Filtering and preprocessing, for example, signal integration to save storage capacity.
- Selection of adequate sampling and storage rate to allow prolonged multi-channel recordings.

- Reliability of software algorithms for signal processing and signal averaging.
- Efficient routines for detection (and correction) of artifacts and outliers.

General Issues

- Improvement of feasibility of recordings and reliability of measurement.
- Recording of events and accompanying behavioral data.
- Definition of adequate baselines (sleep baselines, rest periods).
- Definition of appropriate summary statistics (data reduction).
- Accounting for the confounding of effects and unwanted variance, for example, due to change in posture, physical activity, eating and drinking, circadian rhythms, and variation in ambient temperature.

Posture (supine, sitting, standing) and physical activity will certainly affect many physiological recordings, such as cardiovascular measures. Because physical activity leads to unwanted variance in dependent variables, it has become a major issue in ambulatory monitoring to account for such effects. The automatic detection of posture and motion is based on multichannel accelerometry (Bussmann, Tulen, van Herel, & Stam, 1998; Fahrenberg, Foerster, Smeja, & Müller, 1997; Jain et al., 1996; Veltink & van Lummel, 1994).

The variation in ambient temperature may also constitute a problematic aspect of ambulatory monitoring. Differences between settings, for example, between indoor and outdoor temperatures or seasonal trends, may be important in evaluating change in peripheral circulation (vasoconstriction), electrodermal activity (e.g., Boucsein, 1992; Turpin, Shine, & Lader, 1983), and blood pressure (Giaconi & Ghione, 1992).

The circadian variation in physiological measures has been investigated with particular attention to the ECG (e.g., Myrtek, Brügner, & Fichtler, 1990) and blood pressure (Pickering, 1991). Because circadian effects are often masked by physiological changes due to the subject's daytime activities, it may be necessary to investigate such effects in specific research designs. Differences between workday and non-workday, between daytime and nighttime and, more precisely, between the patterns of decrease at night and increase in the morning, appear to be an important aspect in recovery from occupational load and possibly in development of hypertension (Haynes, Gannon, Orimoto, O'Brien, 1991; Hocking-Schuler & O'Brien, 1997; Pickering, 1991; Steptoe, Roy, Evans, & Snashall, 1995).

The decomposition of HR changes is an important issue in psychophysiological methodology and the advantages and problems of this approach are discussed in other chapters of this volume. One aspect of this approach is the estimation of *additional HR,* that is, HR changes which are nonmetabolic in nature. Such additional HR may be indicative of emotional activation and mental effort.

Myrtek et al. (1988) have developed an algorithm to assess additional HR on-line, based on recordings of the ECG and accelerometry (discussed later). Wilhelm and Roth (1996) proposed a bicycle ergometer calibration to predict the actual HR from minute ventilation. This method was applied in ambulatory monitoring to estimate the additional heart rate caused by flying phobia.

The estimation of additional blood pressure is, likewise, an important issue in ambulatory assessment methodology since blood pressure depends, among other conditions, on posture and physical activity. However, the automatic readings are usually taken at intervals of 15, 30, or 60 minutes making the decomposition of blood pressure variability more difficult than the decomposition of continuously recorded HR. Further studies that include continuous monitoring of posture and motion are necessary to develop more practical methods (Fahrenberg, 1996b; Tuomisto, Johnston, & Schmidt, 1996).

An abundance of methodological studies exist especially in ambulatory monitoring of the ECG and BP that cannot be reviewed here (cf. Pickering, 1991). Such investigations have paved the way for gradual progress in the standardization of these methodologies, although progress in device technology, for example, real-time analysis of the ECG or continuous noninvasive finger arterial blood pressure recordings, should lead to further methodological advances.

Some Practical Suggestions

Hardware/software systems for ambulatory physiological measurement have attained a high degree of complexity, although a number of rather simple devices, often based on algorithms that remain obscure ("black box" type), also exist. The software for parameterization of biological signals is rarely provided by the manufacturers of recording systems, except for a few options for preprocessing (e.g., filtering). The appropriate software must be developed as add-on programs by the user. An investigator, therefore, will largely depend on technical assistance with the consequence that projects in engineering psychophysiology usually require a team of co-workers.

Programming appears to be very demanding on team resources because it takes a great deal of time and skill to develop software for the parameterization of biosignals. This involves formatting data, handling long records and data reduction, defining and testing algorithms, graphic display allowing for visual inspection, detecting artifacts, and interactive editing. The frustrating experience of programming appears to have prevented a considerable number of researchers who have obtained such multichannel recording systems from taking full advantage of such technology. Signal processing can be handled by multipurpose data analysis software like LABVIEW (National Instruments Corp., Austin, TX) and MATLAB (The Math Works Inc., Natick, MA). Software packages are also specifically designed for psychophysiological research. Such developments, like BIO (Foerster, 1997), may be advantageous in the parameterization of certain signal features and may ease data handling, editing, processing, and control procedures because these procedures are specifically supported by tailor-made routines.

Cross-laboratory cooperation in this respect appears to be uncommon because of the specific nature and features of the recording systems and incompatible computer systems. Consequently, an exchange of software appears to be restricted to specific recorder/analyzer and computer systems (e.g., Vitaport user's homepage: www.GWDG.DE/~psyweb/vitaport/welcome.html).

Recording methods differ widely in their feasibility and reliability. There are many aspects of these differences such as ease and unobtrusiveness in application of electrodes and sensors, occurrence of artifacts, precision in parameterization, reactivity (i.e., method-dependent effects), and face validity and reasonableness (i.e., the subject's attitude). Based on the previously cited issues and evidence from multichannel recordings, a tentative ranking of suitability of physiological measures for field studies can be proposed. First, measure the ECG (HR and other derived parameters), accelerometry, EDA and temperature, followed by blood pressure measurement, respiration rate, EOG, and EMG (see also Myrtek, Brügner, & Müller, 1996b). Other measures have shown to be more reactive and more easily disturbed by movement, respiration, talking, and other sources of unwanted variance. Of course, averaging techniques, pattern recognition, or concurrent recordings of movement to detect artifacts may assist in improving reliability. Advanced researchers who make use of modern technology and software development make recordings in up to 16 or 20 channels. However, routine application of such recordings will probably require several years of development and expertise.

The reliability of measures is one issue, and the valid interpretation of findings is another since physiological measures have to be evaluated in the context of regulatory controls and situational demands. HR and blood pressure, thus, are only two aspects of hemodynamic regulation, and their evaluation should relate to physiological concepts like sympathetic-vagal balance or energy expenditure and to behavioral concepts like additional HR, overload, or optimal level of psychophysiological arousal.

PSYCHOLOGICAL DATA

Methods of recording task demands and performance data have a long history in human engineering (e.g., time and motion studies). Similar assessments have been conducted in differential and clinical psychology. Event recorders for the timed registration of stimuli and responses, "beeper" studies in which a programmable wristwatch prompts the subject to respond to a questionnaire, self-ratings on diary cards, and an electronic data logger have all been used for this purpose. Especially the advent of pocket-sized (handheld, palm-top) computers has eased the acquisition of data considerably (for a review, see Fahrenberg, 1996a). There are various sources of data: behavior (performance) measurement, behavior observation, self-reports on experiences, subjective state, and so on, and, possibly, environmental aspects, like ambient noise or temperature, which can be recorded together with physiological measures (see Table 5.3).

Software to facilitate the use of handheld PCs in field studies has been developed in many institutions, more or less specific to certain studies. Flexible software systems suited to a variety of applications are still an exception (e.g., Behavior Observation System, Noldus Information Technology, AG Wageningen, NL; MONITOR for various types of self-reports, developed for use with the PSION 3 series, Brügner, 1998; in-field performance testing, cf. Buse & Pawlik, 1996; Fahrenberg, Brügner, Foerster, & Käppler, 1999).

The application of a programmable pocket PC in ambulatory assessment has many advantages as compared to conventional paper and pencil methods:

- Alarm functions for prompting the observer/subject at predefined intervals and a built-in reminder signal.
- Reliable timing of input, delay of input, and duration of input.
- Flexible layout of questions and response categories.
- Branching of questions and tailor-made sequential or hierarchical strategies.
- Concealment of previously recorded responses from the observer/ subject.
- Overcoming the recall error.

The recall error or retrospection effect has been a methodological issue in several studies. Even if subjects are told to monitor their stress, mood, and symptoms several times daily, such questionnaire reports are often done from memory. Such ratings are, of course, less accurate. A specific retrospection effect may exist when subsequent events and experiences systematically influence and even distort the subjective evaluation and weighting of previous states (DeLongis, Hemphill, & Lehman, 1992; Hedges, Jandorf, & Stone, 1985). Retrospective ratings may indicate more negative mood and unease than is to be expected from the actual ratings averaged across the day, that is, a *negative retrospection bias* (Käppler, Becker, & Fahrenberg, 1993; Fahrenberg et al., 1999).

A higher reliability and ecological validity can be generally assumed with computer-assisted recordings than with paper and pencil questionnaires and diaries that lack flexibility and exactness when timing responses.

The recording of posture and motion is another basic issue in the methodology of behavior observation and performance measurement. Calibrated piezoresistive sensors allow the separation of a DC-component, indicating degree of inclination, that is, body position, and an AC-component, representing acceleration, that is, movement. Multichannel accelerometry can be used for the measurement of

TABLE 5.3
Potential Contents of a Computer-Assisted Protocol

Objective Setting Features (Observables)

Location (place) and time

Persons and objects

Task and stimulus conditions, load measures, task demands

Ambient variables (e.g., noise)

Performance and Behavior Data (Observables, Measures)

Time and motion

Specific behavior categories

Social interaction

Objective test scores

Self-measurements

Self-Report (Ratings)

Subjective state

Physical symptoms

Evaluation of settings

Actions, goals, coping process

Comments on tasks, performance

Further Aspects

Specific events

Artifacts

Reactivity and compliance

posture, specific movements, tremor, gait, and other aspects of motor behavior. Moreover, an automatic detection of posture and motion during ambulatory monitoring can be achieved if an individual's characteristic patterns (a standard protocol for walking, standing, sitting) are obtained at the beginning of the assessment (Fahrenberg et al., 1997; Veltink & van Lummel, 1994). Based on such multisite accelerometry and ambulatory recordings, a continuous record of posture and movement can be obtained rather unobtrusively. In addition to this, accelerometry may be used as a method for the estimation of gross physical activity and energy expenditure (Patterson et al., 1993).

ISSUES IN METHODOLOGY

Acceptance, Compliance, and Reactivity

Ambulatory assessment by means of a pocket PC or recorder depends on the favorable attitude and appropriate motivation of the participating subjects. It is essential that the equipment is readily accepted and that good compliance to instructions is established. A comprehensive postmonitoring interview is recommended in order to obtain information on these essential aspects.

The acceptance of the handheld computer in field studies and of ambulatory monitoring in general appears to be much higher than initially expected by investigators (Fahrenberg & Myrtek, 1996). The awareness of being equipped with a new electronic device can be a motivating condition to participants. Furthermore, motivation can stem from the participation in a diagnostic procedure or research project and the willingness to learn about oneself: Everybody is his own blood pressure (or stress) researcher. However, it is understandable that at some workplaces and for some subjects the attitude may be quite different. Certain environments will be less suited to ambulatory monitoring, and some settings will only allow for a registration of the ECG and actometer recordings that are hidden from the public. Intermittent automatic blood pressure measurements, beeper signals, and the instruction to type a response on a handheld computer may be problems in some instances because they distract the participant or elicit positive or negative responses and comments from colleagues. A specific advantage of a computer-assisted method of obtaining self-reports and physiological measurements is the exact protocol of the subject's compliance and the timing of data acquisition.

Ethical issues specific to ambulatory monitoring studies have hardly been discussed in the literature (Schuler, 1982). Appropriate data confidentiality is a concern because ambulatory assessment may violate privacy more easily than alternative methods. Furthermore, persons not being assessed may become involved in the observation and evaluation of settings when these are being carried out. Obtaining the subject's informed consent before the recording starts is essential, but may be problematic because the exact course of daily activities and events cannot be anticipated. Moreover, it may be more difficult to explain the essential hypothesis of the investigation, variables, and methods of analysis in ambulatory studies than in the laboratory. Thus, the postmonitoring interview is well suited to recall specific events and discuss problematic aspects.

Reactivity means that the method of observation and measurement itself is a source of unwanted variance due to specific interactions such as awareness, adaptation, sensitization, and coping tendencies (Haynes & Horn, 1983; Stern, 1986; Watson & Pennebaker, 1991). Such reactivity may include a bias common to a class of methods or specific to one method. Obviously, nonreactive methods are an exception in psychological assessment and physiological measurement, which is in accordance with Heisenberg's uncertainty principle: The attempt to measure a phenomenon may distort the phenomenon.

Three aspects of reactivity, however, appear to be specific to ambulatory assessment. Subjects may: (a) tend to avoid certain settings during the recording in order to prevent their being monitored there; (b) tend to unintentionally or deliberately manipulate the recording systems, shift settings of the PC, and may even try to get access to the program; and (c) try to test their capacities or the equipment by unusual patterns of behavior, exercise, or vigorous movements.

At present, only a postmonitoring interview can reveal such effects, but this evaluation can be rather difficult. Little is known about whether it is advisable to include a few control items and questions in the computer-assisted questionnaire that directly refer to actual (non-) compliance, noncompliance, and method-dependent effects, for example, the subject's being irritated by the beeps or having specifically changed the usual routine.

Cost-Benefit

Ambulatory assessment methodology has certain advantages compared to stationary equipment. However, the recorder/analyzer will usually be more costly. The expectation is that less expensive devices, suited for multichannel physiological recordings, and pocket-PCs will be developed. Thus, a more extensive use is foreseeable as is already the case in single purpose ECG recorders or blood pressure recorders. The higher ecological validity of the ambulatory assessment will encourage such applications.

There are some investigations that use assessment of hormones in saliva, especially cortisol, and immune function, especially secretory immunoglobulin A, to obtain indicators of arousal, strain, or overload (e.g., Kirschbaum et al., 1992). These methods have been used more recently due to the availability of appropriate assay methods and reduced costs. The subject's compliance concerning adherence to the time schedule in using the salivettes, difficulties of adequate freezing, and impurities in the probe appear to be crucial aspects. Besides the compliance of subjects and the reliability of this method, the cost-benefit aspect can be questioned. Seemingly, there is no methodological study indicating what incremental validity may be obtained by a few data points of hormonal or neuroimmunological assays as compared to a continuous recording of the ECG and appropriate decomposition of HR changes. Moreover, reliable baselines of HR can be easily obtained in 24 hour recordings, and metabolically caused HR changes can be largely partialled out by referring to concurrent recordings of physical activity. At present, HR appears to be the most reliable, valid, and least expensive measure for indicating important changes in sympathetic and vagal influences on the cardiovascular system and associated changes in emotional activation and mental effort (see Myrtek et al., 1996a, 1996b).

EXEMPLARY INVESTIGATIONS
USING AMBULATORY ASSESSMENT METHODOLOGY
IN APPLIED ENVIRONMENTS

A number of studies were selected from recent research publications to illustrate the potentialities of ambulatory monitoring. Only short reviews can be provided here, and the reader is referred to the original articles for further details and discussion of findings. In the present context, such abstracts may be useful in order to indicate the variety of applications of ambulatory assessment methodology and to underline specific issues like multimodal assessment, on-line signal processing and feedback, interactive monitoring and estimation of nonmetabolic changes in HR.

Psychophysiological Assessment of Workload
and Stress Among Air Traffic Controllers

In ambulatory workload studies in operational environments, it is essential, but often quite difficult, to adequately measure variations in task demands. The choice of demand measures depends on the nature of the task and on practical as well as theoretical considerations. Many operational tasks have multiple demand characteristics. Furthermore, as is illustrated in the following example, the nature of the demand characteristics and hypotheses about their relationship with specific psychophysiological processes need to be carefully considered.

In a study of workload and stress among 18 military air traffic controllers (ATCs), Wientjes, ter Maat, and Gaillard (1994) employed a special purpose ambulatory 8-channel digital physiological recorder (see Wientjes, Spiekman, Benschop, & Hoogeweg, 1994) in addition to paper-and-pencil questionnaires and observational methods. On the basis of extensive consultation of ATC experts, two workload measures were collected that were hypothesized to reflect two distinct aspects of the demand characteristics of the ATC task: (a) the number of aircraft that were simultaneously under control per time unit, and (b) the number of potential conflict situations per time unit (i.e., when, mostly due to miscommunication, an aircraft under control was potentially in conflict with another aircraft and special measures were necessary to prevent accidents).

The first measure was considered to primarily reflect variations in momentary workload, whereas the latter measure was hypothesized to additionally reflect a demand component that might be associated with stress. Measures of both task demands were sampled each 5 minutes by the investigator who was closely observing the ATC. The investigator entered these data, in addition to other information (i.e., codes reflecting relevant events), on a handheld keyboard that conveyed the data via an infrared transmitter interface to the digital recorder. The measurement sessions encompassed twenty-four 5-minute blocks (i.e., 2 hours of measurements) per subject. The psychological measures included subjective workload ratings and measures of mood. The physiological measures included HR and transcutaneous pCO_2 ($ptcCO_2$), an indirect measure of arterial CO_2 levels (Garssen, Buikhuisen, Hornsveld, Klaver, & van Doornen, 1994). It should be noted that under nonstressful conditions, $ptcCO_2$ varies very little between individuals. However, during states of stress, ventilation may increase beyond metabolic demands (i.e., due to hyperventilation), which leads to a decrease in $ptcCO_2$ (see Wientjes, 1992).

The results confirmed the hypothesized difference between the two demand measures. HR increased monotonically when the number of aircraft increased, but $ptcCO_2$ did not change in response to variations in workload. However, $ptcCO_2$ did decrease when there were more than two potential conflicts per 5-minute block. These situations occurred in about 15% of the working time. HR also increased when there were potential conflicts, but this measure did not reliably discriminate between variations in workload and variations in the number of potential conflicts. Among the psychological measures, the subjective workload ratings did not differentiate between the two types of demand. However, the mood measures did. The scores for anger and fatigue were higher in blocks when there were two or more potential conflicts than in other blocks. It was concluded that variations in the workload of ATCs are associated with normal, adaptive variations in psychophysiological activation (probably reflecting the increased mental load). On the other hand, negative emotional reactions and maladaptive physiological responses may apparently be observed when the task includes aspects that are difficult to control and when air traffic safety may be compromised.

Evaluation of Stress Tolerance: Physiological Responses, Subjective Reactions, and Performance During Exposure to an Acute Stressor Task

Increasingly, operational tasks are performed in potentially stressful environments. This has generated a demand for research that may help to evaluate whether or not individuals are able to maintain high levels of performance under adverse conditions. It has long been recognized that stress may negatively affect performance, due to enhanced and dysfunctional levels of activation, subjective strain, negative emotions, and coping strategies that are oriented to ward self-protection rather than toward execution of the task (Gaillard & Wientjes, 1994). Hence, in order to capture all relevant aspects of the coping process, evaluation of individual differences in stress tolerance should employ multimethod assessment, preferably in a realistic task environment.

In an ambulatory evaluation of individual differences in stress tolerance, performance, subjective responses, minute ventilation (MV), and HR were assessed during an acute stressor task (Wientjes, Gerrits, Langefeld, & Gaillard,

1997). Seventy-two subjects participated in the study (all military personnel). A few weeks before the ambulatory measurements took place, the subjects filled in a series of questionnaires measuring personality characteristics and coping styles.

The stressor task was crossing a 60-meter long rope bridge across an 80-meter deep ravine. The rope bridge consisted of two ropes, a stepping rope and a supporting rope that was held tightly in both hands. During the crossing, the subjects were secured via a line attached to the supporting rope. Because the crossing was performed quite slowly, while stepping sideways on the stepping rope, the physical demands of the task were modest. However, the rope bridge was considered by experts to be an intense acute stressor. MV was measured via calibrated inductive plethysmography (Respitrace, Inc.) employing two bands with inductive coils around the chest and abdomen (see Wientjes, 1992). All equipment, including the ECG amplifier and the Respitrace device, was integrated in the ambulatory digital recorder (see Wientjes et al., 1994) carried by the subjects in a backpack. In addition to the physiological equipment, paper-and-pencil questionnaires were administered in order to assess the emotional response. The measurement sessions also included a precrossing baseline period. During the crossing, the investigators closely monitored the subjects and entered codes reflecting relevant events (such as the beginning and end of the crossing) on a handheld keyboard. The keystrokes were conveyed telemetrically to the digital recorder. Performance was evaluated in terms of the time needed for the crossing.

There was a wide range of variation in all measures, including crossing time. Compared to the baseline, MV, HR, and feelings of strain all increased substantially during the crossing. There were several correlations among the various measures: Performance was negatively correlated with HR during baseline, as well as with HR during the crossing. Furthermore, performance was negatively correlated with feelings of strain during the crossing and positively correlated with energetic feelings. However, notwithstanding the shared variance, each individual measure appeared to represent a somewhat different aspect of the response to the stressor. Because it had been postulated that stress tolerance should be evaluated in terms of a combination of physiological, subjective, and performance measures, rather than in terms of single measures, a cluster analysis was performed. The results indicated that two groups of subjects could reliably be distinguished. The majority of the subjects (about 70%) was characterized by good performance, low HR, low scores for feelings of strain, and high scores for energetic feelings. The remaining group showed the reverse pattern. Interestingly, subsequent analyses showed that the physiological and subjective group differences had already emerged during the prestressor baseline. It was concluded that stress tolerance was, in this particular stressor task, associated with the ability to optimally prepare for successful coping with the task demands by adaptively regulating physiological and psychological arousal levels. Contrary to expectations, the two groups did not differ with respect to any of the measures of personality or coping style that had previously been determined. Hence, it might be that the value of such questionnaires in predicting stress tolerance is more limited than is often suggested, and that practice tests encompassing assessment of performance as well as subjective and physiological responses in realistic task environments are indispensable when screening individualsfor stress tolerance.

Interactive Monitoring Studies of Emotional, Mental, and Physical Workload Components

Myrtek and coworkers developed a methodology to measure nonmetabolic HR (additional HR) based on ambulatory recordings of the ECG and physical activity. Triaxial actometer devices placed on the chest and thigh were used to measure physical activity. A carefully designed and tested algorithm accounted for activity-related changes in HR and ST-segment, which were recorded with five chest electrodes (Myrtek et al., 1988, 1996a). An emotional event was postulated if HR in a given minute was at least 3 beats per minute faster than the average of the previous minutes with no or only a small increase in physical activity (£10 units). The parameter adapts to conditions of increased HR due to physical activity. Because this analysis is conducted online, such events of nonmetabolic heart rate changes could be fed back to the participant by a beeper signal that also can be activated at random. The subject was instructed to report, by use of a pocket computer, self-ratings on mood, activities, and setting variables. Such interactive monitoring and contingency analysis of emotionally induced ECG changes proved to be applicable as a matter of routine. More than 1,000 subjects have been studied with this method (Myrtek et al., 1996b).

Such studies have included the effects of television viewing on school children, interoception of emotional arousal, and contingencies between ECG change and cardiac perception in patients. Moreover, in ergonomic studies the new methodology was applied to differentiate between emotional, mental, and physical workload components. Train drivers on high speed tracks and mountain tracks, white- and blue-collar workers, and students, the latter three groups at work and during leisure time, have served as participants as well (Myrtek et al., 1994). The conclusion from this unusually extensive research is contained in the statement by Myrtek et al. (1996b): "The results show that heart rate is an indicator of total workload, physical activity an indicator of the physical workload component, heart rate variability an indicator of the mental, and nonmetabolic heart rate an indicator of the emotional workload component. P-wave amplitude was identified as a general indicator of sympathetic activation but not as a contributing factor in the differing workload components. Very inconsistent results were seen for T-wave amplitude, making the interpretation of T-wave equivocal" (p. 302).

Psychophysiological Assessment of Human Reliability in a Simulated Complex System

Rau (1996) conducted a psychophysiological study that is noteworthy in its theoretical concepts and methodology. In her research on operators in an electroenergy network, she made reference to the concept of *human reliability* as proposed by Nebylizyn and to the concept of *decision latitude* as proposed by Karasek. The ability to perform one's job in accordance with the objective task demands and, simultaneously, to maintain an optimal level of psychophysiological arousal are the essential constructs here. Such research requires a multimethod assessment, preferably by ambulatory monitoring technology.

The study was carried out at simulated computer workplaces on a 110-kV electroenergy network with the same technical equipment and information transfer as in real life. The motivation of the operators was high because this 3-hour training was required for certification. Objective methods of task analysis were used to define typical tasks and performance data were recorded accordingly. HR, blood pressure, and physical activity were measured by use of a multichannel monitoring system. The operators, concurrent with blood pressure measurements, answered questions concerning subjective strain, emotion, perceived control, and success, by means of a handheld computer. Such a multilevel approach appears to be especially suited to the assessment of the psychophysiological arousal and recovery within the framework of demand and performance.

The study revealed marked increases in strain with increasing job demand and decision latitude. The operator in the dyadic team having decision authority showed a higher psychophysiological arousal level (HR, systolic blood pressure, self-report data concerning control and success) than the co-operator. Case studies were also reported in order to illustrate the advantages of this methodology for understanding the particular sequence of effects and their interactions. The design of complex systems may profit from such multilevel assessments in order to reduce strain, occurrence of errors, and boredom-induced decrements in human reliability.

CONCLUSION AND PERSPECTIVES

The exemplary investigations showed that computer-assisted ambulatory monitoring in applied environments has many advantages in the valid assessment of workload. However, only few examples are given in the next part of this book.

In conclusion, computer-assisted ambulatory assessment in applied environments is a newly emerging methodology. Progress is apparent not only in instrumentation but in assessment strategies. Ambulatory assessment, like any other method, has problematic aspects; a particular concern is how to account for multiple effects in the recordings. However, the benefits are evident:

- Recording of relevant data in natural settings;
- Real-time measurement of variations in workload and task demands, performance, and behavioral and physiological changes;
- Interactive real-time assessment by automatically prompting the subject to respond to questions or instructions;
- Real-time assessment and feedback by reporting physiological changes to the subject either with or without information as to which changes actually occur;
- Concurrent assessment of variations in workload and task demands, performance, and psychological and physiological changes (events, episodes);
- Correlation and contingency (symptom-context) analysis across systemic levels, as suggested in triple-response models (multimodal assessment);
- Ecological validity of findings and suitability for direct application.

The benefits of ambulatory monitoring for diagnosis and management of cardiovascular diseases, and of more recent applications, for example, polysomnograms and recordings of sleep apnea, are beyond doubt. Substantial findings were also obtained by computer-assisted ambulatory assessment in applied environments. Some of the findings substantiate theories and evaluations that rely on laboratory observations. It is evident, however, that other research findings challenge previously held views. Ambulatory assessment has also been especially fruitful in generating new research questions addressing, for example, the issue of behavior consistency or the discrepancies (response fractionation) between subjectively reported workload and actual measurements of objective workload and physiological change. Thus, research findings in relevant fields support further development and application of ambulatory assessment methodology. It may be expected that as ambulatory methodologies continue to improve, research will leave the laboratory to be performed in the field.

REFERENCES

Amlaner, C.J., & Macdonald, P.W. (Eds.). (1980). *A handbook of biotelemetry and radio tracking.* Oxford, England: Pergamon.

Berntson, G.G., Bigger, J.T., Eckberg, D.L., Grossman, P., Kaufmann, P.G., Malik, M., Nagaraja, H.N., Porges, S.W., Saul, J.P., Stone, P.H., & van der Molen, M.W. (1997). Heart rate variability: Origins, methods, and interpretive caveats. *Psychophysiology, 34,* 623–648.

Boucsein, W. (1991). Arbeitspsychologische Beanspruchungsforschung heute—eine

Herausforderung an die Psychophysiologie [Research on work stress: A challenge to psychophysiology]. *Psychologische Rundschau, 42,* 129–144.

Boucsein, W. (1992). *Electrodermal activity.* New York: Plenum.

Brügner, G. (1998). MONITOR: Ein flexibles Programm zur Datenerhebung mittels Pocket-PC [MONITOR: A flexible software for data acquisition]. *Zeitschrift für Differentielle und Diagnostische Psychologie, 19,* 145–147.

Buse, L., & Pawlik, K. (1996). Ambulatory behavioral assessment and in-field performance testing. In J.Fahrenberg & M.Myrtek (Eds.), *Ambulatory assessment: Computer-assisted psychological and psychophysiological methods in monitoring and field studies* (pp. 29–50). Se-attle: Hogrefe & Huber.

Bussmann, J.B.J., Tulen, J.H.M., van Herel, E.C.G., & Stam, H.J. (1998). Quantification of physical activities by means of ambulatory accelerometry. *Psychophysiology, 35,* 488–496.

Cacioppo, J.T., & Tassinary, L.G. (1990). *Principles of psychophysiology.* New York: Cambridge University Press.

Christie, M.J., & Todd, J.L. (1975). Experimenter-subject-situation interaction. In P.H. Venables & M.J.Christie (Eds.), Research psychophysiology (pp. 50–68). London: Wiley.

DeLongis, A., Hemphill, K.J., & Lehman, D.R. (1992). A structured diary methodology for the study of daily events. In F.B.Bryant, J.Edwards, R.S.Tindale, E.J.Posavac, L.Heath, E.Henderson, & Y.Suarez-Balcazar (Eds.), *Methodological issues in applied social psychology: Social psychological applications to social issues* (Vol. 2, pp. 83–109). New York: Plenum.

de Vries, M.W. (Ed.). (1992). *The experience of psychopathology. Investigating mental disorders in their natural settings.* Cambridge, MA: Cambridge University Press.

Fahrenberg, J. (1996a). Ambulatory assessment: Issues and perspectives. In J.Fahrenberg & M.Myrtek (Eds.), *Ambulatory assessment: Computer-assisted psychological and psychophysiological methods in monitoring and field studies* (pp. 3–20). Seattle: Hogrefe & Huber.

Fahrenberg, J. (1996b). Concurrent assessment of blood pressure, heart rate, physical activity, and emotional state in natural settings. In J.Fahrenberg & M.Myrtek (Eds.), *Ambulatory assessment: Computer-assisted psychological and psychophysiological methods in monitoring and field studies* (pp. 165–187). Seattle: Hogrefe & Huber.

Fahrenberg, J., Brügner, G., Foerster, F., & Käppler, C. (1998). Ambulatory assessment of diurnal changes with a hand-held computer: Mood, attention, and morningness-eveningness. *Personality and Individual Differences, 26,* 641–656.

Fahrenberg, J., Foerster, F., & Müller, W. (1996). Laboratory and field studies for improvement of ambulatory monitoring methodology. In J.Fahrenberg & M.Myrtek (Eds.), *Ambulatory assessment: Camputer-assisted psychological and psychophysiological methods in monitoring and field studies* (pp. 237–255). Seattle: Hogrefe & Huber.

Fahrenberg, J., Foerster, F., Smeja, M., & Müller, W. (1997). Assessment of posture and motion by multi-channel piezoresistive accelerometer recordings. *Psychophysiology, 34,* 607–612.

Fahrenberg, J., & Myrtek, M. (Eds.). (1996). *Ambulatory assessment: Computer-assisted psy-chological and psychophysiological methods in monitoring and field studies.* Seattle: Hogrefe & Huber.

Fahrenberg, J., & Myrtek, M. (in press). Ambulantes Monitoring und Assessment [Ambulatory monitoring and assessment]. In F.Rösler (Ed.), *Enzyklopädie der Psychologie. Biologische*

Psychologie. Band 4- Grundlagen und Methoden der Psychophysiologie. Göttingen, Germany: Hogrefe.

Foerster, F. (1997). *Programm-Paket BIO* [Computer software Bio]. Freiburg i. Br.: Universität, Forshungsgruppe Psychophysiologie [Available from the Psychophysiology Research group]. University of Freiburg, Germany.

Gaillard, A.W.K., & Wientjes, C.J.E. (1994). Mental load and work stress as two types of energy mobilization. *Work & Stress, 8,* 141–152.

Garssen, B., Buikhuisen, M., Hornsveld, H., Klaver, C., & van Doornen, L. (1994). Ambulatory measurement of transcutaneous pCO_2. *Journal of Psychophysiology, 8,* 231–240.

Giaconi, S., & Ghione, S. (1992). Seasonal and environmental temperature effects on arterial blood pressure. In T.F.H.Schmidt, B.T.Engel, & G.Blümchen (Eds.), *Temporal variations of the cardiovascular system* (pp. 344–350). Berlin, Germany: Springer.

Hancock, P.A., & Meshkati, N. (Eds.). (1988). *Human mental workload.* Amsterdam: Elsevier.

Hankins, T.C., & Wilson, G.F. (1998). A comparison of heart rate, eye activity, EEG, and subjective measures of pilot mental workload during flight. *Aviation, Space, and Environmental Medicine, 69,* 360–367.

Haynes, S.N., Gannon, L.R., Orimoto, L., & O'Brien, W.H. (1991). Psychophysiological assessment of poststress recovery. *Psychological Assessment, 3,* 356–365.

Haynes, S.N., & Horn, W.F. (1983). Reactivity in behavioral observation: A review. *Behavioral Assessment, 4,* 369–385.

Hedges, S.M., Jandorf, L., & Stone, A.A. (1985). Meaning of daily mood assessments. *Journal of Personality and Social Psychology, 48,* 428–434.

Hinton, J.W., & Burton, R.F. (1997). A psychophysiological model of psystress causation and response applied to the workplace. *Journal of Psychophysiology, 11,* 200–217.

Hocking-Schuler, J.L., & O'Brien, W.H. (1997). Cardiovascular recovery from stress and hypertension risk factors: A meta-analytic review. *Psychophysiology, 34,* 649–659.

Hugdahl, K. (1995). *Psychophysiology.* Cambridge, MA: Harvard University Press.

Jain, A., Martens, W.L.J., Mutz, G., Weiß, R.K., & Stephan, E. (1996). Towards a comprehensive technology for recording and analysis of multiple physiological parameters within their behavioral and environmental context. In J.Fahrenberg & M.Myrtek (Eds.), *Ambulatory assessment: Computer-assisted psychological and psychophysiological methods in monitoring and field studies* (pp. 215–235). Seattle: Hogrefe & Huber.

Jorna, P.G.A.M. (1992). Spectral analysis of heart rate and psychological state: A review of its validity as a workload index. *Biological Psychology, 34,* 237–257.

Käppler, C., Becker, H.-U., & Fahrenberg, J. (1993). Ambulantes 24-Stunden-Monitoring als psychophysiologische Assessmentstrategie: Reproduzierbarkeit, Reaktivität, Retrospektionseffekt und Bewegungskonfundierung [Ambulatory 24-hour monitoring used as psychophysiological assessment strategy: Reproducibility, reactivity, retrospection effect and physical activity]. *Zeitschrift für Differentielle und Diagnostische Psychologie, 14,* 235–251.

Kirschbaum, C., Read, G.F., & Hellhammer, D.H. (Eds.). (1992). *Assessment of hormones and drugs in saliva in biobehavioral research.* Seattle: Hogrefe & Huber.

Littler, W.A. (Eds.). (1980). *Clinical and ambulatory monitoring.* London: Chapman & Hall.

Luczak, H. (1987). Psychophysiologische Methoden zur Erfassung psycho-physischer Beanspruchungszustände [Psychophysiological methods for the assessment of psychophysiological load]. In U.Kleinbeck & J.Rutenfranz (Eds.), *Enzyklopädie der Psychologie. Arbeitspsychologie* (pp. 185–259). Göttingen, Germany: Hogrefe.

Manzey, D. (1998). Psychophysiologie mentaler Belastung [Psychophysiology of mental load]. In F.Rösler (Ed.), *Enzyklopädie der Psychologie. Biologische Psychologie. Band 5. Ergebnisse und Anwendungen der Psychophysiologie* (pp. 799–864). Göttingen, Germany: Hogrefe.

Martin, L., & Venables, P. (Eds.). (1980). *Techniques in psychophysiology.* Chichester, England: Wiley.

Miles, L.E., & Broughton, R.J. (Eds.). (1990). *Medical monitoring in the home and work environment.* New York: Raven Press.

Mulder, L.J.M. (1992). Measurement and analysis methods of heart rate and respiration for use in applied environments. Biological *Psychology, 34,* 205–236.

Myrtek, M., Brügner, G., & Fichtler, A. (1990). Diurnal variations of ECG parameters during 23-hour monitoring in cardiac patients with ventricular arrhythmias or ischemic episodes. *Psychophysiology, 27,* 620–626.

Myrtek, M., Brügner, G., Fichtler, A., König, K., Müller, W., Foerster, F., & Höppner, V. (1988). Detection of emotionally induced ECG changes and their behavioral correlates: A new method for ambulatory monitoring. *European Heart Journal, 9* (Suppl. N), 55–60.

Myrtek, M., Brügner, G., & Müller, W. (1996a). Interactive monitoring and contingency analysis of emotionally induced ECG changes: Methodology and applications. In J. Fahrenberg & M.Myrtek (Eds.), *Ambulatory assessment: Computer-assisted psychological and psychophysiological methods in monitoring and field studies* (pp. 115–127). Seattle: Hogrefe & Huber.

Myrtek, M., Brügner, G., & Müller, W. (1996b). Validation studies of emotional, mental, and physical workload components in the field. In J.Fahrenberg & M.Myrtek (Eds.). *Ambulatory assessment: Computer-assisted psychological and psychophysiological methods in monitoring and field studies* (pp. 287–304). Seattle: Hogrefe & Huber.

Myrtek, M., Deutschmann-Janicke, E., Strohmaier, H., Zimmermann, W., Lawrenz, S., Brügner, G., & Müller, W. (1994). Physical, mental, emotional, and subjective workload components in train drivers. *Ergonomics, 37,* 1195–1203.

Nelson, R.O., & Hayes, S.C. (Eds.). (1986). *Conceptual foundations of behavioral assessment.* New York: Guilford.

Patry, J.L. (Ed.). (1982). *Feldforschung. Methoden und Probleme sozialwissenschaftlicher Forschung unter natürlichen Bedingungen* [Field research. Methods and issues in social science research in naturalistic settings]. Bern, Switzerland: Huber.

Patterson, S.M., Krantz, D.S., Montgomery, L.C., Deuster, P.A., Hedges, S.M., & Nebel, L. E. (1993). Automated physical activity monitoring: Validation and comparison with physiological and self-report measures. *Psychophysiology, 30,* 296–305.

Pawlik, K. (1988). "Naturalistische" Daten für Psychodiagnostik: Zur Methodik psychodiagnostischer Felderhebungen ["Naturalistic" data in psychological assessment: On the methodology of psychodiagnostic field surveys]. *Zeitschrift für Differentielle und Diagnostische Psychologie, 9,* 169–181.

Pickering, T.G. (1991). *Ambulatory monitoring and blood pressure variability.* London: Science Press.

Rau, R. (1996). Psychophysiological assessment of human reliability in a simulated complex system. *Biological Psychology, 42,* 287–300.

Richter, P., & Hacker, W. (Eds.). (1998). *Belastung und Beanspruchung. Streß, Ermüdung und*

Burnout im Arbeitsleben [Stress and strain: Load, exhaustion and burnout at the workplace]. Heidelberg, Germany: R.Asanger.

Rösler, F. (Ed.). (in press). *Grundlagen und Methoden der Psychophysiologie* [Basics and methods of psychophysiology]. *Enzyklopädie der Psychologie. Biologische Psychologie. Band 4.* Göttingen, Germany: Hogrefe.

Rosenthal, R., & Rosnow, R.L. (Eds.). (1969). *Artifact in behavioral research.* New York: Aca-demic Press.

Schandry, R. (1988). *Lehrbuch Psychophysiologie* (2. Aufl.) [Textbook of psychophysiology]. Weinheim, Germany: Psychologie Verlags Union.

Schneiderman, N., Weiss, S.M., & Kaufmann, P.G. (1989). *Handbook of research methods in cardiovascular behavioral medicine.* New York: Plenum.

Schuler, H. (1982). Ethische Probleme [Ethical issues]. In J.L.Patry (Ed.), *Feldforschung* (pp. 341–364). Bern, Switzerland: Huber.

Shapiro, D., Jamner, L., Jane, J., Light, K.C., Myrtek, M., Sawada, Y., & Steptoe, A. (1996). Blood pressure publication guidelines. *Psychophysiology, 33,* 1–12.

Stemmler, G. (1992). *Differential psychophysiology: Persons in situations.* Berlin, Germany: Springer.

Stemmler, G. (1996). Strategies and designs in ambulatory assessment. In J.Fahrenberg & M. Myrtek (Eds.), *Ambulatory assessment: Computer-assisted psychological and psychophysiological methods in monitoring and field studies* (pp. 257–268). Seattle: Hogrefe & Huber.

Stemmler, G. (in press). Methodische Konzepte. Grundlagen psychophysiologischer Methodik [Basic methodological concepts in psychophysiology]. In F.Rösler (Ed.) *Enzyklopädie der Psychologie. Biologische Psychologie. Band 4. Grundlagen und Methoden der Psychophysiologie.* Göttingen, Germany: Hogrefe.

Stemmler, G., & Fahrenberg, J. (1989). Psychophysiological assessment: Conceptual, psychometric, and statistical issues. In G.Turpin (Ed.), *Handbook of clinical psychophysiology* (pp. 71–104). Chichester, England: Wiley.

Steptoe, A., Roy, M.P., Evans. O., & Snashall, D. (1995). Cardiovascular stress reactivity and job strain as determinants of ambulatory blood pressure at work. *Journal of Hypertension, 13,* 201–210.

Stern, E. (1986). *Reaktivitätseffekte in Untersuchungen zur Selbstprotokollierung des Verhaltens im Feld* [Reactivity in investigations based on self-report data in field studies]. Doctoral dissertation, University of Hamburg. Frankfurt, Germany: Lang.

Suen, H.K., & Ary, D. (1989). *Analyzing quantitative behavioral observational data.* Hillsdale, NJ: Lawrence Erlbaum Associates.

Tuomisto, M.T., Johnston, D.W., & Schmidt, T.F.H. (1996). The ambulatory measurement of posture, thigh acceleration, and muscle tension and their relationship to heart rate. *Psychophysiology, 33,* 409–415.

Turpin, G. (1990). Ambulatory clinical psychophysiology: An introduction to techniques and methodological issues. *Journal of Psychophysiology, 4,* 299–304.

Turpin, G., Shine, P., & Lader, H. (1983). Ambulatory electrodermal monitoring effects of ambient temperature, general activity, electrolyte media, and length of recording. *Psychophysiology, 20,* 219–224.

Veltink, P.H., & van Lummel, R.C. (Eds.). (1994). *Dynamic analysis using body fixed sensors.* Second World Congress of Biomechanics. The Hague, The Netherlands: Mc Roberts bv.

Veltman, J.A., & Gaillard, A.W.K. (1996). Physiological indices of workload in a simulated flight task. *Biological Psychology, 42*, 323–342.

Watson, D., & Pennebaker, J.W. (1991). Situational, dispositional, and genetic bases of symptom reporting. In J.A.Skelton & R.T.Croyle (Eds.). *Mental representation in health and illness* (pp. 60–84). New York: Springer.

Webster, J.G. (Ed.). (1988). *Encyclopedia of medical devices and instrumentation* (Vol. 1). New York: Wiley.

Wientjes, C.J.E. (1992). Respiration in psychophysiology: Methods and applications. *Biological Psychology, 34*, 179–203.

Wientjes, C.J.E., Gerrits, J.L.T., Langefeld, J.J., & Gaillard, A.W.K. (1997). *Stress tolerance, temperament, and coping.* (Rep. No. TM 97–A082 [Dutch]). Soesterberg, The Netherlands: TNO Human Factors Research Institute.

Wientjes, C.J.E., Spiekman, L.W.M., Benschop, J.F., & Hoogeweg, F. (1994). *Development and evaluation of the Stressomat: equipment, software, and manual* (Rep. No. TNO-TM 1994 Λ-54 [Dutch]). Socsterberg, The Netherlands: TNO Human Factors Research Institute.

Wientjes, C.J.E., ter Maat, R., & Gaillard, A.W.K. (1994). *Workload and stress at the CRC/ MilATCC: A test of the Stressomat.* (Rep. No. TNO-TM 1994 A-56 [Dutch]). Soesterberg, The Netherlands: TNO Human Factors Research Institute.

Wientjes, C.J.E., Veltman, J.A., & Gaillard, A.W.K. (1996). Cardiovascular and respiratory responses during a complex decision-making task under prolonged isolation. *Advances in Space Biology & Medicine, 5*, 133–155.

Wilhelm, F., & Roth, W.T. (1996). Ambulatory assessment of clinical anxiety. In J. Fahrenberg & M.Myrtek (Eds.), *Ambulatory assessment: Computer-assisted psychological and psychophysiological methods in monitoring and field studies* (pp. 317–345). Seattle: Hogrefe & Huber.

Wilson, G.F (1992). Applied use of cardiac and respiration measures: Practical considerations and precautions. *Biological Psychology, 34*, 163–178.

Zeier, H. (1994). Workload and psychophysiological stress reactions in air traffic controllers. *Ergonomics, 37*, 525–539.

II

APPLICATIONS

Chapter 6

A Psychophysiological Approach to Working Conditions

Gijsbertus Mulder
Lambertus J.M.Mulder
Theo F.Meijman
Johannes B.P.Veldman
Arie M.van Roon
University of Groningen, The Netherlands

In a recent review of the literature on mental workload Manzey (1998) gives a survey of the psychophysiological variables that have been proposed as indices of the amount of workload. Among them are spontaneous EEG activity in the different frequency bands, event-related brain potentials, and in particular the amplitude of the P300 component, cardiovascular indices, such as heart rate (HR), heart rate variability (HRV), and indices related to respiration and the oculomotor system. In this chapter we discuss the use of HRV, blood pressure, and catecholamines in assessing the load imposed upon workers in jobs where the load is mainly on perceptual and cognitive subsystems in the human information processing system. Our approach combines behavioral and psychophysiological studies. Most of the work starts in laboratory environments to explore psychophysiological measures and is followed by studies in the field. The results of these studies usually lead to further laboratory studies, including detailed analysis and modeling followed again by new applications.

We believe that research in the field of mental workload requires a sound theoretical model of human information processing, a model that is detailed enough to specify the information processing transactions involved in the task. Secondly, a thorough understanding of the control mechanisms underlying the particular psychophysiological indices of study is required. Computer simulation of both task performance and physiological control mechanisms are characteristic of this approach.

In this chapter, we discuss the use of cardiovascular and beta-adrenergic indices in assessing workload and the proper way to interpret changes in these parameters. Cardiovascular reactivity (e.g., HR and blood pressure) can be recorded noninvasively and important indices can be derived that can also be predictive and diagnostic for cardiovascular risks associated with performance of mental loading tasks. We first discuss the concept and measurement of mental effort. The remainder of the chapter focuses on research methodology using HRV, on baroreflex blood pressure control and its simulation, and finally on a discussion of the use of catecholamines. We hope that the discussion is detailed enough for the reader to consider our approach in their own work.

THE CONCEPT OF MENTAL EFFORT

On the Usefulness of Measuring Psychophysiological Activity During and After Work

The main reason to measure physiological activity during and after mental work is to assess the costs involved in performing mental tasks and to measure the duration of the imposed effects upon the task performer. In discussing the effects of mental work upon the task performer, it is useful to introduce the concept of *mental effort.* We distinguish between two types of effort: *computational* and *compensatory* effort.

Everything that makes a task more attention-demanding makes it more effortful. A task can also become very effortful if the task performer is very tired, works under conditions of extreme environmental noise, has been deprived of sleep, and so on. We believe that these two aspects of effort become manifest in different ways and in different psychophysiological indices and as a consequence should also be measured and approached in different ways.

Computational Effort

The first type of effort is related to the processing complexity of tasks. This type of mental effort was first discussed and specified by Kahneman (1973). Kahneman considered effort to be a nonspecific input to an information processing structure (a computational mechanism or a processor). Effort corresponds to what a subject is doing and is merely determined by the intrinsic demands of the task. In order to know how complex the task is and which processes are involved, cognitive task analysis is required (Card, Moran & Newell, 1983; Lamain, in press). In such an analysis the Executive-Process/Interactive-Control (EPIC) architecture of Meyer and Kieras (1997) is used to describe the task in terms of the processing load of the task on the three processors of EPIC (i.e., the perceptual, cognitive, and motor processors).

In our laboratory work, we searched for psychophysiological indices more or less uniquely related to perceptual, cognitive, and motor operations (Mulder et al., 1995). In order to validate psychophysiological measures, we have used, to a large extent, visual and memory search tasks designed by Shiffrin and Schneider (1977) and Schneider and Shiffrin (1977). In these tasks, the participants could either use automatic detection or controlled search. If automatic detection is used, performance is virtually independent of load (the number of items on the display or in the memory set or their product). In contrast, controlled search is heavily dependent on load. In consistent mapping conditions, in which targets (members of the memory set) and nontargets are kept separate, target stimuli or features of target stimuli acquire an automatic attention response. The subject can simply wait for the occurrence of one of the learned attention responses, check the detected item, and then respond. In varied mapping conditions, controlled search is used, a slow, effortful, serial item-by-item comparison process of all displayed items occurs to all the items in working memory. In some versions (counting versions), the subject had to keep a running mental count of the number of targets present in working memory (Aasman, Mulder, & Mulder, 1987). In other versions (selective attention versions), participants had to search in the visual display for stimuli with a certain size, color, or location (Wijers, 1989). We believe this type of task is prototypical for many tasks outside the laboratory in which controlled and automatic processes are present in different combinations.

A most surprising finding was the clear reduction in HRV during the performance of effortful mental tasks. Actually, we measured the reduction of HRV in a special frequency band of the cardiac interval signal (the mid-frequency or 0.10 Hz component). G.Mulder (1980) hypothesized that this reduction is due to the emergence of a phylogenetic old emotional response pattern, the defense reaction. During the defense reaction, the sensitivity of the baroreflex is reduced (as is discussed later). This type of finding suggests that emotional and cognitive systems be closely coupled. Also important was the observation that effortful processing, as indexed with the 0.10 Hz component, was not entirely determined by the intrinsic demand of the task as Kahneman hypothesized, but was clearly subject controlled. If the task became too difficult and/or exceeded the current capacities of the subject, the suppression of the 0.10 Hz component was less obvious (Aasman et al., 1987; Vicente, Thornton, & Moray, 1987). In other words, the task was performed with less effort and, as a consequence, errors became more likely.

We also concluded that the suppression of the 0.10 Hz component was very diagnostic (Mulder & Mulder, 1981): The index seems to indicate that the subject heavily uses attention-demanding control operations (so called *resource-limited*) in working memory (rehearsal, search, planning, etc.). If, on the other hand, the task required operations that were limited by the data (so-called *data-limited processing*), the 0.10 Hz component was unaffected. As Manzey (1998) correctly observes, the sensitivity of the 0.10 Hz component to the cognitive demand of tasks could not always be replicated. As far as we can see there are two important reasons involved here: The demands of the task and the differences between them should be relatively high. In addition, the respiration frequency of the subject should, in all conditions, be above the frequency of the 0.10 Hz component (Althaus, Mulder, Mulder, Van Roon, & Minderaa, 1998; Mulder, Van Roon, Veldman, Elgersma, & Mulder, 1995).

Compensatory Effort

Mental effort is also involved if participants perform tasks in adverse conditions. Under these conditions the other aspect of mental effort (compensatory effort) becomes important. In this respect, Broadbent's two arousal mechanisms are relevant (Broadbent, 1971): a *lower* mechanism concerned with well established (automatic) processes, which is affected by noise and sleeplessness, and an *upper* mechanism, which monitors and alters the parameters of the lower mechanism. Inefficiency of the lower mechanism will not become manifest in performance provided that the upper mechanism remains efficient. In that case, performance is hardly affected by stress, but has become less efficient because it can only be achieved at higher costs. We (G.Mulder, 1986) hypothesized that this aspect of effort would mainly manifest itself in another set of psychophysiological indices, that is, in changes in muscular activity (e.g., forearm flexor EMG), beta-adrenergic changes in cardiovascular parameters, and increased levels of adrenaline.

A second aspect of our laboratory research has therefore been to determine the sensitivity of a number of psychophysiological indices to this type of compensatory effort. The same experimental tasks mentioned before were now carried out under conditions of sleep deprivation and after fatiguing work conditions (Wiethoff, 1997). Such studies can be either performed in normal working conditions or in laboratory situations. Results of such studies confirm these predictions (see Waterink, 1999, and Wiethoff, 1997, for reviews).

Two important concepts are associated with the previously mentioned forms of effort: *effectiveness* and *efficiency.* Effectiveness refers to the quality of performance, whereas efficiency describes the relation between the quality of performance and the amount of effort invested in it. Efficiency decreases if more effort must be invested for the same quality of performance.

USING HRV

The fastest changes in HR are attributed to the pattern of respiration and will mainly be found in the frequency range between 0.15 and 0.40 Hz. Frequencies, corresponding to a rhythm with a duration of about 10 seconds (0.10 Hz component) are related to basic properties (eigen frequency) of the baroreflex, a reflex that regulates short time blood pressure. Lower frequencies are related to both adaptations to the task situation and to the control of body temperature.

Based on these mechanisms G.Mulder (1980) and L.Mulder (1988) defined the low (0.02–0.06 Hz), the mid- (0.07–0.14 Hz) and the high frequency band (0.15–0.40 Hz) of HRV. As mentioned before, we proposed to use the power in the mid-frequency band as an index of task-related mental effort. The main reason to use this mid-frequency band and not the power in the high frequency band was the interaction with respiration in this latter band. The HRV index of mental effort is defined as the rest-task difference of variability in the mid-frequency band.

Two types of effects can be distinguished, using this approach in a pre-post experimental paradigm. First, in order to reach the same task performance, mental effort after the work period has to be higher, resulting in a larger HRV index. Second, after the work, the subject is not prepared or able to deliver the same amount of effort in the task as before the working period (Meijman & Kompier, 1998). This situation results in decreased task performance in combination with less effort and, thus, a lower HRV index. By plotting (normalized) task performance on the x-axis and the HRV effect on the y-axis a mental efficiency diagram can be constructed. An example is given in Fig. 6.1, which illustrates the effect of working conditions of driving examiners. During a day off, mental effort spent in a standard task is low, whereas reaction times are at an intermediate level.

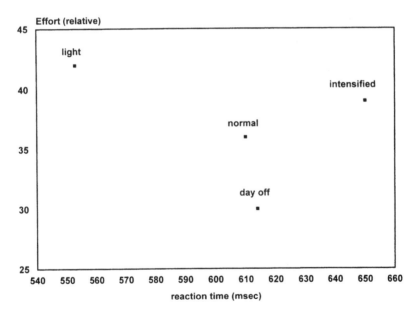

FIG. 6.1. Mental efficiency of driving examinators. Mental effort (vertical) and reaction time (horizontal) during a laboratory test task after three working sessions and during a day off obtained from 30 driving examiners. Mental effort scale: percentage decrease in HRV (mid-frequency band) as compared to a preceding rest. The higher the score, the higher the invested effort. Light: 9 examinations per day; Normal: 10; Intensified: 11.

After a working day of 9 examinations (light), mental effort in the same standard task is high as is the performance level Measurements after the two other working conditions, 10 examinations (normal) and 11 examinations (intensified), show that mental effort is decreased, whereas performance is diminished. However, the effort level is higher than during the day off. The data illustrate that subjects are no longer able (or motivated) to spend enough effort in the standard task for an adequate task performance.

Comparable data are obtained in studies on city busdrivers and in studies after heavy clerical work (Meijman & Kompier, 1998). Veldman (1992) used about the same procedure studying the effects of environmental noise on the efficiency of task performance, although he used blood pressure and heart rate as effort indices and combined measures of reaction times and errors for task performance. From several studies using this approach (Meijman, 1989), it can be concluded that HRV estmates can serve as an indication of a changed pattern of invested mental effort due to preceding mental work. The combination of the HRV index and performance measures in a mental efficiency diagram improves the insight into the strategies used by participants in such an experimental paradigm.

In the second paradigm, we measure mental effort during normal daily work. In this situation, task-related mental effort is hard to assess: Workload is changing continuously and a period of a few minutes of constant workload (and constant effort) are necessary for the computation of spectral measures (Mulder, 1980). This period can be shortened to about 30 seconds when spectral profile techniques are applied (Mulder, 1992; Veltman & Gaillard, 1996; Wiethoff, 1997). However, in such a case, numerous repetitions of similar situations will be required to get reliable estimates by averaging across these events. This approach uses psychophysiological measures as an indirect way to determine moments in which the real tasks are extremely loading or information processing load is moderate or low. HRV is used here as such an exploratory tool. With the use of a moving window monitoring changes from moment to moment in the mid-frequency band, the periods mentioned earlier are detected and with the use of simultaneous video recording the nature of these task elements can be identified (Wiethoff, 1997; Wildervanck, Mulder, & Michon, 1978; Van Westrenen, 1999). For example, Van Westrenen (1999) applied this technique in an attempt to identify moments of intense mental loading in the daily work of river pilots. River pilots navigate ships through the Nieuwe Waterweg to the port of Rotterdam. He measured HR and HRV and found six points in the route that contained difficult and even crucial decisions. At these points, HRV was suppressed.

SIMULATING BAROREFLEX CONTROL
OF BLOOD PRESSURE

One of the main research issues in understanding HRV as an index of mental effort is the background of this reduction in HRV The main question is how the decrease in HRV can be explained, that is, which mechanisms are responsible? The pattern during task related mental effort can largely be described by the defense reaction. Do all cardiovascular reactions (including changes in HRV, blood pressure variability, and baroreflex sensitivity) fit into this defensive response pattern?

The Baroreflex Model

We believe that an answer to these questions is of fundamental importance in interpreting the findings in both laboratory and applied studies. A model of the processes involved should be available in order to determine the nature of activation in autonomic space (e.g., Berntson, Cacioppo, Quigley, & Fabro, 1994). Such a model requires an adequate representation of the baroreflex blood pressure control subsystem. Increased activation of the baroreceptors in the vessel wall of the aortic arch and the carotid sinus is accompanied by an increase in vagal and a decrease in sympathetic activity, whereas decreased activation does the reverse (baroreflex).

As we mentioned before, mental workload temporarily evokes the defense reaction, characterized by an increase in HR and blood pressure, a decrease in HRV and blood pressure variability, and a decrease in the sensitivity of the baroreflex. Basic properties of the baroreflex are described very well in the literature (Karemaker, 1987) and several simulation models are available (Saul, Berger, Chen, & Cohen, 1991; Ten Voorde, 1992; Wesseling & Settels, 1993). We have decided on an extended version of the Wesseling & Settels model (Van Roon, 1998). Van Roon has implemented a complete simulation setup for the model, making the assumption that the main differences between rest and task are caused by both vagal and sympathetic gain changes, which have to be estimated from the HR and blood pressure data, including its variability measures.

Van Roon applied this approach to a series of laboratory studies mainly involving memory search and counting tasks. Experimental results (rest-task differences) can be summarized as follows: HR increases (5%-10%), blood pressure increases (5–10 mmHg), and HRV as well as blood pressure variability decrease in all frequency bands. In general, it seems that the pattern can be described by the occurrence of a defense reaction during task performance.

Van Roon found even with simple, initial simulations that blood pressure changes with the magnitude found in the experimental data can only be obtained by a clear increase in sympathetic activation. Furthermore, a simultaneous decrease of HRV and blood pressure variability in the mid-frequency band can only be obtained when both vagal and sympathetic gain are affected. However, as is discussed later, not all effects of mental tasks could be simulated by changes in the gain of the vagal and sympathetic component of the control system. We found that, in certain circumstances (e.g., tasks lasting for an extensive period of time), there is a direct change in autonomic activation, bypassing the baroreflex (for details, see van Roon, 1998).

Applying the Simulation Model

In this section we discuss two examples. In the first study, we attempt to simulate the differences between rest and task conditions in a simple short-lasting laboratory task. In the second (field) study, we investigated the development of visual fatigue in the course of the working day. Both simulations aim at a description of the effects of mental work in terms of changes in autonomic activity. It is the task of the simulation model to estimate these changes in autonomic activity using measures of HR, blood pressure, and its variability measures.

The first study was performed at the University of Bonn, Germany. The purpose of the experiment was to study the cardiovascular effects of vagal blockade during the performance of a mental loading task. In this chapter, however, only data of the preblockade part of the experimental session are described. A comparison is made between a baseline (rest) period (5 minutes) and a task period (5 minutes) in which participants had to perform a memory search and counting task. The results of this study are summarized in Fig. 6.2. For details, see van Roon (1998).

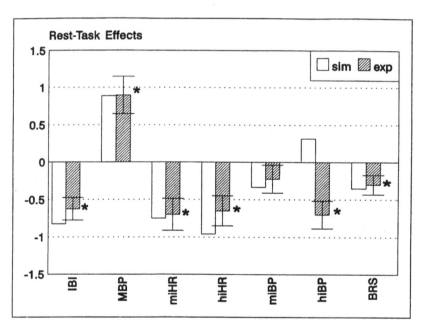

FIG. 6.2. Simulation of the effects of mental task load. Experimental and simulated effects of the rest-task difference in the memory search task (including counting). IBI: interbeat interval; MBP: mean blood pressure; miHR & hiHR, respectively: mid- and high-frequency band of heart rate; miBP & hiBP, respectively: mid- and high-frequency band of MBP; BRS: baroreflex sensitivity (transfer gain of changes in IBI/changes in MBP). Asterisks indicate significant experimental effects; intervals are mean +/− s.e. Scale for IBI: 0.5 corresponds with 50 ms effects; scale for MBP: 0.5 corresponds with 5 mmHg effect. Other variables are effect values after logarithmic transformation (nat. log).

The figure shows the rest-task differences of both the experimental and simulated data. Experimental rest-task effects can be characterized by a decrease in interbeat interval time (higher HR), an increase in blood pressure, a decrease in HRV (mid- and high-frequency band), and a decrease in blood pressure variability. Also, baroreflex sensitivity was significantly smaller. Blood pressure variability in the mid-frequency band is the only variable in which the rest-task difference is nonsignificant. The simulated data are obtained by a vagal gain decrease from 1.0 to 0.56 and a sympathetic gain decrease from 1.0 to 0.74. (Note that a sympathetic gain decrease results in an increase in sympathetic activation.) For all but one of the variables, the simulated data well resemble the experimental data.

In addition, the pattern of rest-task effects can be characterized as the appearance of the defense reaction: an increase in blood pressure and HR, a decrease in variability, and a decrease in baroreflex sensitivity. This clear pattern explains the excellent fit of the simulated data: A strong vagal gain decrease (of about 50%) and a sympathetic gain change (about 25%) which strengthens this effect is all that has to be changed in the model in order to fit the experimental data. This example study shows, like other comparable studies that have been simulated, that the computational effects of mental task load can be largely described by a change in both vagal and sympathetic gain. In the terms of Berntson, Cacioppo, and Quigley (1991), reciprocal autonomic activation occurs.

The aim of the second experiment was to investigate the pattern of cardiovascular effects during a working day; participants were required to perform a word processing and editing task during several hours. Because of the impossibility of measuring blood pressure at the finger during the typing work (and for such a long time) a number of short-lasting test measurements were included using a selective attention task. The selective attention task consisted of short lasting (3 minutes), alternating auditory and visual tasks (Veldman, Mulder, van Roon, van der Veen, & Mulder, 1998). The main aim in the present discussion of this experiment is to see whether the cardiovascular changes during the day can also be described by a pattern of autonomic gain adaptations estimated by the baroreflex simulation model. The results of this study are summarized in Fig. 6.3.

The figure shows the differences from the baseline of both the experimental and simulated data. Experimental effects show a specific pattern as a function of time: Blood pressure increases in the first part of the session and decreases thereafter. Interbeat interval time (IBI) increases in the first part of the session, while it decreases in the last part of the session; baroreflex sensitivity (BRS) follows the IBI pattern exactly. HRV, however, follows a different time course. A simple conclusion from the experimental data has to be that the cardiovascular data during long-term task performance is quite different from a pattern that mostly occurs during short-lasting laboratory tasks. The simulation results show (see Fig. 6.3) that part of the data can be surprisingly well simulated. This holds, in particular, for the pattern of changes. The estimates of vagal and sympathetic gain parameters (last panel of Fig. 6.3) show that in the prework test measurements

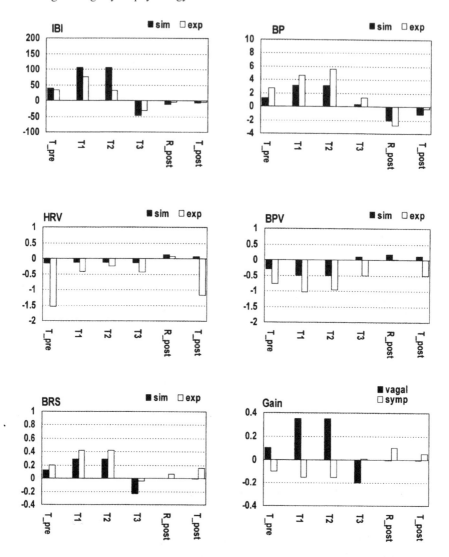

FIG. 6.3. Simulation of the effects of long-lasting clerical work. The pictures show differences with preceding baseline of the pre-work task (T_pre), three intermediate test periods (T1, T2, and T3) and post work rest/task (R_post, T_post), respectively. The first five panels show the experimental (exp) and simulated (sim) data of interbeat interval time (IBI), mean blood pressure (MBP), variability in the mid-frequency band of heart rate (HRV) and blood pressure (BPV), and finally baroreflex sensitivity (BRS). Scale for IBI: ms; Scale for MBP: mmHg. Other variables are effect values after logarithmic transformation (nat. log). The last panel shows the vagal and sympathetic gain changes during the session as estimated by the simulation model. Units: fractional changes in relation to the preceding baseline (rest). Note: A decrease of sympathetic gain corresponds to an increase in sympathetic activity.

and the first intermediate test measurements there is a vagal gain increase in combination with a sympathetic decrease, whereas during the third intermediate measurement there is a vagal gain decrease without a sympathetic change. (Note that a sympathetic gain decrease results in an increase of sympathetic activation.) In both the postwork rest and task situations vagal gain does not differ from baseline, whereas sympathetic gain is minimally increased. The changes in the mid-frequency band of HRV cannot be described adequately by these simulations. Direction of changes is not different, but the magnitude of the effects is. It can be concluded that, despite the long duration of the session (and thereby a long time distance between baseline and test measurements), the baroreflex simulation approach appears to be valuable in this case for determining autonomic gain changes during the session as a function of time.

In this section we have seen that effects of short-lasting mental workload (5 minutes) are reflected in HRV as well as in other variables, such as HR, blood pressure, and baroreflex sensitivity. Simulation reveals that autonomic gain changes can adequately describe the pattern that occurs, which can be characterized as a defense reaction. If a task lasts much longer the subject has an additional problem: how to keep the cardiovascular system in a state appropriate for task performance. Simulation studies indicate that this state-regulating process modulates the same variables but in a different way. Again, the whole pattern of changes can readily be described by autonomic gain changes, except for HRV Experimentally, in the prework test the same effects on HRV are found as in short-lasting tasks. On other variables (HR, BRS), however, the effect is opposite. Simulation shows that in such a case autonomic gain changes are not enough to describe the complete measured cardiovascular pattern. More detailed simulation, including parameter estimation of direct central (hypothalamic) influences will be necessary. In order to understand this type of change it will be necessary to move to another index of a change in the psychophysiological state of the operator: the level of catecholamines.

CATECHOLAMINES, EFFORT, AND STATE CONTROL

Human beings react with increased sympathoadrenal activity when they are confronted with challenging events or situations demanding active involve' ment. The secretion in the peripheral blood of the catecholamines adrenaline and noradrenaline are important parts of this reaction. Sympathetic innervation via the splanchnic nerve stimulates the adrenal medulla. This gland secretes adrenaline and a small fraction of the circulating noradrenaline in the blood. The latter hormone is secreted mainly at the presynaptic endplates of the (postganglionic) sympathetic nerve fibers. It has been shown that the excretion rates in urine are substantially and reliably correlated with the amount of the catecholamines circulating in peripheral blood (Moleman, Tulen, Blankenstijn, Man in't Veld,

& Boomsma, 1992). Therefore, these urinary excretion rates can be used as indicators of the sympathoadrenal activation. When we speak of adrenaline and noradrenaline levels in this chapter, we refer to the urinary excretion rates of peripheral circulating catecholamines.

The activation of adrenaline and noradrenaline is differentially related to the mobilization of resources needed in the adaptation of the organism to situational demands. Elevated noradrenaline levels in blood and urine predominantly characterize the mobilization of physical resources. When the adaptation to the situational demands mainly depends on the mobilization of mental resources, elevated levels of adrenaline are observed in blood as well as in urine (Frankenhaeuser, 1975, 1980; Dimsdale & Moss, 1980). Mental effort is the general concept denoting the mobilization of mental resources in order to adapt to changing demands. Therefore, adrenaline has been used as a psychophysiological indicator of mental effort.

Two aspects of the concept of mental effort were distinguished earlier. A task is said to be *mentally effortful* when it relies on attention-demanding mental operations. The other aspect of mental effort is related to the self-regulating mechanisms of the organism in the adaptation to changing situational demands (Hockey, 1986,1993). Due to factors like circadian rhythmicity, sleep loss, drug use, and, most commonly, fatigue, the actual psychophysiological state of the organism may not be the optimal one for meeting the demands placed upon it. In order to react properly under these circumstances, the actual, suboptimal state of the organism has to be adapted. Effort is used here to denote the compensatory mechanisms that are involved in such state-control processes. We hypothesize that these state-regulating mechanisms, denoted by the concept of compensatory effort, are stressful. On the hormonal level, they are characterized by elevated rates of adrenaline and cortisol; on the subjective level, they are characterized by feelings of tension and anxiety.

On the Use of Catecholamines in Workload Studies

The question now is whether changing adrenaline levels are indicators of both aspects of mental effort or of just one aspect, and if so, of which; it may even be that these hormonal changes have nothing to do with effort expenditure. In order to investigate this question, we have performed a number of studies.

Habituation and Workload

The first study was done in the field with 30 driving examiners (Meijman, Mulder, van Dormolen, & Cremer, 1992), who were investigated during a session of several memory search tasks they performed after their working day. Three conditions were distinguished. In an *intensified* working day, examiners were productive during 86% of their working time. Apart from the scheduled rest pauses there were no rest pauses between successive examinations in this intensified condition. In a *normal* working day, examiners were productive during 80% of their working time. In addition to the scheduled rest pauses, there were short breaks of about 3 minutes between successive examinations. In a nonintensified or *light* working day, examiners were productive during 76% of their working time. In addition to the scheduled rest pauses, there were intermittent breaks between successive examinations of about 7 minutes.

In all three conditions the examiners performed the same memory search tasks between 4:15 p.m. and 5:45 p.m., immediately after the end of the working day. The order of the three conditions was balanced over the 30 participants according to a Latin square design. At least 1 week separated the conditions. Urinary excretion rates of catecholamines, adrenaline, and noradrenaline were measured at the start and end of the memory search tasks session. In addition, several performance measures were taken in order to study the changes in mental efficiency under the various levels of workload. What is of interest now are the urinary excretion rates of both catecholamines measured at the end of the memory search task session, broken down to *order* (first time studied, second time studied, and third time studied) and to *work load condition* (light, normal, and intensified).

With respect to noradrenaline, no differences were found between the three order and the three work load conditions. The results with respect to adrenaline are summarized in Fig. 6.4.

Figure 6.4 shows that in all three work load conditions adrenaline decreased from the first time that the subject was studied to the third time. In all three order conditions, however, the highest levels of adrenaline were observed during the memory search task performance after the intensified working day. Both effects were statistically significant. Analysis of the performance data proved that reaction times were significantly worse after the intensified working day, the condition with the highest adrenaline rates. Details on these latter results can be found in Meijman et al. (1992).

These results may lead us to various conclusions. Apparently, a habituation resulted from the repetition of the same task situation. The challenging character or newness was gradually lost, and consequently a decrease of the adrenaline reaction was observed. On the other hand, the elevated adrenaline levels after the intensified working day may indicate that the subjects had to invest compensatory effort in performing the memory search tasks. This was confirmed by the performance data and the data on the subjective estimations of effort expenditure during the memory search tasks. The expenditure of compensatory effort in this condition, however, did not result in better performances compared to the other conditions. Noradrenaline levels were not related to these two psychological aspects of the task performance. This is in accordance with the theory that this hormone is not highly related to the mobilization of mental/psychic resources.

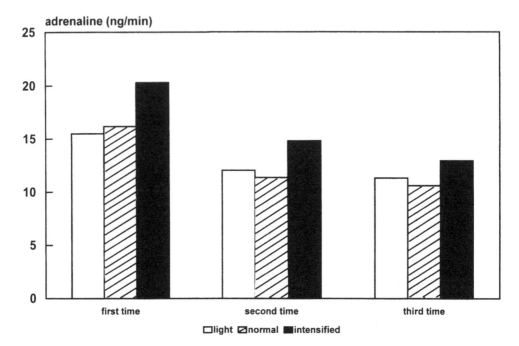

adrenaline (ng/min)

□light ⊠normal ■intensified

FIG. 6.4. Mean urinary adrenaline excretion rates during sessions of memory search after a working day, differentiated to order of the study and to preceding work load; data of 30 driving examiners. Light: 9 examinations per day; Normal: 10; Intensified: 11. Data are from the same study described in Fig. 6.1.

Mobilization and Compensation

Elevated adrenaline levels may be observed as part of two different reaction patterns. The first might be described as a general mobilization of (mental) resources or an *orienting reaction,* which is shown by the organism in the confrontation with new and/or challenging situations. Such mobilization is a prerequisite for adequate performance when needed in such situations, but which do not necessarily require any other state regulating processes. It is this aspect that is stressed by the toughness model of Dienstbier (1989, 1991). The orienting reaction will fade away by habituation. The second reaction pattern might be described as a *compensatory reaction,* which is shown by the organism in the confrontation with situational demands requiring an adaptation of the actual psychophysiological state in order to cope properly with the demands.

Frankenhaeuser (1989) and Frankenhaeuser and Lundberg (1985) made the distinction between *effort without distress* and *effort with distress,* denoting these different reaction patterns. On the subjective level, the former reaction is characterized by feelings of activation and the latter by feelings of tension or anxiety. Moreover, on the hormonal level, the latter reaction is characterized by elevated levels of cortisol in addition to the elevated levels of adrenaline.

In a field study of city bus drivers, we were able to differentiate between these two reaction patterns. The study shows that elevated levels of adrenaline may correlate either with feelings of activation or with feelings of tension or anxiety, depending on the psychophysiological state of the subject (Meijman & Kompier, 1998). During the morning hours of the working day, the increase in adrenaline levels is correlated (r=.40) with feelings of activation and not with feelings of anxiety or tension. It seems to be part of a general activation syndrome, indicating the mobilization of resources in order to perform the daily work: *effort without distress.* During the afternoon, after several hours of work, we may suppose that the psychophysiological state of the subject has changed. It could be considered suboptimal due to fatigue resulting from the first working hours. Elevated adrenaline levels correlated, under these circumstances, with increased (subjective) effort investment in order to maintain a (minimum) performance level under increased task load (r=.55) and correlated with feelings of anxiety or tension (r=.35), not with feelings of activation: *effort with distress.*

It may be concluded that the increase in adrenaline levels may have very different meanings depending on the psychophysiological state of the subject. When this state is suboptimal and the subject is forced to perform at a certain level, state regulating processes are needed, that is, compensatory effort has to be invested. Increased adrenaline secretion is a necessary part of this process that must be considered stressful. A similar conclusion may be drawn from studies of the aftereffects of intensified workload. Rissler and Elgerot (1978) report on such aftereffects, indicated by elevated levels of adrenaline, in the evening (after work) during a 13-Week period of overtime work and up to 4 weeks after this period.

In the study on driving examiners described earlier, a similar phenomenon was observed. During the evening, after the intensified working day, considerably elevated adrenaline excretion rates were measured, which did not return to baseline levels. These results are presented in Fig. 6.5. The baseline curve in Fig. 6.5 is based on the mean values at each sampling time of 3 separate days off. These values did not deviate from each other significantly.

After the light and normal working days, the adrenaline excretion rates returned to the baseline levels during the evening. After the intensified working day, however, they remained elevated over the 3 after work (4:00 p.m., 8:00 p.m., and 10:30 p.m.) sampling times. It might be concluded that the examiners remained in a state of overactivation after the intensified working day, as did the office workers in the Rissler and Elgerot (1978) study after working overtime for several consecutive weeks. As a result, the examiners reported sleeping problems at the end of the week with 5 intensified working days, in particular complaints like "staying awake for half an hour or longer," "tossing and turning," and "restless sleep." Measured with a standardized subjective sleep quality scale, developed

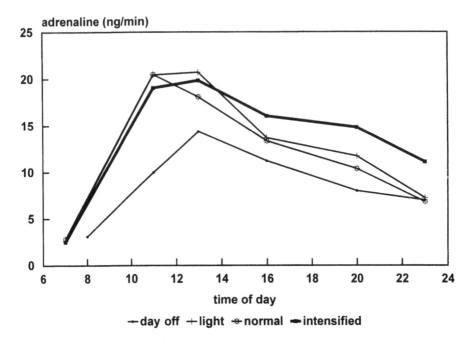

FIG. 6.5. Mean urinary adrenaline secretion at various times during the day off and three working days differing in intensity of work; data of 30 driving examiners. Data are from the same study described in Figs. 6.1 and 6.4.

by Mulder-Hajonides van der Meulen (see Meijman, de Vries-Griever, de Vries, & Kampman, 1985) these complaints were significantly higher compared to the self-reported sleep quality at the end of the light working week and the normal working week. In addition, considerably elevated feelings of tension or anxiety were measured during the Saturday (first day off) after the intensified working week, in particular during the first part of that day.

Some Conclusions

Humans, like most other mammals studied in the various biosciences, respond with a sympathoadrenal reaction to challenging situations. The two catecholamines, adrenaline and noradrenaline, which are part of this reaction, seem to have quite different functions. The (peripheral) secretion of noradrenaline plays a role in the mobilization of physical resources. The secretion of adrenaline plays a role in the mobilization of mental resources. Recently, it has become clear how adrenaline is able to do this. Peripheral adrenaline has an effect on the vagus nerve, which terminates in the nucleus of the tractus solitarius (NTS) in the medulla. NTS then sends outputs to the locus coeruleus (LC), which releases noradrenaline in widespread areas of the forebrain (LeDoux, 1996).

Elevated peripheral adrenaline levels may be part of two different reaction patterns that have quite different psychological or emotional connotations. The physiological function of the sympathoadrenal reaction is always the same: It plays a role in the mobilization of resources. But, depending on the psychophysiological state of the subject and the stringency of the situational demands requiring state changes, such mobilization may have very different meanings. Challenging situations demand an active mental involvement of the subject that provokes the adrenaline reaction. It becomes less pronounced or even fades away when such (same or similar) situations are repeated, provided they were mastered successfully in the past. In the studies by Ursin, Baade, and Levine (1978) the repetition of a parachute jump, after a first successful jump, provoked less pronounced adrenergic reactions. In our data from the driving examiner study, a similar phenomenon was found. Without further complications, the elevation of adrenaline rates may be interpreted, on the psychological level, as an indicator of *effort without distress* (Frankenhaeuser, 1989). On the subjective level, such elevations are accompanied by feelings of (positive) activation.

However, when subjects are forced to compensate for their apparently inadequate state in the confrontation with situational demands the elevation of peripheral adrenaline rates is embedded in a syndrome that is stressful. In such cases, the elevation of adrenaline rates is accompanied by feelings of tension and anxiety. The important thing for theories on work and health is that such compensatory mechanisms may be mobilized more or less chronically during work, depending on the general condition of health and wellbeing of the subject. Ursin (1980, 1988) pointed to the adverse effects of such *sustained* (sympathetic) *activation.* Our data suggest that such a phenomenon might apply in regular working situations.

FINAL CONCLUSIONS

In this chapter, we have identified the effects of working conditions on the psychophysiological state of the worker. The load of the work and its associated health effects can only be determined by recording psychophysiological measures in addition to performance and subjective measures. We found two major effects of work: one related to the cognitive complexity of the task (computational effort), the other to effects of the working conditions on the state of the worker (compensatory effort). With respect to the first type of effort, some scientists have complained (Luczak, 1987; see also Manzey, 1998) that HR and HRV are not diagnostic enough since they fail to distinguish between cognitive and emotional load. We believe that such a distinction is completely artificial. Cognitive and emotional systems are closely coupled in the brain and, as a consequence, the distinction between cognition and emotion is a typical example of a "Descartes type of error" (Damasio, 1994).

In studies in the field, HR and blood pressure (not always available continuously) are very often used. We showed that the sensitivity of the baroreceptor reflex is an important way by which the worker can change his cardiovascular state in order to cope with the complexity of the task. We also showed that computer simulation is necessary to understand the nature of the changes in the underlying physiological subsystems. The nature of the changes is expressed in terms of changes in the gain of the vagal and sympathetic control of cardiovascular parameters.

In this respect, it is important to mention the results of a study of Backs, Ryan, and Wilson (1994), also referred to in Manzey (1998). In this study, the authors argued that the failing diagnosticity of tonic HR and the 0.10 Hz component can be explained by the fact that in both indices sympathetic and parasympathetic effects are reflected. The authors argue that only in the sympathetic parts of HR and HRV perceptual-cognitive load is reliably present. In order to demonstrate this, these authors used principal component analysis. Hopefully, we have shown that with computer simulation of the dynamic changes in cardiovascular system parameters the contribution of the two components of the baroreflex can be differentially estimated. In short-lasting cognitive tasks, computational effort can be adequately described in terms of changes in the baroreflex sensitivity (defense reaction). We also showed that the simulation of cardiovascular effects in extreme working conditions requires an extra mechanism; we suggest a direct central effect on the rostral ventral medulla, which controls sympathetic output (Van Roon, 1998). During these conditions, we may observe the cardiovascular effects of compensatory effort. Moreover, we suggested that this particular change in state could also be nicely observed by measuring urine levels of catecholamines. It appeared that adrenaline reflects almost directly the change in the state of the worker, but adrenaline is insensitive to the complexity of a task or its pacing (i.e., it only reflects compensatory effort). Peripheral noradrenaline is not sensitive to the mental state of the subject.

Measuring changes in the sensitivity of the baroreflex requires continuous recording of blood pressure. This is not always possible. For that reason, we introduced the pre-post paradigm enabling the researcher to monitor the effects of work under controlled conditions.

We think that the time has now come to assess the effects of mental work in more absolute values that can be compared across a wide range of work situations. In this respect, we propose that the researcher attempt to simulate the effects on the cardiovascular system in terms of the absolute changes in gain of the controlling mechanisms. If strong sympathetic changes become apparent, the next step is to assess these effects in the secretion rates of adrenaline.

Finally, we know that for many situations such an approach is not always possible. Monitoring during-task performance the 0.10 Hz component of the cardiac interval signal can be a first step in identifying task situations that require adaptations of the cardiovascular system. Cognitive task analysis and further refinement should then follow.

ACKNOWLEDGMENT

Work on this chapter was supported by the Netherlands Concerted Research Action "Fatigue at Work," by the Netherlands Organisation for Scientific Research (NWO).

REFERENCES

Aasman, J., Mulder, G., & Mulder, L.J.M. (1987). Operator effort and the measurement of heart-rate variability. *Human Factors, 29,* 161–170.

Althaus, M., Mulder, L.J.M., Mulder, G., Van Roon, A.M., Minderaa, R.B. (1998). The influence of respiratory activity on the cardiac response pattern to mental effort. *Psychophysiology, 35,* 420–430.

Backs, R.W., Ryan, A.M., & Wilson, G.F. (1994). Psychophysiological measures of workload during continuous manual performance. *Human Factors, 36,* 514–531.

Berntson, G.G., Cacioppo, J.T., & Quigley, K.S. (1991). Autonomic determinism: The modes of autonomic control, the doctrine of autonomic space, and the laws of autonomic constraint. Psychological *Review, 98*(4), 459–487.

Berntson, G.G., Cacioppo, J.T., Quigley, K.S., Fabro, V.T. (1994). Autonomic space and psychophysiological response. *Psychophysiology, 31,* 44–61.

Broadbent, D.E. (1971). *Decision and stress.* London: Academic Press.

Card, S.K., Moran, T.P., & Newell, A.L. (1983). *The psychology of human-computer interaction.* Hillsdale, NJ: Lawrence Erlbaum Associates.

Damasio, A. (1994). *Descartes' error: Emotion, reason, and the human brain.* New York: Grosset/Putnam.

Dimsdale, J.E., & Moss, J. (1980). Plasma catecholamines in stress and exercise. *Journal of the American Medical Association, 243,* 340–342.

Dienstbier, R.A. (1989). Arousal and physiological toughness: implications for mental and physical health. *Psychologial Review, 96,* 84–100.

Dienstbier, R.A. (1991). Behavioral correlates of sympathoadrenal reactivity: The toughness model. *Medicine and Science in Sports and Exercise, 23,* 846–852.

Frankenhaeuser, M. (1975). Experimental approaches to the study of catecholamines and emotion. In L.Levi (Ed.), *Emotions—their parameters and measurement* (pp. 212–225). New York: Raven.

Frankenhaeuser, M. (1980). Psychobiological aspects of life stress. In S.Levine & H.Ursin (Eds.), *Coping and health* (pp. 102–123). New York: Plenum.

Frankenhaeuser, M. (1989). A biopsychosocial approach to work life issues. *International Journal of Health Services, 19,* 747–758.

Frankenhaeuser, M., & Lundberg, U. (1985). Sympathetic-adrenal and pituitary-adrenal response to challenge. In P.Pichot, P.Berner, F.Wolf, & K.Thau (Eds.), *Psychiatry* (Vol. 2, pp. 53–68). New York/London: Plenum.

Hockey, G.R.J. (1986). A state control theory of adaptation and individual differences in stress management. In G.R.J.Hockey, A.W.K.Gaillard, & M.G.H.Coles (Eds.), *Energetics and human information processing.* Dordrecht, The Netherlands: Martinus Nijhoff Publishing.

Hockey, G.R.J. (1993). Cognitive-energetical control mechanisms in the management of work demands and psychological health. In A.Baddeley & L.Weiskrantz (Eds.), *Attention:*

Selection, awareness, and control A tribute to Donald Broadbent. Oxford, England: Clarendon.

Kahneman, D. (1973). *Attention and effort.* Englewood Cliffs, NJ: Prentice Hall.

Karemaker, J.M. (1987). Neurophysiology of the baroreceptor reflex. In R.I.Kitney & O. Rompelman (Eds.), *The beat-by-beat investigation of cardiovascular function* (pp. 27–49). Oxford, England: Clarendon.

Lamain, W. (in press). *Cognitive task analysis for camputer supported tasks.* Unpublished doctoral dissertation, University of Groningen, The Netherlands.

Luczak, H. (1987). Psychophysiologische Methoden zur Erfassung psychophysischer Beanspruchungszustände [Psychophysiological methods for psychophysical mental states]. In I.U.Kleinbeck & J.Rutenfranz (Eds.), *Arbeitspsychologie. Enzyklopädie der Psychologie.* Göttingen, Germany: Hogrefe.

LeDoux, J. (1996). *The emotional brain.* New York: Simon and Schuster

Manzey, D. (1998). Psychophysiologie mentaler Beanspruchung [Psychophysiology of mental load]. In F.Rösler (Ed.), *Ergebnisse und Anwendungen der Psychophysiologie. Serie I. Biologische Psychologie.* Enzyklopädie der Psychologie. Göttingen, Germany: Hogrefe.

Meyer, D.E., & Kieras D.E. (1997). A computational theory of executive cognitive processes and multiple task performance: Part 1. Basic mechanisms. *Psychological Review, 104,* 3–65.

Meijman, T.F. (1989). *Over Vermoeidheid* [On fatigue: Studies of the aftereffects of workloads]. Doctoral dissertation, University of Groningen, The Netherlands.

Meijman, T.F., de Vries-Griever, A.H.G., de Vries, G., & Kampman, R. (1985). *The evaluation of the Groningen Sleep Quality Scale* (Heymans Bulletin 1720 HB 0767). Groningen, The Netherlands: Department of Experimental and Work Psychology.

Meijman, T.F., & Kompier, M. (1998). Busy business: How city bus drivers cope with time pressure, passengers, and traffic safety. *Journal of Occupational Health Psychology, 3*(2), 109–121.

Meijman, T.F., Mulder, G., van Dormolen, M., & Cremer, R. (1992). The workload of driving examiners: A psychophysiological field study. In H.Kragt (Ed.), *Enhancing industrial performance.* London: Taylor & Francis.

Moleman, P., Tulen, J.H.M., Blankenstijn, J.P., Man in't Veld, A.J., & Boomsma, F., (1992). Urinary excretion of catecholamines and their metabolites in relation to circulating catecholamines: Six-hour infusion of epinephrine and norepinephrine in healthy volunteers. *Archives of General Psychiatry, 49,* 568–572.

Mulder, G. (1980). *The heart of mental effort: Studies in the cardiovascular psychophysiology of mental work.* Doctoral dissertation, University of Groningen, The Netherlands.

Mulder, G. (1986). The concept and measurement of mental effort. In G.R.J.Hockey, A.W. K.Gaillard, & M.G.H.Coles (Eds.), *Energetics and human information processing* (pp. 175–198). Dordrecht, The Netherlands: Martinus Nijhoff.

Mulder, G., & Mulder, L.J.M. (1981). Task-related cardiovascular stress. In J.Long & A. Baddeley (Eds.), *Attention and performance* IX. Hillsdale, NJ: Lawrence Erlbaum Associates.

Mulder, G., Wijers, A.A., Lange,J.J., Buijink, B.M., Mulder, L.J.M., Willemsen, A.T.M., & Paans, A.M.J. (1995). The role of neuroimaging in the discovery of processing stages. A review. *Acta Psychologica, 90,* 63–79.

Mulder, L.J.M. (1988). *Assessment of cardiovascular reactivity by means of spectral analysis.* Doctoral dissertation, University of Groningen, The Netherlands.

Mulder, L.J.M. (1992). Measurement and analysis methods of heart rate and respiration for the use in applied environments, *Psychophysiology, 34,* 205–236.

Mulder, L.J.M., Van Roon, A.M., Veldman, J.B.P., Elgerma, A.F., Mulder, G. (1995).

Respiratory pattern, invested effort, and variability in heart rate and blood pressure during the performance of mental tasks. In M.Di Rienzo, G.Mancia, G.Parati, A.Pedotti, & A. Zanchetti (Eds.), *Computer analysis of cardiovascular signals.* Amsterdam: IOS Press.

Rissler, A., & Elgerot, A. (1978). Stress *reactions related to overtime at work.* Stockholm, Sewden: Department of Psychology, University of Stockholm.

Saul, J.P., Berger, R.D., Chen, M.H., & Cohen, R. (1991). Transfer function analysis of autonomic regulation: II. Respiratory sinus arrythmia. *American Journal of Physiology, 256,* H153–H161.

Shiffrin, R.M., & Schneider, W. (1977). Controlled and automatic human information processing: II. Perceptual learning, automatic attending and a general theory. *Psychological Review, 84,* 127–190.

Schneider, W., & Shiffrin, R.M. (1977). Controlled and automatic human information processing: I. Detection, search, and attention. *Psychological Review, 84,* 1–66.

TenVoorde, B.J. (1992). *Modelling the baroreflex.* Master's thesis, Vrije University, Amsterdam, The Netherlands: CopyPrint 2000.

Ursin, H. (1980). Expectancy, activation and somatic health. A new psychosomatic theory. In S.Levine & H.Ursin (Eds.), *Coping and health* (pp. 2–26). New York: Plenum.

Ursin, H. (1988). Expectancy and activation: An attempt to systematize stress theory. In D. Hellhammer, I.Florin, & H.Weiner (Eds.), *Neurobiological approaches to human disease.* Toronto: Hans Huber.

Ursin, H., Baade, E., & Levine, S. (1978). *Psychobiology of stress. A study of coping men.* New York: Acade.

Van Roon, A.M. (1998). *Short-term cardiovascular effects of mental work.* Doctoral dissertation, University of Groningen, The Netherlands.

van Westrenen, F. (1999). *The river-pilot at work.* Doctoral dissertation, Delft University of Technology, The Netherlands.

Veldman, J.B.P. (1992). *Hidden effects of noise.* Doctoral disseration, University of Groningen, The Netherlands.

Veldman, J.B.P., Mulder, L.J.M., Van Roon, A.M., vander Veen, F.M., & Mulder, G. (1998). Test measurements are a powerful tool in determining cardiovascular effects of long lasting mental work. *Journal of Psychophysiology, 12,* 338–352.

Veltman, J.A, & Gaillard, A.W.K. (1996). Physiological indices of workload in a simulated flight task. *Biological Psychology, 42,* 323–242.

Vincente, K.J., Thornton, D.C., & Moray, N. (1987). Spectral analysis of sinus arrhythmia: A measure of mental effort. *Human Factors, 29,* 171–182.

Waterink, W. (1999). *Facial muscle activity as an index of energy mobilization during processing of information processing: An EMG study.* Doctoral dissertation, Catholic University of Brabant, Tilburg, The. Netherlands.

Wesseling, K.H., & Settels, J.J. (1993). Circulatory model of baro- and cardio-pulmonary reflexes. In M.Di Rienzo, G.Mancia, G.Parati, A.Pedotti, & A.Zanchetti (Eds.), *Blood pressure and heart rate variability.* Amsterdam: IOS Press.

Wiethoff, M. (1997). *Task analysis is heart work. The investigation of heart rate variability: A tool for task analysis in cognitive work.* Doctoral dissertation, University of Groningen, The Netherlands.

Wildervanck, C., Mulder, G., & Michon, J.A. (1978). Mapping mental load in car driving. *Ergonomics, 21,* 255.

Wijers, A.A. (1989). *Selective visual attention.* Doctoral dissertation, University of Groningen, The Netherlands.

Chapter 7

The Validity of Factor Analytically Derived Cardiac Autonomic Components for Mental Workload Assessment

John K.Lenneman
Richard W.Backs
Central Michigan University

Heart rate has probably been used to assess mental workload in the field more often than any other psychophysiological measure. Heart rate has often been found to be a sensitive measure of mental workload. O'Donnell and Eggemeier (1986) define *sensitivity* as the "capability of a technique to detect changes in the amount of workload imposed on task performance" (p. 42–2). Heart rate typically increases as mental workload increases and decreases as mental workload decreases. However, heart rate does not always differ between conditions that impose different amounts of mental workload according to other measures such as task performance or subjective reports. For example, heart rate has not been found to differ between day and night conditions of flight (Lewis, Jones, Austin, & Roman, 1967; Roscoe, 1978) or as the angle of approach during landing increases (Roscoe, 1975).

Further, heart rate does not provide diagnostic information about the source of mental workload. O'Donnell and Eggemeier (1986) define *diagnosticity* as the "capability of a technique to discriminate the amount of workload imposed on different operator capacities or resources" (p. 42–3). Backs (1995) offers two reasons for heart rate's limited diagnostic utility. First, heart rate is affected by physical demands that may be independent of mental workload. Second, heart rate does not provide information about the separation of the sympathetic and parasympathetic nervous system activity.

Knowledge of the underlying activity of the autonomic nervous system (ANS) may increase both sensitivity and diagnosticity in mental workload assessment beyond that obtainable with heart rate. Uncovering the pattern of sympathetic and parasympathetic activity elicited during task performance may facilitate inferences about the nature of the cognitive demands imposed by a task. For example, Obrist (1981) discussed two different mechanisms that are responsible for heart rate change. The first mechanism, cardiac-somatic coupling, produces a parasympathetic response that occurs during passive coping. The second mechanism, cardiac-somatic uncoupling, produces a sympathetic response that occurs during active coping. The same heart rate can result from either mechanism. Information about the activity of the sympathetic and parasympathetic nervous systems must be obtained to identify the mechanism responsible for observed heart rate. In turn, knowledge of the mechanism would permit inferences about the psychological processes (e.g., active vs. passive coping) used to perform a task.

We believe that the Obrist type of cardiac mechanism has been subsumed by the more general construct of the "mode of autonomic control" (Berntson, Cacioppo, & Quigley, 1991). The mode of autonomic control refers to the neurogenic pattern of sympathetic and parasympathetic activity responsible for heart rate. Traditionally, the sympathetic and parasympathetic branches of the ANS were thought to have reciprocal effects on heart rate; that is, heart rate change was due to the activation of one branch with concomitant inhibition of the other branch. However, the "doctrine of autonomic space" (Berntson et al., 1991, p. 459) posits that the modes of autonomic control are multidimension-ally determined instead of merely reciprocally coupled. There are eight possible modes of autonomic control of heart rate in autonomic space (Berntson et al., 1991): the traditional coupled reciprocal modes, the uncoupled modes, and the coupled nonreciprocal modes (see Table 7.1). With coupled nonreciprocal modes of control, heart rate may not change, even though the underlying sympathetic and parasympathetic activity may change greatly, because the effects of the two branches may cancel each other.

Identifying the mode of autonomic control for heart rate can increase sensitivity and diagnosticity. Sensitivity may be increased when tasks that differ in mental workload according to other indicators, but not heart rate, differ in autonomic space. For example, the tasks may elicit different coupled nonreciprocal control modes (i.e., coactivation vs. coinhibition). Diagnosticity would also be increased by allowing one to form inferences about the psychological processes that occur during the performance of a task. For example, if the heart rate observed during a task was produced by uncoupled sympathetic activation that represented cardiac-somatic uncoupling, then it may be inferred that performance of the task required effortful, active coping.

Several cardiovascular measures reflect activity of one branch of the ANS more than the other. However, careful consideration must be given to the measures that will be used to assess the activity of the ANS branches. In the present study, *heart period* (the time between successive heart beats) was used instead of heart

TABLE 7.1
Possible Modes of Autonomic Control for Heart Rate

Control Mode	Sympathetic Response	Parasympathetic Response
Reciprocally Coupled Modes		
Sympathetic Activation	⇑	⇓
Parasympathetic Activation	⇓	⇑
Nonreciprocally Coupled Modes		
Coactivation	⇑	⇑
Coinhibition	⇓	⇓
Uncoupled Modes		
Sympathetic Activation	⇑	–
Sympathetic Inhibition	⇓	–
Parasympathetic Activation	–	⇑
Parasympathetic Inhibition	–	⇓

rate (the number of beats per minute) because it has superior statistical properties (Berntson, Cacioppo, & Quigley, 1995). Other measures used were the high-frequency (respiratory sinus arrhythmia or RSA) and low-frequency (0.1 Hz or blood pressure band) components of heart period variability (HRV), residual heart period, and pre-ejection period (PEP). RSA is considered to be a good measure of parasympathetic (specifically, vagal) activity (Berntson et al., 1997). PEP (Cacioppo et al., 1994) and residual heart period (Grossman & Svebak, 1987) are considered to be measures of sympathetic activity. The low-frequency HRV is considered to be a measure of both sympathetic and parasympathetic activity (Berntson et al., 1997). RSA, low-frequency HRV, and residual heart period are all obtained from additional analyses of heart period, whereas PEP is obtained from the impedance cardiogram.

A multivariate approach of analyzing the psychophysiological measures obtained from heart period has been proposed as a method for obtaining cardiac autonomic information (Backs, 1995, 1998). The multivariate approach attempts to improve the sensitivity and diagnosticity of heart rate by identifying the neurogenic activity of the sympathetic and parasympathetic nervous systems responsible for the observed heart rate in a task. Principal components analysis (PCA) was used in the present study to extract information about the sympathetic and parasympathetic nervous systems common to RSA, low-frequency HRV, residual heart period, and heart period. Details of how the components were derived are presented in the Method section.

PCA of the measures results in two components that are used to compute autonomic component scores that may reveal the activity of the sympathetic and parasympathetic nervous systems within the "doctrine of autonomic space" (Berntson et al., 1991). However, the use of PCA to extract cardiac autonomic information can only be validated through examination of the extent to which the PCA method is consistent with better methods of obtaining the same information.

The present study was designed to assess the validity of the PCA method. The first approach to assessing the validity was to determine whether the autonomic space created by the PCA components was consistent with the autonomic space created by PEP and RSA. Because PEP and RSA are considered to be more direct sympathetic and parasympathetic measures, respectively (e.g., Cacioppo et al., 1994), similarity of the autonomic spaces would indicate that PCA is a valid method of obtaining cardiac autonomic information. The second approach was to examine correlations between the sympathetic component and PEP and between the parasympathetic component and RSA. If the PCA components are valid, then they should correlate significantly with the more direct measures of autonomic function.

METHOD

Twenty male university students, in good health and free of medications that affect the cardiovascular system, participated in the study. Electrocardiogram (EKG), basal thoracic impedance, and the first derivative of pulsatile changes in transthoracic impedance (dZ/dt) were obtained via a Minnesota Impedance Cardiograph Model 304B. A tetrapolar montage of Mylar band electrodes was used in the acquisition of the EKG and impedance cardiogram (ZKG) (Sherwood et al., 1990). Respiration (the sum of abdominal and thoracic circumference) was obtained using the Grass Model 8–16 E polygraph.

Each subject participated in five tasks: cold pressor, exercise, illusion, memory, and tracking. For the cold pressor task, subjects submerged their left hand to the wrist for 2 minutes in a bucket containing ice water maintained at a temperature of 3 to 4° C. For the exercise task, subjects were asked to pedal a mechanically braked bicycle ergometer at a steady rate of 50 revolutions per minute under two exercise loads, 8 and 25 W. For the illusion task, the only requirement was to experience the illusion, which was a moving pattern of dots that formed various geometric shapes. For the memory task, subjects counted each member of a one or three letter target set. Letters were presented visually for 0.5 seconds every 2 seconds, and subjects verbally reported their count for each member of the target set at the end of the trial. For the tracking task, subjects were required to maintain the position of a moving cursor upon a fixed target. Four different tracking tasks were formed from the factorial combination of two levels of order-of-control (velocity and acceleration) and two levels of disturbance (sum of 11 cosine waves that were low-pass filtered with a break frequency of 0.06 or 0.25 Hz). Subjects were allowed to practice each condition of the memory and the tracking tasks for 1 minute before beginning the actual experiment. A 2 minute resting baseline condition preceded each of the tasks; that is, a baseline was collected before the cold pressor, the four tracking tasks, the two memory tasks, the visual illusion task, and the two exercise tasks.

The MXedit (Delta-Biometrics, Inc.) software package was used to determine the heart period variability in the low (0.06–0.11 Hz) and RSA (0.12–0.40 Hz) frequency bands from the heart period data according to the method of Porges and Bohrer (1990). A bandwidth of 0.12–1.0 Hz was used to determine RSA for the 25 W exercise task. Residual heart period was calculated as the difference between the actual heart period and the heart period predicted from RSA by linear regression (Grossman & Svebak, 1987). EKG and ZKG were sampled at 500 Hz and heart period and PEP were obtained using the software described by Kelsey and Guethlein (1990).

PCA was used to extract information about the sympathetic and parasympathetic nervous systems common across the four cardiovascular measures obtained with heart period. A correlation matrix for heart period, RSA, low-frequency HRV, and residual heart period was computed for each task and baseline and then pooled to yield a single correlation matrix. PCA was performed on the pooled correlation matrix. One component was expected to be defined by the marker variable for sympathetic activity (residual heart period) and the other was expected to be defined by the marker variable for parasympathetic activity (RSA). The first two components were rotated to simple structure using varimax. The component scores for a particular subject in a task were computed as the weighted sum of the subject's standardized score for each cardiovascular measure multiplied by the standardized scoring coefficient.

RESULTS

SAS Proc Factor was used to compute the component loadings and standardized scoring coefficients for each of the cardiac measures (SAS Institute, Inc., 1989). Component 1 was identified as the parasympathetic component because of the large weight for RSA. Component 2 was identified as the sympathetic component because of the large weight for residual heart period. The standardized scoring coefficients for Components 1 and 2 that were used to produce the parasympathetic and sympathetic component scores, respectively, for each subject within a task are shown in Table 7.2.

Difference scores from baseline-to-task were computed for the PCA components, PEP, and RSA. A positive difference score means that a task elicited increased heart period (i.e., slower heart rate) that could be caused by sympathetic

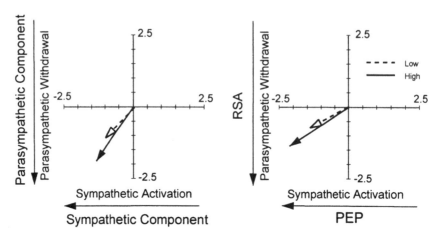

FIG. 7.1. An autonomic space representation of cardiac control modes using the group mean autonomic component (left panel) and PEP and RSA (right panel) difference scores for the low and high intensity exercise tasks.

TABLE 7.2

The Rotated Factor Patterns and Standardized Scoring Coefficients for RSA, Low Frequency HRV, Heart Period, and Residual Heart Period

Cardiovascular Measure	Rotated Factor Pattern		Standardized Scoring Coefficients	
	Component 1	Component 2	Component 1	Component 2
RSA	.96	–.14	.41	–.07
Low frequency HRV	.88	–.26	.39	–.17
Heart period	.74	.50	.36	.46
Residual heart period	–.16	.93	–.03	.77
% variance explained before rotation	55.84	29.57		

RSA = Respiratory Sinus Arrhythmia; HRV = heart rate variability.

withdrawal and/or parasympathetic activation. A negative difference score means that a task elicited decreased heart period (i.e., faster heart rate) that could be caused by sympathetic activation and/or parasympathetic withdrawal.

Autonomic modes of control are represented graphically in Figs. 7.1 through 7.6 by a vector in autonomic space from the origin (i.e., zero after baseline correction) to a point determined by the sympathetic and parasympathetic change from baseline-to-task. PEP and RSA have been converted to z-scores in the figures to make the display more directly comparable with the PCA components. Change parallel to the positive diagonal suggests a coupled reciprocal mode of control, whereas change parallel to one axis suggests an uncoupled mode of control. Separate Hotelling's T^2s were used to test the significance of the vector in autonomic space for the PCA components and for PEP and RSA. Then, t-tests were used to test the significance of the difference from baseline-to-task for each PCA component, PEP and RSA, and heart period.

Consistency Between the PCA Components and PEP and RSA

The autonomic space created by the PCA components was consistent with the autonomic space created by PEP and RSA for the exercise tasks (see Fig. 7.1). Both the PCA component (T^2 [2, 18]=86.32 and 217.91, p<.001) and the PEP and RSA (T² [2, 18]=82.82 and 185.74, p<.001) vectors were significantly different from the origin for low and high intensity exercise. In addition, heart period, the sympathetic and parasympathetic components, PEP, and RSA were each significantly different from baseline for the two exercise tasks (Table 7.3). The vectors indicate that the exercise tasks elicited reciprocally coupled sympathetic activation and parasympathetic withdrawal. This consistency between vectors indicates that the PCA component scores are valid indicators of cardiac autonomic information for the exercise task.

The autonomic space created by the PCA components was consistent with the autonomic space created by PEP and RSA for the tracking task and for the high load condition of the memory task (see Figs. 7.2 through 7.4). Both the PCA component and the PEP and RSA vectors were significantly different from the origin for velocity and acceleration order-of-control (PCA: T^2 [2, 18] = 12.84 and 14.39, p<.01; PEP and RSA: T^2 [2, 18]=8.85 and 8.56, p<.05), for low and high disturbance (PCA: T^2 [2, 18]=11.48 and 14.57, p<.02; PEP and RSA: T^2 [2, 18]=7.58 and 9.37, p<.05), and for high memory load (PCA: T^2 [2, 18]=19.34, p<.01; PEP and RSA: T^2 [2, 18]=11.74, p<.05). The parasympathetic component and RSA were significantly different from baseline-to-task for all conditions; however, heart period was only significantly different from baseline-to-task for the acceleration and high disturbance conditions of the tracking task and high load condition of the memory task (Table 7.3). Both vectors indicate that all four conditions of the tracking task and the high load condition of the memory task elicited uncoupled parasympathetic withdrawal. This consistency between vectors indicates that the PCA component scores are valid indicators of cardiac autonomic information in all four conditions of the tracking task and in the high load memory task.

The autonomic space created by the PCA components was consistent with the autonomic space created by PEP and RSA in the illusion task and in the low memory load condition of the memory task (see Fig. 7.5). For these tasks, however, neither vector was significantly different from the origin, and heart

TABLE 7.3

Mean and Standard Deviation Difference Scores for Heart Period, the Autonomic Components, PEP, and RSA for the 10 Tasks

Task	Measure				
	Heart Period (ms)	SC	PC	PEP (ms)	RSA (ln(ms²))
Cold pressor[1]	−63.96*	−0.71*	−0.08	0.36	−0.08
	(84.11)	(1.05)	(0.51)	(7.20)	(1.15)
Low intensity exercise[1,2]	−153.02*	−0.98*	−1.09*	−25.00*	−1.24*
	(90.76)	(1.14)	(0.79)	(15.45)	(1.56)
High intensity exercise[1,2]	−268.02*	−1.32*	−1.91*	−37.70*	−2.32*
	(110.21)	(1.44)	(1.07)	(13.65)	(2.33)
Illusion task	−14.78	−0.08	−0.01	1.31	−0.20
	(53.73)	(0.94)	(0.49)	(4.21)	(1.21)
Low memory	−20.51	0.06	−0.13	1.86	−0.38
	(48.42)	(0.76)	(0.62)	(3.99)	(1.43)
High memory[1,2]	−62.93*	−0.34	−0.36*	−1.83	−0.69*
	(53.70)	(0.82)	(0.55)	(6.12)	(1.12)
Velocity[1,2]	−8.76	0.26	−0.27*	0.21	−0.58*
	(42.38)	(0.63)	(0.35)	(4.45)	(0.87)
Acceleration[1,2]	−31.96*	−0.04	−0.30*	−1.44	−0.67*
	(42.46)	(0.82)	(0.49)	(8.64)	(1.07)
Low disturbance[1,2]	−13.19	0.21	−0.29*	−0.49	−0.60*
	(38.41)	(0.58)	(0.39)	(6.02)	(0.99)
High disturbance[1,2]	−27.54*	0.01	−0.29*	−0.74	−0.66*
	(40.84)	(0.91)	(0.47)	(6.56)	(0.96)

[1]$T^2 p < .05$ for the autonomic components. [2]$T^2 p < .05$ for PEP and RSA.
* $t p < .05$.
SC = Sympathetic Component; PC = Parasympathetic Component; PEP = Pre-ejection Period; RSA = Respiratory Sinus Arrhythmia.
df = 19.

period, the PCA components, PEP, and RSA were not significantly different from baseline-to-task. Both vectors indicate that the illusion task and the low load condition of the memory task elicited no significant change in autonomic activity

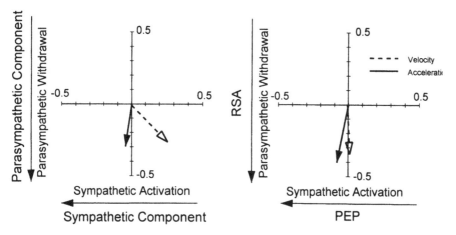

FIG. 7.2. An autonomic space representation of cardiac control modes using the group mean autonomic component (left panel) and PEP and RSA (right panel) difference scores for velocity and acceleration tracking conditions.

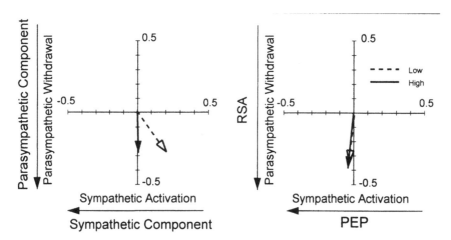

FIG. 7.3. An autonomic space representation of cardiac control modes using the group mean autonomic component (left panel) and PEP and RSA (right panel) difference scores for low and high disturbance tracking conditions.

from baseline-to-task. This consistency between vectors indicates that the PCA component scores are valid indicators of cardiac autonomic information for the illusion task and low memory load condition of the memory task.

Finally, the only task for which the autonomic spaces created by the PCA components and PEP and RSA were not consistent was the cold pressor (see Fig. 7.6). Although heart period was significantly shorter than baseline (i.e., faster

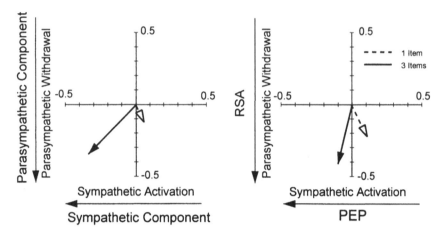

FIG. 7.4. An autonomic space representation of cardiac control modes using the group mean autonomic component (left panel) and PEP and RSA (right panel) difference scores for 1- and 3-item memory load tasks.

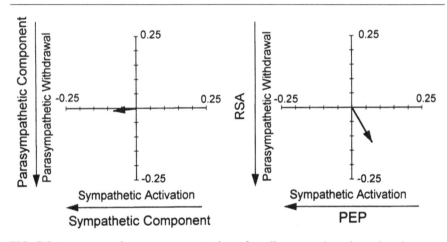

FIG. 7.5. An autonomic space representation of cardiac control modes using the group mean autonomic component (left panel) and PEP and RSA (right panel) difference scores for the visual illusion task.

heart rate, see Table 7.3), only the PCA component vector was significantly different from the origin (T^2 [2, 18]=13.88, $p<.01$). The sympathetic component was significantly different from baseline-to-task, but the parasympathetic component, PEP, and RSA were not (Table 7.3). The PCA components indicate the cold pressor task elicited uncoupled sympathetic activation. However, PEP and RSA indicate that the cold pressor did not elicit any change in activity from baseline-to-task. These results suggest that the PCA component scores may not

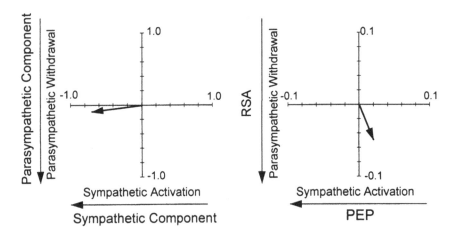

FIG. 7.6. An autonomic space representation of cardiac control modes using the group mean autonomic component (left panel) and PEP and RSA (right panel) difference scores for the cold pressor task. Note the difference in scales between panels.

be valid indicators of neurogenic cardiac autonomic information for the cold pressor task.

Correlation Analyses

If the sympathetic and parasympathetic components correspond to cardiac neurogenic autonomic inputs, then they should correlate with PEP and RSA, respectively. Pearson product-moment correlations were computed for the sympathetic component and PEP and for the parasympathetic component and RSA. The correlations were computed both within tasks and within subjects.

The within-task correlations were computed across subjects for each of the 10 tasks and five baselines. The sympathetic component and PEP were significantly correlated in the cold pressor task (r [18]$=.63$, $p<.05$), but not in the other nine tasks and five baselines ($-.18<r$ [18]$<.29, p>.05$). In contrast, the parasympathetic component and RSA were significantly correlated for all 10 tasks and the five baselines ($.84<r$ [18]$<.97$, $p<.05$).

The within-subject correlations were computed across the ten tasks and five baselines for each of the 20 subjects. The sympathetic component and PEP were significantly correlated for 14 of the 20 subjects ($.68<r$ [13]$<.86$, $p<.05$). The parasympathetic component and RSA were significantly correlated for all 20 subjects ($.65<r$ [13]$<.98$, $p<.05$).

DISCUSSION

The goal of the present study was to examine the validity of PCA of multiple psychophysiological measures computed from heart period for use in mental workload assessment. Validity of the PCA components was examined in two ways: (a) by comparing the autonomic space created from the PCA component scores to that created from PEP and RSA, and (b) by the correlations between the sympathetic component and PEP and between the parasympathetic component and RSA. Both approaches provided some support for PCA component validity.

Although the ideal outcome of the present study would have been that the PCA components veridically reflected cardiac sympathetic and parasympathetic neurogenic activity in all task conditions, it may be that the PCA components can provide useful information for mental workload assessment without achieving this ideal. To have the sensitivity and diagnosticity desired for mental workload assessment there must be the type of psychological-physiological relation that Cacioppo and Tassinary (1990a, 1990b) have termed a *marker*. A marker relation is a one-to-one psychological-physiological mapping obtained in a context-bound environment. Backs (in press) suggested that identification of the modes of autonomic control elicited during tasks that require different cognitive processes would be a useful first step toward establishing marker status for cardiac measures. Thus, a successful outcome of the present study would be that the PCA components led to the same conclusions regarding task effects as did PEP and RSA, the more direct measures of cardiac sympathetic and parasympathetic neurogenic activity. Table 7.4 summarizes the conclusions regard' ing task effects drawn from the PCA components compared to PEP and RSA and to heart period and summarizes the within-task correlations.

Comparison of the autonomic space derived from the sympathetic and parasympathetic components to that derived from PEP and RSA suggests that the PCA components were valid for all task conditions except the cold pressor (Table 7.4). The discrepancy for the cold pressor task illustrates the disadvantage of using residual heart period as the sympathetic marker variable. Residual heart period is the variance in heart period that is unaccounted for by RSA and is affected by all other factors that determine heart period. For cognitive tasks like memory and tracking, where the primary determinants of heart period are probably neurogenic in origin, the PCA components led to the same conclusions as PEP and RSA. However, for tasks like the cold pressor, where heart period may be primarily determined by responses to vasoconstriction, the PCA components dissociate from PEP and RSA.

Changes in autonomic space must be consistent with changes in heart period. Comparison of the autonomic space derived from the sympathetic and parasympathetic components to heart period change from resting baseline also suggests that the PCA components were valid (Table 7.4). All task conditions, except for velocity and low disturbance tracking showed changes in the

TABLE 7.4
Consistencies (+) and Inconsistencies (−) Between the PCA Components and PEP
and RSA, Heart Period, and the Within-Task Correlations

Task	Autonomic Space for PEP and RSA	HP	Within-Task Correlations	
			SC & PEP	PC & RSA
Cold pressor	−	+	+	+
Low intensity exercise	+	+	−	+
High intensity exercise	+	+	−	+
Illusion task	+	+	−	+
Low memory	+	+	−	+
High memory	+	+	−	+
Velocity	+	−	−	+
Acceleration	+	+	−	+
Low disturbance	+	−	−	+
High disturbance	+	+	−	+

SC = Sympathetic Component; PC = Parasympathetic Component; PEP = Pre-ejection Period;
RSA = Respiratory Sinus Arrhythmia.

autonomic space created by the PCA components that were consistent with heart period change. As discussed later, the dissociation between the PCA components and heart period in these tracking conditions may be advantageous for mental workload assessment.

In contrast, the correlation results provided mixed support for the validity of the PCA components. The within-task correlations indicated that only the parasympathetic component was valid (Table 7.4). However, the within-task correlations are not as important to the conclusions about the validity of the PCA components for the purpose of mental workload assessment as are the within-subject correlations. Because mental workload will typically be assessed within an individual (e.g., a pilot), the psychological-physiological relation need only hold within that individual. The correlation between the sympathetic component and PEP across subjects (within a task) is much less important than the correlation within subjects. Therefore, the correlation results support the validity of the PCA method when viewed from the perspective of whether the PCA components would be valuable for mental workload assessment.

More generally, the present study also demonstrated the potential of the autonomic-space approach to increase sensitivity and diagnosticity over using heart period alone for mental workload assessment. Other than the cold pressor, significant change in autonomic space was observed for each task condition for which there was a significant heart period change. The potential for greater sensitivity was evident in the velocity and low-disturbance conditions of the tracking task. In these task conditions, heart period was shorter than baseline (i.e., faster heart rate), but not statistically significant, whereas the change in autonomic space was significantly different from the baseline. The potential for greater diagnosticity was evident in the exercise and tracking tasks, where a coupled reciprocal mode of control was observed during exercise compared to an uncoupled parasympathetic withdrawal mode of control observed during tracking.

CONCLUSIONS

The PCA components examined in the present study were proposed as a method for obtaining cardiac autonomic information in the field. Impedance cardiography and pharmacological blockades can be used to obtain cardiac autonomic information in the laboratory, but these methods are not practical for most field studies. Although ambulatory impedance cardiographs have recently been developed, the technique is difficult to use in the field and is more susceptible to artifact than is the electrocardiogram. One advantage of the PCA method is that it does not require the collection of any more data than are needed to obtain heart period. Another advantage of the PCA method is that it can be applied post hoc to heart period data that have been previously collected (e.g., Backs, Lenneman, & Sicard, 1999; Backs, Wilson, & Hankins, 1995), which could potentially disambiguate instances where other mental workload assessment measures were sensitive but heart period was not. Thus, the PCA method presented in this chapter appears to be a valid and logical choice for obtaining cardiac autonomic information in the field.

REFERENCES

Backs, R.W. (in press). An autonomic space approach to the psychophysiological assessment of mental workload. In R.A. Hancock & P.A. Desmond (Eds.), *Stress, workload, and fatigue.* Mahwah, NJ: Lawrence Erlbaum Associates.

Backs, R.W. (1995). Going beyond heart rate: Modes of autonomic control in the cardiovascular assessment of mental workload. *The International Journal of Aviation Psychology, 5,* 25–48.

Backs, R.W. (1998). A comparison of factor analytic methods of obtaining cardiovascular autonomic components for the assessment of mental workload. *Ergonomics, 41,* 733–745.

Backs, R.W., Lenneman, J.K., & Sicard, J.L. (1999). The use of autonomic components to improve cardiovascular assessment of mental workload in flight simulation. *The International Journal of Aviation Psychology, 9,* 33–47.

Backs, R.W., Wilson, G.F., & Hankins, T.C. (1995). Cardiovascular assessment of mental workload using autonomic components: Laboratory and in-flight examples. In R.S. Jensen & L.A.Rakovan (Eds.), *Proceedings of the Eighth International Symposium on Aviation Psychology* (pp. 875–880). Columbus: The Ohio State University.

Berntson, G.G., Bigger, J.T., Jr., Eckberg, D.L., Grossman, P., Kaufman, P.G., Malik, M., Nagaraja, H.N., Porges, S.W., Saul, J.P., Stone, P.H, & van der Molen, M.W. (1997). Heart rate variability: Origins, methods and interpretive caveats. *Psychophysiology, 34,* 623–648.

Berntson, G.G., Cacioppo, J.T., & Quigley, K.S., (1991). Autonomic determinism: The modes of autonomic control, the doctrine of autonomic space, and the laws of autonomic constraint. *Psychological Review, 98,* 459–487.

Berntson, G.G., Cacioppo, J.T., & Quigley, K.S., (1995). The metrics of cardiac chronotropism: Biometric perspectives. *Psychophysiology, 32,* 162–171.

Cacioppo, J.T., Berntson, G.G., Binkley, P.F., Quigley, K.S., Uchino, B.N., & Fieldstone, A. (1994). Autonomic cardiac control: II. Noninvasive indices and basal response as revealed by autonomic blockade. *Psychophysiology, 31,* 586–598.

Cacioppo, J.T., & Tassinary, L.G. (1990a). Inferring psychological significance from physiological signals. *American Psychologist, 45,* 16–28.

Cacioppo, J.T., & Tassinary, L.G. (1990b). Psychophysiology and psychophysiological inference. In J.T.Cacioppo & L.G.Tassinary (Eds.), *Principles of psychophysiology: Physical, social, and inferential elements* (pp. 3–33). New York: Cambridge.

Grossman, P., & Svebak, S. (1987). Respiratory sinus arrhythmia as an index of parasympathetic cardiac control during active coping. *Psychophysiology, 24,* 228–235.

Kelsey, R.M., & Guethlein, W. (1990). An evaluation of the ensemble averaged impedance cardiogram. *Psychophysiology, 27,* 24–33.

Lewis, C.E., Jr., Jones, W.L., Austin, F., & Roman, J. (1967). Flight research program: IX. Medical monitoring of carrier pilots in combat-II. *Aerospace Medicine, 38,* 581–592.

Obrist, P.A. (1981). *Cardiovascular psychophysiology: A perspective.* New York: Plenum.

O'Donnell, R.D., & Eggemeier, F.T. (1986). Workload assessment methodology. In K.R. Boff, L.Kaufman, & J.P.Thomas (Eds.), *Handbook of perception and human performance: Vol. 11. Cognitive processes and performance* (pp. 42–1–42–49). New York: Wiley Interscience.

Porges, S.W., & Bohrer, R.E. (1990). The analysis of periodic processes in psychophysiological research. In]. T.Cacioppo & L.G.Tassinary (Eds.), *Principles of psychophysiology: Physical, social, and inferential elements* (pp. 708–753). New York: Cambridge University Press.

Roscoe, A.H. (1975). Heart rate monitoring of pilots during steep gradient approaches. *Aviation, Space, and Environmental Medicine, 46,* 1410–1413.

Roscoe, A.H. (1978). Stress and workload in pilots. *Aviation, Space, and Environmental Medicine, 49,* 630–636.

SAS Institute, Inc. (1989). *SAS/STAT user's guide, version 6* (4th ed., Vols. 1–2). Cary, NC: SAS Institute, Inc.

Sherwood, A., Allen, M.T., Fahrenberg, J., Kelsey, R.M., Lovallo, W.R., & van Doornen, L. J.P. (1990). Methodological guidelines for impedance cardiography. *Psychophysiology, 27,* 1–23.

Chapter 8

Alertness in Airline Pilots During Night Flights: Assessment of Alertness Using EEG Measures

Alexander Gundel

Jürgen Drescher

Jacqueline Turowski

DLR-Institute of Aerospace Medicine,
Cologne, Germany

An important issue in the operation of modern civil aircraft is the management of pilot alertness. It is expected that a future alertness management system will be based on a mathematical model for alertness that incorporates several mechanisms for the dynamics of alertness (Folkard & Akerstedt, 1992). This model will be built on an empirical data base. The purpose of the present study is to provide additional knowledge about alertness dynamics that will contribute to an alertness management system in civil aviation.

The operation of long-duration night flights usually conflicts with human circadian regulation, as the trough in several physiological and psychological functions including job performance occurs during flight. Pilots are used to this situation and are able to cope with increased fatigue to some extent. A special situation emerges if two extended night flights follow each other. This automatically involves a sleep deficit and the necessity for daytime sleep between flights. Possibly, a further decrement in alertness is observed during the second night flight.

Alertness or sleepiness is composed of at least three distinct components: (a) a circadian component with low alertness at night corresponding, approximately, with the time of the temperature minimum; (b) a component that depends on the preceding sleep; and (c) a sleep inertia effect with elevated fatigue immediately following sleep (Folkard & Akerstedt, 1992; Folkard, Hume, Minors, Waterhouse, & Watson, 1985; Monk, Moline, Fooksen, & Peetz, 1989).

The sleep-dependent component shows highest alertness immediately after sleep resembling the recovery effect, and then alertness decreases with time. Therefore, this component is sometimes called *time-since-sleep effect.* Sleep inertia and the time-since-sleep component are influenced by quantity and quality of previous sleep. However, the quantitative effect of sleep quality and quantity on these components is less well documented.

In a flight roster with two subsequent night flights, the circadian component of alertness is expected to be the same for both nights if no time zone transition is involved, but the time-since-sleep component may be affected by irregular sleep and work during such a flight schedule. The second flight is preceded by a daytime sleep that is usually shorter and more disturbed than a regular night-time sleep (Akerstedt & Gillberg, 1982; Wegmann et al., 1986), a fact that may reduce alertness during this flight.

The situation of two consecutive night flights is similar to the first two night shifts at the beginning of a shift work schedule as has been investigated in nurses (Folkard, Monk, & Lobban, 1978). Alertness ratings showed that during the second night shift fatigue was actually higher than during the first one. This applied to nurses who also worked during the day. Results for permanent night nurses were different (Folkard et al., 1978).

Our hypothesis is that, due to the reduced quality and quantity of daytime sleep and the affected time-since-sleep component, alertness will be lower during the second flight compared to the outgoing flight. This hypothesis is in congruence with predictions of a mathematical model (Folkard & Akerstedt, 1992). This chapter presents a reanalysis of data from Gundel, Drescher, Maaß, Samel, and Vejvoda (1995).

METHOD

Measurements of alertness could be conducted in a two-pilot crew operating a B767 on a flight roster in which the outgoing flight left Frankfurt, Germany, in the evening bound for Mahe, Seychelles (three-letter codes FRA and SEZ). The return flight was scheduled for the following night. Local time in the Seychelles is 2 hours ahead of Germany, and the consequences of this difference were expected to be marginal. The initial night flight took place approximately 24 hours after the beginning of the last regular night sleep period. There was a possibility of an additional nap during that long waking time.

Alertness is reflected in the central nervous system and physiological measures for reduced alertness or increased drowsiness and sleepiness have been derived from observations of the transition from wakefulness to sleep when the central parameters change completely within minutes (e.g., Rechtschaffen & Kales, 1968). During this transition, increased sleepiness is reflected by rolling eyeballs observed in the electrooculogram (EOG), a disappearance of alpha activity and the occurrence of theta waves in the electroencephalogram (EEG). If these phenomena occur preceding sleep and do not qualify to be scored as sleep stage 1 according to Rechtschaffen and Kales (1968), they are sometimes called *micro-sleep*. With fully open eyes, sleepiness is indicated by the reappearance of alpha activity, in particular in coincidence with eye blinks.

Subjects

Twenty-two male pilots, 11 captains and 11 flight officers, volunteered for this study. They did not receive any compensation for taking part, but regarded their participation as part of their job duties. They expected, however, that the outcome of the study would be for their benefit in regard to an improvement of working conditions, in particular the flight duty regulations. The age range of the pilots extended from 25 to 55 years with a median age of 43 years. There was no medical selection criterion for participation in this study. Subjects had passed their last regular medical checks successfully. Pilots gave their informed consent to this study that complied with the recommendations of the Helsinki Declaration. According to the sleep logs that the pilots kept beginning 3 days before the flights, they did not take any medication during the study.

Study Design

Average flight times were 8:47 hours for the outgoing and 9:37 hours for the homegoing flight. Average takeoffs were at 19:48 and 18:45 UTC (Universal Time Coordinates), respectively. Sleeping times before the flight roster and during layover were subjectively assessed by the pilots using a sleep questionnaire.

EEG results obtained during flight operations are often contaminated by artifacts resulting from various sources, such as head movements, muscle activity, and eye movements. In order to get cleaner EEG recordings, and to have a standardized and controlled situation, hourly experimental phases were introduced during flights starting about 1 hour after takeoff. These phases were staggered for pilot and flight officer.

An experimental phase started with the request to rate mood on a questionnaire. The questionnaire was an adapted and translated German version from that of Bond and Lader (1974). It consists of 16 visual analogue scales representing the three mood factors: alertness, contentedness, and calmness.

The phase ended with a 90-second resting period in order to obtain resting EEG with the subject's eyes closed. For the resting period subjects took a relaxed position in their seat. Relaxation was supported by a neck roll. After taking this position, their eyes were covered by sleeping shades and they closed their eyes. Eyeball movements to the right and left marked the beginning of the 90-second resting period. The mind-set of subjects for this resting period was to relax, not to move, to think of nothing in particular, to have no goal-directed thoughts, and to stay awake. The sleeping shades should ensure approximately the same light conditions during resting phases regardless of the environmental light conditions. Light exposure could not be controlled in this experiment.

There were constraints for the performance of this experiment that resulted from the operational environment. Electrodes for EEG and EOG recordings had to be applied in the cockpit immediately before the airplane left the blocks. A minimal amount of time was allocated for this procedure. To meet this constraint, an elastic headband with integrated Ag/AgCl electrodes was used that was successfully applied during a space experiment with similar constraints (Gundel, Nalishiti, Reucher, Vejvoda, & Zulley, 1993). EEG electrodes were placed according to the international 10–20 system. The electrode positions of this band were C3, C4, Cz (ground electrode), and O2. The band provided clip-on connectors for four disposable electrodes that served as mastoid references (A1, A2) and as EOG derivations from the forehead (above and slightly lateral to the outer canthus of the right and left eyes, respectively). From the set of electrodes, five polygraphic channels, C4-A1, C3-A2, O2-A1, EOG1-A2, and EOG2-A1, were recorded. This setting provided some useful redundancy of EEG and EOG. The recordings were the same as used for sleep recordings and allowed the detection of drowsiness (Rechtschaffen & Kales, 1968).

EEG Analysis

EEG and EOG signals were recorded by an Oxford Medilog 8-channel recorder that allows recording for 24 hours continuously. Before each resting phase, electrode impedances were checked and improved if they were higher than 5 kiloohms.

In our laboratory in Cologne, data were A/D converted with a sampling rate of $256s^{-1}$. Resting phases were visually identified on a computer screen by means of the intended eyeball movements at the beginning of each resting phase and by a protocol that was written by an observer during flights. Using EOG and EEG traces, artifacts that occurred during resting phases were detected and the contaminated intervals rejected.

Air turbulence caused a larger number of artifacts mediated by passive head movements with frequencies around 3 cps. Despite the considerable effort taken during the flights, artifacts and operational considerations resulted in missing values. The number of missing values led to the decision to pool the data obtained in 2-hour intervals beginning with the time period between 19:00 and 20:59 UTC. There were still 31 missing values out of 192 observations.

Analysis of EEG recordings was focused on the bipolar derivaton C4-O2 that was generated by computer after calibration of the signals. This derivation showed fewer artifacts due to head movements than those referenced to a mastoid electrode. Recordings from 16 subjects entered the analysis. Data from the other 6 subjects were incomplete or showed too many artifacts.

Artifact-free 4-second epochs of each 90-second resting period underwent power spectral analysis yielding a frequency resolution of 0.25 cps. Broadband powers were extracted from the power spectra in the frequency bands: delta, 1.5 to 3.5 Hz; theta, 3.5 to 7.5 Hz; alpha, 7.5 to 13.5 Hz; and beta, 13.5 to 32 Hz. They were obtained by averaging the power spectral values in the particular frequency range. Subsequently, these averages underwent a logarithmic transformation to get parameters with a more normal distribution (Gasser, Bächer, & Möcks, 1982).

In addition to broadband powers, peak firequencies in the spectra were determined for the alpha band. A numerical value for the alpha peak frequency was obtained by calculating the frequency at the center of gravity of the spectra in the alpha band; that is, the sum of spectral values multiplied by their frequency was divided by the sum of spectral values.

The first sign of sleepiness in the EEG is the desynchronization of the alpha rhythm. Therefore, an index was derived that describes this desynchronization. For each subject for the two flights separately, EEG power in the alpha band was normalized by the 4-second epoch that showed the highest power in the alpha band. Then the mean squared deviation from the value one (highest power) was calculated using all artifact-free epochs of a resting phase. The resulting index ranges between zero and one, it is one if there is no alpha power at all and zero for maximal alpha power in all 4-second epochs of the resting phase. The index is more sensitive to alpha desynchronization than to the slight reduction of alpha amplitudes shown by subjects who are not completely relaxed.

Statistical Evaluation

The statistical evaluation of EEG data and mood ratings was accomplished by a repeated measure analysis of variance (ANOVA) with repetitions on the factors of flight (outgoing and homegoing flight) and time of day (six time points for EEG data and seven for mood ratings). This design makes an economical use of subjects. Because of the number of missing values it was decided to pool data in 2-hour instead of 1-hour intervals. Remaining missing values were estimated by a procedure created by Yates as described in Kirk (1982). Degrees of freedom were corrected for the replacement of missing values.

It was reasonable to assume that the covariance matrices for the ANOVAs are nonspherical, and, therefore, the degrees of freedom for F-tests had to be corrected. Here a two-step procedure was used (Kirk, 1982). First, the degrees of freedom were corrected according to Geisser and Greenhouse (Kirk, 1982). When these conservatively corrected F-tests failed to show significant results, the less conservative but more difficult to calculate correction of Huynh and Feldt (Kirk, 1982) was applied. Error probabilities obtained after correction are indicated by GG or HF.

Interpretation of data was also based on multiple comparisons of parameter estimates by calculating F-statistics for contrasts. For the factor "time of day" a family of contrasts between the mean obtained for one time point and the average of the means for the other times was calculated. With six contrasts, the probability of one or more type I errors is 0.26 for an error probability of 0.05, and it is 0.06 for an error probability of 0.01. This type of contrast was calculated to estimate the time of the circadian trough. Circadian amplitudes were *compared* by calculating polynomial contrasts for the interaction between "time of day" and "flight." The circadian component is represented by the quadratic term. Differences of parameters between the two flights that occur only at certain time points are described by contrasts between the mean at one time point and the average of the means for the other time points for the interaction of flight and time of day.

Prior to ANOVA, ratings for the 16 mood scales underwent factor analysis (principal components) of the correlation matrix and subsequent orthogonal rotation (varimax) of the factor loadings.

RESULTS

The evaluation of the sleep logs showed that the average total sleep time for the night preceding the outgoing flight was 8:40 hours. Twelve out of 16 subjects additionally napped during the day before the outgoing flight (average nap time 1:18 hours). The average total sleep time during layover was 6:04 hours.

Factor analysis of the mood scales revealed two eigenvalues larger than one. Varimax rotation showed 34.5% of the variance in the factor loading mainly on alertness scales and 12.2% in the factor loading on the scales for the combination of contentedness and calmness, factors that could not be separated in these data. ANOVA of factor scores for alertness showed that the main effects flight and time were statistically significant, $F(1,6)=8.34, p=.011$ and $F(6,90) = 10.87, p=.001$ GG. Figure 8.1 demonstrates the reduction in alertness on the homegoing flight that is almost constant over time. The interaction of flight and time was not significant, $F(6, 62)<1$. Contrasts for the effect of time showed that pilots were most alert at 18:00 and 20:00 hours, $F(1, 15)=17.93, p= .0007$, and $F(1, 15)=10.85, p=.004$. The lowest alertness occurred at 00:00, 02:00 and 04:00 hours, $F(1, 15)=10.68, p=.0052$, $F(1, 15)=19.34, p= .0005$, and $F(1, 15)=10.61, p=.0053$. For the combination of contentedness and calmness, no significant effects were found.

For the six EEG parameters that are broadband power in four bands (delta, theta, alpha, and beta), alpha desynchronization index, and alpha peak frequency, ANOVAs were calculated separately resulting in significant effects for the alpha band only. Figure 8.2 displays mean values for outgoing and homegoing flights for the alpha band parameters. In the other bands (delta, theta, and beta), no significant effects were found. In some individual cases, periods of higher drowsiness were accompanied by theta activity; however, these effects did not reach statistical significance for the group.

For alpha power, the interaction of time of day and flight was almost significant, $F(5, 44)=2.4, p=.052$ HF, whereas the main effects flight, ($F<1$) and time, $F(5, 75)=1.27, p=.2877$, were not significant. Contrasts for the interaction

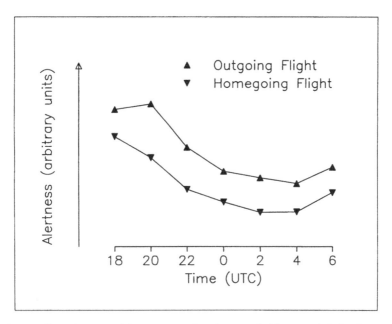

FIG. 8.1 Profiles of averaged factor scores for alertness (arbitrary units) for the outgoing flight and the homegoing flight (UTC: Universal Time Coordinates).

showed that the alpha power was significantly smaller on the return flight at 02:00 hours, $F(1, 15)=7.39$, $p=.0159$. Also for the desynchronization index main effects did not reach statistical significance, $F<1$ and $F(5, 75)=1.11$, $p=.3621$. The interaction of flight by time of day showed a trend, $F(5, 44)=2.04$, $p=.083$ HF. The contrasts for this interaction again showed statistical significance at 02:00 hours, $F(1, 15)=5.31$, $p=.0360$. At this time, alpha activity was more desynchronized on the return flight than on the outgoing flight. At 02:00 hours (UTC) pilots had already flown for about 7 hours on the return flight and for about 8 hours on the outgoing flight. In general, alpha desynchronization displays a remarkable uniformity for the three intervals from 20:00 to midnight (UTC) and was elevated afterwards.

The modulation of alpha peak frequency over time is very marked, $F(5, 75)=6.85$, $p=.0001$ HF. It showed a distinct circadian pattern with a minimum around 02:00 hours (UTC) (Fig. 8.2). Contrasts reveal that alpha peak frequency is lowest for 00:00 hours, $F(1, 15)=7.02$, $p=.0182$, and for 02:00 hours, $F(5, 75)=18.60$, $p=.0006$. Also the interaction of flight and time of day was significant, $F(5, 44)=4.94$, $p=.001$ HF, whereas the effect of flight was not significant, $F(1, 15)=3.00$, $p=.1039$. Contrasts for the interaction of flight and time of day were significant for 00:00 hours, $F(1, 15)=5.69$, $p=.0307$, 02:00 hours, $F(1, 15)=5.75$, $p=.0299$, and 04:00 hours, $F(1, 15)=8.81$, $p=.0096$ (see Fig. 8.2). Polynomial contrasts were significant for the quadratic term for the main effect time, $F(1,$

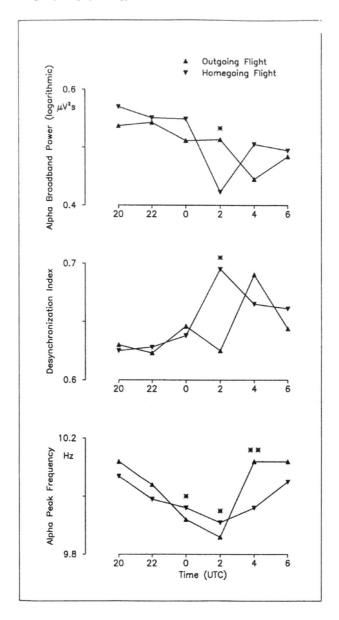

FIG. 8.2 Profiles of three parameters for the EEG alpha band (alpha broadband power, alpha desynchronization index and alpha peak frequency) for the outgoing flight and the homegoing flight. One star denotes an error probability of less than 5%, two stars of less than 1% for the contrasts in the interaction of the factors "time of day" and "flight" (UTC: Universal Time Coordinates).

15)=29.53, p=.0001, and the interaction of time and flight, $F(1, 15)$=6.48, p=.0224. The latter contrast indicated the reduced circadian amplitude on the return flight.

Data were arranged and pooled according to UTC throughout. In a second approach they were arranged and averaged according to time since takeoff. The ANOVA model did not fit the data as well as the data arranged according to UTC. Therefore, it was decided not to pursue this second approach.

DISCUSSION

This study demonstrates an increased sleepiness of pilots during the second of two consecutive night flights. This was found in subjective ratings of alertness and in EEG parameters. Daytime sleep during layover was reduced from normal amounts by about 2 hours, which is in accordance with results from an experiment with displaced sleep (Akerstedt & Gillberg, 1982).

Both flights took place at almost the same time of day, and the circadian phase did not reveal a shift from one night to the next. This could be shown by the alterations in alpha peak frequency that are known to follow a distinct circa' dian pattern (Gundel & Hilbig, 1983). Also, in the subjective ratings of alertness, the usual dip at night occurred at the same time during both flights. The circadian trough occurred around 2:00 hours UTC corresponding to 4:00 hours local time at the home base of the pilots.

The circadian amplitude in peak frequency seemed to be reduced on the return flight. This fact may find a plausible explanation by the reduction of the external zeitgeber strength due to the daytime sleep before the return flight. The daytime sleep automatically reduces exposure to bright daylight, which is believed to be the most potent zeitgeber for the circadian system. Model simulations have shown that the circadian amplitude increases with zeitgeber strength (Gundel & Spencer, 1992).

The subjective alertness ratings showed a marked increase in sleepiness for the entire return flight, compared to the ratings at corresponding times on the outgoing flight. Physiological consequences of this elevated subjective sleepiness could be demonstrated by the resting EEG. Two indicators of increased sleepiness, the reduction of power in the alpha frequency range and the desynchronization of alpha rhythms, occurred predominantly during the return flight at the time of the circadian trough. For about the first 6 hours of both flights, values for these two parameters were very similar, and during the last part of the flights they clearly indicated higher fatigue.

Except for the circadian amplitude differences in alpha peak frequency, the circadian component in alertness was unchanged during both flights. The experiment was not meant to allow a statement concerning the sleep inertia component. In the context of alertness models (Folkard & Akerstedt, 1992), the differences in alertness that were observed must be due to the time-since-sleep component and its modulation by sleep quality and quantity of the daytime sleep during layover.

Subjective ratings and EEG parameters describe different dimensions of the multidimensional alertness phenomenon (Broughton, 1992). Alertness ratings show a smooth time course representing overall assessment of capacities, whereas alpha power in the EEG measures spontaneous dozing off and the susceptibility to sleep.

Sleepiness that may be caused by monotony or that may follow high workload is not part of the alertness model described by Folkard and Akerstedt (1992). It cannot be ruled out that monotony in the second part of the flights may have added to the observed sleepiness.

The findings in the subjective sleepiness assessments are supported by results obtained from night nurses (Folkard et al., 1978). Like night nurses that work on different shifts as opposed to permanent night workers, pilots experience two consecutive night flights as an unusual schedule and are less prone to cope with the situation than permanent night workers.

As a consequence of increased sleepiness, countermeasures have to be considered. Naps are reported to alleviate spontaneous sleepiness (Dinges, Orne, Whitehouse, & Orne, 1987; Minors & Waterhouse, 1989; Mullaney, Kripke, Fleck, & Johnson, 1983; Naitoh, Englund, & Ryman, 1982). Timing and duration of naps, however, seem to be of crucial importance to assess the benefit of a nap for job performance (Dinges, Whitehouse, Orne, & Orne, 1988; Rosa, 1993). This awaits further research. Recently, the use of preplanned naps in three crew operations was discussed in detail as one of the possible measures to cope with the situation of extended night flights (Rosekind, Graeber, Dinges, Connell, Rountree, Spinweber, & Gillen, 1992). The possibility of recommend' ing napping for two crew operations provided napping is compatible with crew management and does not compromise flight safety should also be considered.

ACKNOWLEDGMENTS

The cooperation of the pilots and the airline management is gratefully acknowledged. The work was in part funded by the CEU, Directorate-General for Transport, contract no. C2, B93, B2–7020, SIN 004101.

REFERENCES

Akerstedt, T., & Gillberg, M. (1982). Displacement of the sleep period and sleep deprivation. *Human Neurobiology, 1,* 163–171.

Bond, A., & Lader, M. (1974). The use of analogue scales in rating subjective feelings. *British Journal of Medical Psychology, 47,* 211–218.

Broughton, R.J. (1992). Qualitatively different states of sleepiness. In R.J.Broughton & R. D.Ogilvie (Eds.), *Sleep, arousal, and performance* (pp. 45–59). Boston: Birkhäuser.

Dinges, D.F., Orne, M.T., Whitehouse, W.G., & Orne, E.C. (1987). Temporal placement of a nap for alertness: Contributions of circadian phase and prior wakefulness. *Sleep, 10,* 313–329.

Dinges, D.F., Whitehouse, W.G., Orne, E.C., & Orne, M.T. (1988). The benefits of a nap during prolonged work and wakefulness. *Work and Stress, 2,* 139–152.

Folkard, S., & Akerstedt, T. (1992). A three-process model of the regulation of alertness-sleepiness. In R.J.Broughton & R.D.Ogilvie (Eds.), *Sleep, arousal, and performance* (pp. 11–26). Boston: Birkhäuser.

Folkard, S., Hume, K.I., Minors, D.S., Waterhouse, J.M., & Watson, F.L. (1985). Independence of the circadian rhythm in alertness from the sleep/wake cycle. *Nature, 313,* 678–679.

Folkard, S., Monk, T., & Lobban, M.C. (1978). Short and long-term adjustment of circadian rhythms in "permanent" night nurses. *Ergonomics, 21,* 785–799.

Gasser, T., Bächer, P., & Möcks, J. (1982). Transformations towards the normal distribution of broad band spectral parameters of the EEG. *Electroencephalography and Clinical Neurophysiology, 53,* 119–124.

Gundel, A., Drescher, J., Maaß, H., Samel, A., & Vejvoda, M. (1995). Sleepiness of civil airline pilots during two consecutive night flights of extended duration. *Biological Psychology, 40,* 131–141.

Gundel, A., & Hilbig, A. (1983). Circadian acrophases of powers and frequencies in the waking EEG. *International Journal of Neuroscience, 22,* 125–133.

Gundel, A., & Spencer, M.B. (1992). A mathematical model of the human circadian system and its application to jet lag. *Chronobiology International, 9,* 148–159.

Gundel, A., Nalishiti, V., Reucher, E., Vejvoda, M., & Zulley, J. (1993). Sleep and circadian rhythm during a short space mission. *The Clinical Investigator, 71,* 718–724.

Kirk, R.E. (1982). *Experimental design.* Monterey, CA: Brooks/Cole.

Minors, D.S., & Waterhouse, J. M. (1989). Masking in humans: The problem and some attempts to solve it. *Chronobiology International, 6,* 29–53.

Monk, T.H., Moline, M.L., Fookson, J.E., & Peetz, S.M. (1989). Circadian determinants of subjective alertness. *Journal of Biological Rhythms, 4,* 393–404.

Mullaney, D.J., Kripke, D.F., Fleck, P.A., & Jobnson, L.C. (1983). Sleep loss and nap effects on sustained continuous performance. *Psychophysiology, 20,* 643–651.

Naitoh, P., Englund, C.E., & Ryman, D.H. (1982). *Restorative power of naps in designing continuous work schedules.* San Diego, CA: Naval Health Research Center.

Rechtschaffen, A., & Kales, A. (1968). *A manual of standardized terminology, techniques, and scoring system for sleep stages of human subjects.* UCLA BIS/BRI, Los Angeles.

Rosa, R.R. (1993). Napping at home and alertness on the job in rotating shift workers. *Sleep, 16,* 727–735.

Rosekind, M.R., Graeber, C.R., Dinges, D.F., Connell, L.J., Rountree, M.S., Spinweber, C. L., & Gillen, K.A. (1992). Crew factors in flight operations: IX. Effects of preplanned cockpit rest on crew performance and alertness in long-haul operations. *NASA Technical Memorandum 103884*

Wegmann, H.M., Gundel, A., Naumann, M., Samel, A., Schwarz, E., & Vejvoda, M. (1986). Sleep, sleepiness and circadian rhythmicity in air-crews operating on transatlantic routes. *Aviation Space and Environmental Medicine, 57,* B53–B64.

Chapter 9

The Effect of Naps and Caffeine on Alertness During Sleep Loss and Nocturnal Work Periods

Michael H.Bonnet
Dayton Veterans Administration Medical Center
Wright State University
Kettering Medical Center
Donna L.Arand
Kettering Medical Center

The ability to perform a wide variety of tasks is dependent on the level of alertness. Alertness at a given point in time is related to many variables including the time of day, the length of time awake, and the length and characteristics of the preceding sleep period. This chapter explores attempts to modify alertness and psychomotor performance during nocturnal work periods and sleep deprivation using naps and caffeine.

It has been known for many years that alertness and the ability to perform tasks declines across nocturnal work periods. Because shift work is common in modern society and because almost everyone has the occasional need to remain awake and active during the night, many examinations of nocturnal capabilities and factors that affect those abilities have been undertaken. Such studies have several common methodological procedures. It is typical for studies of sustained operations (SUSOPS) to repetitively measure several aspects of psychomotor performance and alertness over long periods of time to document changes in ability as a function of time of day and degree of sleep deprivation. Tasks typically used include reaction time, memory, coding, decision making, logical reasoning, vigilance, and mood. In addition to measures of subjective alertness, it is now common to measure objective alertness by monitoring the electroenceph-alogram

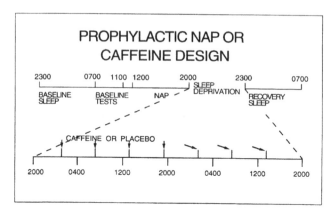

FIG. 9.1. Schematic presentation of events in Study 1.

(EEG). The EEG shows characteristic and easily identifiable changes as subjects fall asleep. The speed with which the transition to sleep occurs is quantified by giving subjects either repeated controlled opportunities to fall asleep in a Multiple Sleep Latency Test (MSLT) or repeated attempts to stay awake in a Maintenance of Wakefulness Test (MWT).

Remaining awake and functional at night or for extended periods of time is important in many fields including military, industrial, medical, and law enforcement. In these settings, individuals typically attempt to maintain alertness by drinking coffee or, in some settings, taking naps. We have conducted a series of controlled laboratory studies in recent years to learn more about the benefits of naps of various length and caffeine in helping to reverse some of the decrements associated with nocturnal work periods and SUSOPS. An initial study examined the ability of naps or caffeine to preserve alertness during an extended work period including two consecutive nights of total sleep deprivation. A second study examined the additive effects of naps and caffeine, and a third study looked at the placement of naps and caffeine in time. All subjects in these studies were 18 to 30-year-old healthy males, usually college students or naval recruits. The time line for a typical study is presented in Fig. 9.1. All studies began with a training period where subjects learned various tasks and then practiced them until they reached an asymptotic level. In the studies to be described, standard test batteries were presented, scored, and stored by computer on a routine schedule. As can be seen in Fig. 9.1, all of the naps began in the afternoon after a full night of sleep. Such naps, taken prior to the beginning of sleep loss, are called prophylactic naps.

STUDY 1

In the initial dual-center study, the effects of 3 levels of prophylactic naps (0 hours, 2–4 hours, and 8 hours) and 3 doses of caffeine (placebo, 150–300 mg, and

400 mg) were compared over a two-night period of sleep loss (Bonnet, Gomez, Wirth, & Arand, 1995). It was hypothesized that caffeine and prophylactic naps would each improve alertness and performance and that the improvements would be related to the dose and time-course of the caffeine or nap. The data were obtained at the VA Hospital in Long Beach, California, and the San Diego Naval Research Center. Subjects at both sites followed the same schedule of MSLTs, performance tests, and caffeine or placebo administration.

Subjects at the Long Beach Center were randomly assigned to one of the nap conditions. Subjects received either no afternoon nap or an available nap time of 2, 4, or 8 hours. Nap onset was varied so that all naps ended at 20:00 hours. Subjects not sleeping between 12:00 and 20:00 hr were allowed to work on homework or perform recreational activities such as playing pool, taking a walk, or watching television. Beginning at 20:00 hr, all subjects followed the same schedule of alternating performance test blocks, MSLT observations, and meals/breaks for 52 hours before being allowed a night of recovery sleep scheduled at their normal sleep time. During the sleep loss period, subjects were administered placebo capsules every 6 hours starting at 01:30 hr.

Subjects at the San Diego Center were randomly assigned to one of the placebo/caffeine conditions. Caffeine was administered in capsule form or matched placebo. One group received the placebo throughout the study. This group was the primary control group, and data from this group and the no nap placebo group at the Long Beach center were compared to insure compatibility between centers. Caffeine conditions consisted of a single administration of 400 mg caffeine at 01:30 hr each night or 150 mg or 300 mg caffeine administered every 6 hours starting at 01:30 hr on the first night of sleep loss. As with the Long Beach site, beginning at 20:00 hr, all subjects followed the same schedule of alternating performance test blocks, MSLT observations, and meals/breaks for 52 hours before being allowed a night of recovery sleep scheduled at the normal sleep time.

All subjects at both centers were assigned their own room for the course of the study. Each room contained a standard hospital bed and furniture including a desk with an Apple IIE or IIGS computer. Subjects participated in the study in groups of 1 to 4 individuals. Subjects completed all tests and questionnaires at their individual computer workstation in their room under technician observation. Nonstartling procedures, such as calling the subject's name, were used by the technicians to awaken faltering subjects. Meals and breaks were scheduled in another area of the laboratory, which was also within technician observation. Caffeinated beverages were not available.

Performance and mood were assessed with an extensive battery of psychomotor and subjective report measures that are described elsewhere in detail (Bonnet et al., 1995). Alertness was assessed with the MSLT. To help reduce between-subject variance, scores on all of these measures were calculated as percentage changes from performance or alertness levels attained on the baseline day in the laboratory (preceding the prophylactic nap when given).

Four-channel sleep electromyogram (EOG) and EEG recordings (Left Eye-A2, Right Eye-A2, C3-A2, OZ-A1) were made during nocturnal sleep periods, naps, and MSLT evaluations. All sleep recordings were scored for sleep stages in 30-second epochs using standard criteria (Rechtschaffen & Kales, 1968). Objective alertness was assessed during the study by the use of the MSLT. This test, in which participants are asked to fall asleep in a bed under sleep-inducing conditions, has been shown to be extremely sensitive to sleep loss (Carskadon & Dement, 1979, 1981) and is currently used as a standard clinical test (Carskadon, 1994). For the clinical test, patients are considered to have "normal" alertness when wakefulness is maintained for 10 minutes, and patients are considered to be incapacitated by sleepiness when they fall asleep in less than 5 minutes. Seventeen MSLT evaluations were made during the study proper. The first occurred at 10:00 hours on the baseline day. The remaining 16 MSLT tests began at 22:00 hours that night and continued at 3-hour intervals until 19:00 hours 2 days later. The score for the MSLT was the time from the beginning of the nap to the onset of Stage 2 (or in very rare instances REM) sleep. The MSLT should not be confused with a nap because it is designed to assess ability to fall asleep without allowing sleep to accumulate.

The study included 140 subjects. In order to establish comparability between the placebo groups across both centers, separate comparisons of age, weight, and performance for placebo groups were conducted. These ANOVAs revealed no significant main or interaction differences between placebo groups across study centers on any measure except for two mood scales, which will not be reported here. Therefore, the placebo groups were combined (total $n=27$) for subsequent analyses. Similarly, performance, mood, and sleepiness did not differ significantly between the 2- and 4-hour nap groups and between the caffeine 150 and 300 mg repetitive-dose groups. Therefore, to simplify interpretation and maximize group differences, these groups were combined for subsequent analyses yielding a 2- to 4-hour nap group (total $n=60$) and a 150–300 mg repetitive-dose caffeine group (total $n=17$). There were 24 subjects in the 8 hour nap group and 12 subjects in the single-dose 400-mg caffeine group. Total sleep time during the naps in the two nap groups was 157 minutes and 375 minutes.

Consistent differences were found for psychomotor performance, mood, and physiological measures. Because of space limitations, only physiological data from the MSLT is presented here. Additional performance and subjective data can be found in other publications (Bonnet et al., 1995; Bonnet & Arand, 1994a, 1994b). Baseline sleep latencies during the MSLT ranged from 12 to 16 minutes in the various groups. Multiple sleep latency observations were analyzed with a mixed ANOVA with effects for group (between subjects) and time of test (repeated measures), and a significant group by time interaction was found, $F(24, 812)=3.18, p<.001$. Pairwise comparisons (see Fig. 9.2) revealed significant group differences through 01:00 hours on the second night of sleep loss and at 16:00 and 22:00 hours on the final evening. Latency values for the 8-hour nap group

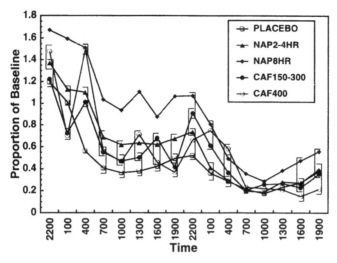

FIG. 9.2. MSLT observations (proportion of prestudy baseline latency to Stage 2 sleep) for nap and caffeine groups during sleep loss. The presence of brackets indicates overall statistically significant differences between groups at that time point. Data points within brackets indicate nonsignificant differences. From Bonnet et al. (1995). Copyright 1995 by *Sleep* journal. Reprinted with permission.

diverged most and were generally longer (i.e., subjects were less sleepy) than those for other groups except between 16:00 and 22:00 hours on the second night and day of sleep loss. For subjects who received a single 400-mg dose of caffeine, mean sleep latency values initially increased significantly (equal to that of the 8-hour nap group) at 04:00 hours, but then returned to placebo levels at 07:00 hours. Results for the 2- to 4-hour nap group and 150–300 mg caffeine group were intermediate. Although the caffeine 150–300 mg group was significantly sleepier than the placebo group prior to the initial caffeine administration, alertness was greater than placebo at 04:00 hours (after caffeine) and remained improved (compared to the placebo) until 04:00 hours on the second night of sleep loss with one exception. Similarly, MSLT latencies were longer than the placebo in the 2- to 4-hour nap group from 04:00 hr on the first night of sleep loss until 16:00 hr on the first day of sleep deprivation. The 150–300 mg caffeine group and the 2- to 4-hour nap group differed significantly at only three points during the study, with the nap group having longer latencies than the caffeine group at 01:00 (prior to caffeine use) and 19:00 hrs on the first day of sleep loss and the 150–300 mg caffeine group having longer latencies at 01:00 hours on the second night of sleep loss.

The data demonstrated a dose-related increase in alertness and performance for both prophylactic sleep and dose of caffeine. As a rough generalization, 2- to 4-hours of prophylactic nap was similar to 150–300 mg of caffeine in the magnitude of improvement. The 8-hour prophylactic nap provided the greatest increase in alertness and performance compared to all other nap and caffeine conditions. The dose response effect of caffeine was most evident at the 04:00 hours MSLT on the first night of sleep loss. At that point, latency was equal in the 2- to 4-hour nap group and the 150–300 mg caffeine groups and equal in the 8-hour nap group and 400-mg caffeine group.

The beneficial effect of both naps and caffeine seemed most predominant during the first night of deprivation. Subjects in all conditions were relatively incapacitated on the second night. Five minute latencies to Stage 1 sleep, typically used as a criterion for pathological sleepiness, correspond to a level of about 0.55 in Fig. 9.2. Using that criterion, the placebo group was pathologically sleepy by 04:00 hours on the first night of sleep loss, whereas both caffeine groups reached that criterion by 07:00 hours on the first night. The 2- to 4-hour nap group maintained alertness until 01:00 hours on the second night, and the 8-hour nap group reached the criterion for pathology at 04:00 hours on the second night.

Time-course effects of caffeine are clearly evident in the data (see Fig. 9.2). In general, 400 mg of caffeine caused a clear spike in alertness for about 6 hours following the initial administration. This period of action corresponds roughly with the work schedule period in a previous study showing the beneficial impact of caffeine on an all-night work shift (Sugerman & Walsh, 1989). As one might expect, there was a return to baseline level of function, usually without dropping below placebo levels, as the caffeine action diminished. However, the time-course effects of caffeine are in contrast to the steady state effect of the prophylactic naps, which seemed to show relatively slow changes over time (i.e., if the naps were a drug, they would have a long half-life).

Summary

Both prophylactic naps and caffeine helped to maintain alertness and performance. In the real world, however, naps and caffeine have separate advantages and disadvantages that help dictate their use. Prophylactic naps clearly had the advantage of a long-lasting effect and could probably be used with some frequency without the development of tolerance, dependency, withdrawal or side effects. On the other hand, naps must be planned and may consume a substantial amount of time. Caffeine clearly can be used when time is insufficient for a nap, but carries the potential risks of most pharmacological interventions. One strategy that may be superior to either the use of caffeine or prophylactic naps may be the use of both. This was tested in the second study.

STUDY 2

The beneficial effect of both prophylactic naps and caffeine in maintaining alertness and performance means that it may be practical to decrease the length of prophylactic nap and the amount of caffeine used by combining naps with caffeine during the work period. The second study reports the effects of the use of caffeine following a 4-hour prophylactic nap in comparison to a 4-hour prophylactic nap without caffeine over a 27-hour continuous operation (Bonnet & Arand, 1994b).

Three consecutive nights and two days were spent in the laboratory (usually Thursday night through Sunday morning). The initial night was a baseline sleep night scheduled according to the subject's habitual bedtime and wake time. On the following morning, subjects completed baseline testing and had their baseline nap latency test between 08:00 and 12:00 hours. Subjects were allowed to leave the laboratory until 15:00, when they returned to be readied for a 4-hour nap that began at 16:00 hours and ended at 20:00 hours. Beginning at 20:00 hours, all subjects followed the same schedule of alternating performance test blocks, MSLT observations, meals, and breaks for 27 hours before being allowed a night of recovery sleep scheduled at their normal sleep time. All subjects received pills at 01:30, 07:30, 13:30, and 19:30 hours. For all subjects, the pills received at 13:30 and 19:30 hours were placebos. For 12 subjects randomly assigned to the caffeine group, the pills administered at 01:30 and 07:30 hours contained 200 mg of Eleveine, a sustained release formulation of caffeine. For another 12 subjects assigned to the placebo group, all pills were placebo.

Objective alertness (MSLT) was measured 10 times during the study, including a baseline observation prior to the 4-hour nap and every 3 hours during the study starting at 22:00 hours and ending at 22:00 hours on the next evening. The resulting MSLT data were not normally distributed. As a result, the data underwent a log transformation before calculation of the percentage of baseline and ANOVA. ANOVA for the MSLT (terms for group, time of test, and interaction) revealed a significant group×time interaction, $F(8, 176)=2.92$, $p< .01$. Neuman-Keuls pairwise comparisons revealed significantly longer latencies in the caffeine group as compared to the placebo group at seven of the nine test points. These data are plotted in Fig. 9.3 with significant differences designated. In terms of time effects, MSLT values fell rapidly from baseline levels to 70% of baseline levels in the placebo group between 04:00 and 07:00 hours and then remained at 70% to 90% of baseline for the remainder of the study. In contrast, in the caffeine group values slowly declined across the study, but never fell significantly below the baseline level (the lowest level being 92% of baseline at 19:00 hours).

Summary

The data from this study showed that the combination of a 4-hour prophylactic nap and the use of caffeine during a 24-hour continuous operation resulted in significantly improved alertness as compared to a 4-hour prophylactic nap by itself. Of equal importance, the data from the nap plus caffeine group indicated that, at least under the acute conditions of this study, alertness could be maintained at baseline levels for 24 hours without a nocturnal sleep period. In this study,

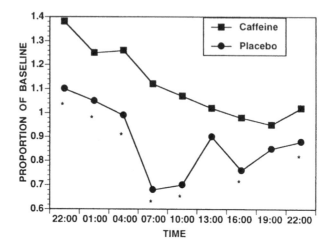

FIG. 9.3. MSLT values (proportion of prestudy baseline latency to Stage 1 sleep) for caffeine and placebo groups in Study 2. Significant differences are noted (*) in the figure. Data from Bonnet and Arand (1994b).

no evidence of negative effects from caffeine use, such as increased tremor, were found.

STUDY 3

Study 2 suggests that performance across a nocturnal shift and the following day can be maintained at a high level by taking a prophylactic nap prior to the work shift and using caffeine during that work shift. In hospitals and some other applied work settings involving sustained nocturnal work periods, employees are allowed to nap during the night until their presence is required. As might be expected, a debate concerning work schedules has begun in hospitals, where it has been common practice for residents to work 36-hour shifts and obtain short periods of sleep during the night (Godlee, 1992; Vassallo, Chana, Clark, Smith & Wood, 1992). Little empirical work has compared various nocturnal work and nap schedules, but a fair amount is known about the impact of reduced sleep and short nocturnal naps on the ability to maintain alertness and performance.

It is well known that significant sleep reduction will lead to a decline in the ability to perform many tasks. Empirical studies of young adults suggest that alertness and performance deficits can be measured on the day following a reduction in sleep to 4 to 6 hours per night on an acute basis (Carskadon, Harvey, & Dement, 1981; Dinges et al., 1997; Rosenthal, Roehrs, Rosen, & Roth, 1993), and these deficits accumulate under chronic conditions (Carskadon & Dement,

1981; Dinges et al., 1997). A number of studies have examined sleep, mood, and performance in physicians at various levels of training (reviewed in Bonnet, 1994; Leung & Becker, 1992). Of 11 studies that examined performance or mood in physicians who had slept an average of 2.8 hours during an on-call night as compared with recent baseline sleep amounts of 7.1 hours, nine studies reported significantly worse performance on at least one test.

In hospital on-call situations, it is common for physicians to attempt to sleep for periods of time during the night when their immediate presence is not required. The amount of sleep possible is widely variable and the ambiguity of the situation, which presumably conflicts the physician's desire and need for sleep with patient care responsibilities, is less than ideal. This work/rest schedule is by nature ill-defined and physicians are, therefore, unlikely to have the option of using caffeine to maintain alertness because caffeine might interfere with the ability to fall asleep if an opportunity should arise. The standard on-call/sleep schedule is also less than ideal because it is the rule that physicians are not awakened until needed. Performance immediately after arousal from sleep will frequently be poor because it can take up to 30 minutes to overcome the negative effects of waking out of deep sleep (*sleep inertia*; Dinges, 1989; Stampi, 1989). During the night, circadian effects add to sleep deprivation to increase sleep inertia. This reasoning indicates that individuals allowed relatively short intervals of sleep across an all-night work period should be less alert and should perform more poorly than individuals who have had naps at earlier times, but empirical comparisons of such schedules have not been done.

Therefore, in the third study, alertness following a combination of prophylactic nap with nocturnal caffeine use was compared with alertness after a schedule allowing four nocturnal naps. A group of normal young adults participated in two (counterbalanced) continuous work periods (Bonnet & Arand, 1994a). In one schedule, subjects were allowed to take four 1-hour naps during the nocturnal part of a 24-hour work period. In the other schedule, subjects took a prophylactic 4-hour nap before the 24-hour operation and received caffeine during the night as in Study 2. The empirical intent of this design was to compare a simulated on-call schedule allowing naps during the night to the empirical schedule that was shown to be most effective in maintaining alertness and performance during the night. It was hypothesized that a prophylactic nap and caffeine would allow maintenance of the baseline level of alertness and performance across the night and following day. It was also hypothesized that naps during the night would permit normal circadian drops in temperature, alertness, and performance.

The study involved spending 3 consecutive nights and 2 days in the laboratory (usually Thursday night through Sunday morning) for 2 consecutive weeks. The initial night was a baseline sleep night scheduled according to the subject's habitual sleep/wake time. On the following morning, subjects completed baseline testing and had their baseline nap latency test between 08:00 and 12:00 hours. For 1 week, subjects were allowed to leave the laboratory until 15:00 hours, when they returned to the lab to be readied for a 4-hour nap that began at 16:00 hours and ended at 20:00 hours. For the other (counterbalanced) week, subjects returned to the lab at 19:00 hours. Beginning at 20:00 hours, all subjects followed the same schedule of alternating performance test blocks, MSLT observations, and meals/breaks for 24 hours before being allowed a night of recovery sleep scheduled at their normal sleep time. In one group, subjects did not have a nap between 16:00 and 20:00 hours, but were allowed to stay in bed for 1 hour on each of their first four (nocturnal) MSLT observations. Subjects in the other group were awakened at sleep onset as in previous studies. For the 1 hour and 4 hour naps, subjects were required to remain in bed for the allotted time even if unable to sleep. All subjects in both conditions received pills at 01:30, 07:30, 13:30, and 19:30 hours. For subjects in the 4-hour afternoon nap group, the pills administered at 01:30 and 07:30 hours contained 200 mg of Eleveine. All other pills were placebos. In the group with four 1 -hour naps during the night, all pills were placebos. Subjects were unaware of when caffeine might be administered and reported no consistent expectations about afternoon or evening naps being "better."

Data from the 4-hour nap and the four 1-hour naps are shown in Table 9.1. Total sleep time in the two 4-hour nap conditions did not differ, but there were significant differences in the distribution of sleep stages. These differences reflected the timing and length of the nap sleep periods. Fifty-eight minutes of wake during sleep in the 4-hour nap was offset by four 12-minute sleep onset latencies (total of 47 minutes) in the 1-hour naps. *Slow wave sleep* (Stage 3 plus Stage 4) was increased in the nighttime naps as compared to the prophylactic nap. Despite the 4 hours of available sleep time for both groups, the recovery

TABLE 9.1
Sleep Values for the 4-Hour Prophylactic Nap and the Sum of the Four 1-Hour Nocturnal Naps

	4-Hour Nap		4 × 1-Hour Naps		t	p
Total sleep (min)	167	(57)	174	(29)	0.33	NS
Stage 1%	6.7	(3.7)	10	(4.5)	3.22	0.01
Stage 2%	33	(16)	21	(10)	3.14	0.01
Stage 3%	7.2	(4.9)	7.6	(4.9)	0.30	NS
Stage 4%	15	(9.0)	32	(13)	5.55	0.01
Stage REM%	12	(8.2)	17	(8.7)	2.41	0.04
Sleep latency (min)	10	(7.1)	12*	(6.7)	1.03	NS

*The total time spent falling asleep in the four naps was 47 minutes or an average of 12 minutes per nap.

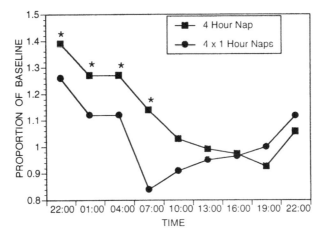

FIG. 9.4. MSLT values (proportion of prestudy baseline latency to Stage 1 sleep) for 4-hour prophylactic nap and 4×1-hour nap groups in Study 3. Significant differences are noted (*) in the figure. Data from Bonnet and Arand (1994a).

night of sleep demonstrated decreased sleep latency, increased sleep efficiency, and increased Stage 4 sleep. These changes are usually seen following sleep loss (Bonnet, 1994). However, there were no statistically significant differences in the recovery night sleep between the prophylactic 4-hour nap and the four 1-hour nocturnal nap conditions.

Objective alertness (MSLT) was measured 10 times during the study, including a baseline observation prior to the 4-hour nap and every 3 hours during the study starting at 22:00 hours and ending at 22:00 hours on the next evening. The data underwent a log transformation before calculation of the percentage of baseline and ANOVA. ANOVA for the MSLT (terms for group, time of test, and interaction) revealed a significant Group x Time interaction, $F(8, 88)= 3.156$, $p<.01$. These data are plotted in Fig. 9.4 with significant differences, as determined by Neuman-Keuls pairwise comparisons, designated. Nap latencies were significantly longer at 22:00, 01:00, 04:00, and 07:00 hours in the 4-hour nap group as compared to the 4×1-hour nap group. The subjects having four naps during the course of the night were pathologically sleepy at 07:00 hours, but their alertness improved to nearly normal levels by the end of the study. Although alertness was significantly greater in the 4-hour prophylactic nap group through 07:00 hours, alertness continued to decline in that group across the day. In summary, the third study has shown that alertness across the night shift was significantly improved following a prophylactic nap and caffeine use as compared to a series of nocturnal naps.

CONCLUSION

The data from all three studies indicated that individuals who take a prophylactic nap and use caffeine during their work shift will have significantly increased objective alertness during the nocturnal work shift. It would be predicted that the potential for a catastrophic mistake would be the greatest when an employee was awakened and immediately forced to make an important decision. Although such a situation was not explicitly measured in Study 3, it is common that individuals allowed to sleep during work situations will be awakened only when an emergency occurs. Several studies have indicated that performance is worse immediately after awakenings during nocturnal sleep periods (Dinges, 1989; Stampi, 1989). Sleep inertia effects are increased after awakenings from slow wave sleep (Bonnet, 1983) and have been found in many tasks including simple and choice reaction time (Feltin & Broughton, 1968; Scott & Snyder, 1968), short term memory (Bonnet, 1983; Stones, 1977), and letter substitution (Dinges, Orne, & Orne, 1985). In the current study, 60% of nocturnal naps (vs. 10% of prophylactic naps) ended in slow wave sleep. This figure is probably a low estimate for the real world where on-call situations and other stresses typically cause chronic as opposed to acute sleep deprivation. The high percentage of awakenings from deep sleep implies that individuals in nocturnal on-call situations will frequently suffer from the maximum negative consequences of sleep inertia, sleep deprivation, and declining circadian rhythm.

Study 3 indicated that individuals who nap during the night in their work situation and who are awakened frequently or accumulate less than 4 hours of sleep will be less alert than individuals who take an afternoon nap and remain awake during the night. In terms of practical use, the results suggest that taking a nap prepares individuals to deal with the sleep loss that occurs during nocturnal work periods and that caffeine probably reverses some of the normal circadian decline in nocturnal alertness. However, this strategy also has limitations, and extending work periods into the afternoon or evening would have required additional sleep or caffeine administration.

In summary, these studies have demonstrated the beneficial effect of obtaining additional sleep prior to periods of sleep loss and the additive benefits of caffeine use. Subjects for the studies were healthy young adults who did not report significant daily caffeine use and who were not sleep deprived prior to the beginning of each study. In the real world, of course, it is common that shift workers will suffer from both chronic sleep loss and tolerance to caffeine. Such limitations could reduce the beneficial impact of both prophylactic naps and caffeine in applied settings. On the other hand, knowledge of the ability of naps and caffeine to produce baseline levels of alertness across nocturnal work periods when used correctly should allow planners to develop improved sustained work schedules.

ACKOWLEDGMENTS

This work was performed at the Long Beach Veterans Administration Medical Center and the San Diego Naval Health Research Center supported by a Merit Review Grant from the Department of Veterans Affairs, the Sleep-Wake Disorders Research Institute, and the Naval Medical Research and Development Command, Department of the Navy, Bethesda, Maryland under Research Work Unit 61153N MR 04101.03–6003. The views presented in this paper are those of the authors. No endorsement by the Department of the Navy has been given or should be inferred.

REFERENCES

Bonnet, M.H. (1983). Memory for events occurring during arousal from sleep. *Psychophysiology, 20*(1), 81–7.

Bonnet, M.H. (1994). Sleep deprivation. In M.Kryger, T.Roth, & W.C.Dement (Eds.), *Principles and practice of sleep medicine* (pp. 50–68). Philadelphia: Saunders.

Bonnet, M.H., & Arand, D.L. (1994a). The impact of naps and caffeine on extended nocturnal performance. *Physiology & Behavior, 56*(1), 103–109.

Bonnet, M.H., & Arand, D.L. (1994b). The use of prophylactic naps and caffeine to maintain performance during a continuous operation. *Ergonomics, 37*(6), 1009–1020.

Bonnet, M.H., Gomez, S., Wirth, O., & Arand, D.L. (1995). The use of caffeine versus prophylactic naps in sustained performance. *Sleep, 18,* 97–104.

Carskadon, M.A. (1994). Measuring daytime sleepiness. In M.Kryger, T.Roth, & W.C.De-ment (Eds.), *Principles and practice of sleep medicine* (pp. 961–966). Philadelphia: Saunders.

Carskadon, M.A., & Dement, W.C. (1979). Effects of total sleep loss on sleep tendency. *Perceptual and Motor Skills, 48*(2), 495–506.

Carskadon, M.A., & Dement, W.C. (1981). Cumulative effects of sleep restriction on daytime sleepiness. *Psychophysiology, 18*(2), 107–13.

Carskadon, M.A., Harvey, K., & Dement, W.C. (1981). Sleep loss in young adolescents. *Sleep, 4*(3), 299–312.

Dinges, D.F. (1989). Napping patterns and effects in human adults. In D.F.Dinges & R.J. Broughton (Eds.), *Sleep and alertness: Chronobiological, behavioral, and medical aspects of napping* (pp. 171–204). New York: Raven Press.

Dinges, D.F., Orne, M.T., & Orne, E.C. (1985). Assessing performance upon abrupt awakening from naps during quasi-Continuous operations. *Behavior, Research Methods, Instruments, & Computers, 17,* 37–45.

Dinges, D.F., Pack, F., Williams, K., Gillen, K.A., Powell, J.W., Ott, G.E., Aptowicz, C., & Pack, A.I. (1997). Cumulative sleepiness, mood disturbance, and psychomotor vigilance performance decrements during a week of sleep restricted to 4–5 hours per night. *Sleep, 20,* 267–277.

Feltin, M., & Broughton, R.J. (1968). Differential effects of arousal from slow wave sleep and REM sleep. *Psychophysiology, 5,* 231.

Godlee, F. (1992). Juniors' hours: Is the end in sight? *British Medical Journal, 305,* 937–940.

Leung, L., & Becker, C.E. (1992). Sleep deprivation and house staff performance. Update 1984–1991. *Journal of Occupational Medicine, 34,* 1153–60.

Rechtschaffen, A., & Kales, A. (1968). *A manual of standardized terminology, techniques, and scoring systems for sleep stages of human subjects.* Washington, DC: Public Health Service, U.S. Government Printing Office.

Rosenthal, L., Roehrs, T.A., Rosen, A., & Roth, T. (1993). Level of sleepiness and total sleep time following various time in bed conditions. *Sleep, 16,* 226–232.

Scott, J., & Snyder, F. (1968). "Critical reactivity" (Pieron) after abrupt awakenings in relation to EEG stages of sleep. *Psychophysiology, 4,* 370.

Stampi, C. (1989). Ultrashort sleep/wake patterns and sustained performance. In D.F. Dinges & R.J.Broughton (Eds.), *Sleep and alertness: Chronobiological, behavioral, and medical aspects of napping* (pp. 139–170). New York: Raven Press.

Stones, M.J. (1977). Memory performance after arousal from different sleep stages. *Psychophysiology, 68,* 177–181.

Sugerman, J.L., & Walsh, J.K. (1989). Physiological sleep tendency and ability to maintain alertness at night. *Sleep, 12,* 106–112.

Vassallo, D.J., Chana, J., Clark, C.L., Smith, R.E., & Wood, R.F. (1992). Introduction of a partial shift system for house officers in a teaching hospital. *British Medical Journal, 305,* 1005–1008.

Chapter 10

Studying Pharmacological Performance Enhancement With Behavioral, Subjective, and Electroencephalographic Measures

John A.Caldwell, Jr.
J.Lynn Caldwell
U.S. Army Aeromedical Research Laboratory,
Fort Rucker, Alabama

Effective sustained operations are difficult because personnel, unlike machines, need periodic sleep for the restitution of both the body and brain (Horne, 1978). Although it is possible for organizations to function continuously throughout the day and night while adequate staffing is available, downsizing and cutbacks have made it difficult to ensure proper sleep and rest among the personnel involved in around-the-clock operations.

Ultimately, sleep deprivation limits the effectiveness of individuals who are unable to obtain adequate sleep on a daily basis. Humans who are deprived of sleep experience increased mental lapses, reductions in cognitive abilities, decrements in alertness, and decreases in motivation (Krueger, 1989). No amount of training or experience with sleep deprivation enables someone to overcome these effects.

Because it is often impossible to obtain adequate sleep in situations that require sustained work, effective countermeasures to minimize the mood and performance impact of sleep deprivation must be explored. The research described here examines the use of behavioral, subjective, and electrophysiological measures to assess the efficacy of three basic approaches. The first is to use stimulants to sustain alertness and performance in personnel required to work extended periods with no opportunity for sleep. The second is to use short prophylactic naps (with or without a medication) to sustain performance in circumstances where sleep opportunities exist, but are limited to short durations. The third is to use hypnotics to indirectly aid performance by improving sleep quality in situations where 6 to 8 hours of sleep are possible, but environmental or other factors impair sleep initiation and maintenance. Because there was a particular interest in applying these countermeasures in aviation scenarios, the majority of studies involved simulated or actual flight performance measures in addition to more standard laboratory tests.

BACKGROUND

Estimating the efficacy of any strategy to sustain aviator performance in the operational environment is a demanding task. As Angus, Pigeau, and Heslegrave (1992) indicate, few sustained operations studies examine the interactions between the operational environment and the individual's reactions to that environment They also point out that laboratory tests can be difficult to generalize to the real world because they may not be sufficiently difficult or interesting, may be administered too infrequently to be reliable, may be given in durations that are too brief to yield sensitivity, or may be inappropriate to the stressor under investigation. Caldwell (1995) suggested that many of the commonly used laboratory tests also suffer from a lack of sensitivity because of overtraining and a general lack of complexity. To address many of these concerns, Angus et al. (1992) favor conducting sequences of field and laboratory experiments with carefully selected tests and schedules that closely mimic operational demands, and Caldwell (1995) emphasizes the importance of collecting physiological measures along with both performance and cognitive measures in order to improve the accuracy and generalizability of research findings. A psychophysiological approach not only provides a well-rounded evaluation of promising countermeasures (and other variables), but is less susceptible to motivation, practice, and other extraneous factors.

The studies described here employ a variety of tests designed to approximate the multitasking demands of the operational environment; they utilize test schedules that evaluate performance and alertness throughout the entire period of continuous wakefulness, employ a multidimensional approach that integrates behavioral, cognitive, questionnaire, and electroencephalographic assessments, and, in most cases, include measures of flight performance from a specially instrumented, high-fidelity, full-visual UH-60 helicopter simulator (and in one case, actual in-flight performance in a UH-60 aircraft). They also utilize subjects drawn from the population to which the final results will be applied (aviators). In several cases, a similar experimental design was used across studies to facilitate comparisons of different countermeasures for sleep loss. This approach maximizes the similarities between research and operational situations while lending more objectivity, sensitivity, reliability, and validity to findings that will be incorporated into operational policies.

General Study Designs

Four investigations are discussed. The first is a helicopter simulator study of the performance-sustaining effects of dextroamphetamine (Dexedrine®) given to sleep-deprived pilots. The second is an actual in-flight study of the sustaining effects of dextroamphetamine designed to validate findings from the simulator investigation. The third is a study of the alertness-sustaining potential of prophylactic naps (induced with zolpidem tartrate vs. placebo) in aviators subjected to prolonged work periods. The fourth is a study of the potential hangover effects of the sleep medication triazolam (Halcion®) given to pilots prior to a full and an interrupted night of sleep. Each is a counterbalanced, double-blind, placebo-controlled, within-subjects study. After training and adaptation, subjects completed a series of control and test days. Although control days (washout periods for drug effects and/or sleep-deprivation effects) are not presented, they were examined before performing the test-day analyses to minimize the possibility of preexisting systematic differences prior to the experimental interventions.

Testing Schedules

In the two dextroamphetamine studies and the zolpidem-napping study, a test schedule was selected that would permit continuous evaluations of aviator performance and alertness throughout most of the circadian cycle. It is known that alertness suffers the most around 05:00 hours when the body's core temperature is at its lowest point (Åkerstedt, 1995). Conversely, during daytime hours, alertness and performance tend to improve as a function of normal circadian influences. When evaluating countermeasures for use in around-the-clock operations, it is necessary to ensure assessments are conducted at several time points during the day. In addition, because personnel who are involved in real-world sustained operations are required to work prolonged periods with only minimal rest breaks, the schedules used in these studies placed continuous demands on participants. In three investigations, subjects were tested on performance abilities, psychological mood, and electrophysiological measures of arousal at 4-hour intervals from 01:00 hours until approximately 20:00 hours. In the triazolam study, the test schedule was less demanding since the objective of this study was to examine hangover effects of a sleep medication after either a full or an interrupted night of sleep rather than alertness and performance sustainment.

Dependent Measures

Each study employed a variety of related assessments. In the dextroamphetamine and triazolam studies, flight performance was objectively measured by computer in a UH-60 helicopter simulator or aircraft while subjects flew a standardized course; central nervous system (CNS) function was measured by a resting eyes-open/eyes-closed electroencephalograph (EEG); mood and/or sleepiness were measured by self-report instruments such as the Profile of Mood States (POMS) or the Stanford Sleepiness Scale; and vigilance and cognition were measured with the Walter Reed Performance Assessment Battery (PAB), the Synthetic Work Battery (SYNWORK), or the Multi-Attribute Task Battery (MATB).

In the zolpidem/napping study, flight performance data were not collected, but CNS activity (resting eyes-open/eyes-closed EEG), mood (POMS and visual analog scales [VAS]), vigilance and cognitive performance (MATB and SYNWORK), and physiological sleepiness (the repeated test of sustained wakefulness [RTSW]) were assessed. In all of the studies, sleep architecture was measured by polysomnography since overall sleep quality during postdeprivation periods has a significant impact on subsequent arousal and performance.

SIMULATOR STUDY OF DEXTROAMPHETAMINE

This study investigated the efficacy of using dextroamphetamine (Dexedrine®) as a countermeasure to sleep loss during sustained operations in which it is impossible for personnel to gain any restorative sleep for up to 40 hours. Although, previously published reports have proven that amphetamines are effective for enhancing performance and alertness (see Caldwell et al., 1994, for a review), there had been no controlled aviation studies. However, there was evidence that amphetamines would effectively counter sleep loss and fatigue in aviation personnel.

Stimulants have been used in several recent military operations in which aviators were engaged in continuous operations that included sleep deprivation. Dextroamphetamine was administered to EF-111A Raven crews during an Air Force strike on Libya in April 1986 (Senechal, 1988) and to F-15C pilots flying combat air patrol missions in Operation Desert Shield/Storm (Cornum, 1992; Emonson & Vanderbeek, 1995). The medication was found to be both safe and beneficial in terms of overcoming fatigue without producing unwanted side effects.

Thus, there was good reason to believe that dextroamphetamine would be an effective countermeasure for sleep loss in sustained aviation operations. However, because there were no controlled aviation studies, 12 subjects (6 males and 6 females) were exposed to 2 periods of 40 hours of continuous wake-fulness during which they flew a helicopter simulator and completed EEG and mood evaluations. During the final 23 hours of one period, subjects were administered three separate 10-mg doses of dextroamphetamine (at 00:00, 04:00, and 08:00 hours) and, during the final hours of the other period, subjects were administered placebo. Test sessions occurred at 01:00, 05:00, 09:00, 13:00, and 17:00 hours. As shown in Fig. 10.1, each session began with a 1-hour flight in a UH-60 simulator consisting of hovers, low-level navigation, upper-altitude airwork

Time	Sunday	Monday	Tuesday	Wednesday	Thursday	Friday	Saturday
00-01				DEX/PBO		DEX/PBO	
01-02				Simulator		Simulator	
02-03				EEG		EEG	
03-04				Minisim POMS/Synw		Minisim POMS/Synw	
04-05				DEX/PBO		DEX/PBO	
05-06				Simulator		Simulator	
06-07				EEG		EEG	
07-08		Wake up	Wake up	minisim POMS/Synw	Wake up	Minisim POMS/Synw	Wake up Breakfast
08-09		Testdose Breakfast	Breakfast	DEX/PBO Breakfast	Breakfast	DEX/PBO Breakfast	Release
09-10		Simulator	Simulator	Simulator	Simulator	Simulator	
10-11		EEG	EEG	EEG	EEG	EEG	
11-12		Minisim	Minisim POMS/Synw	Minisim POMS/Synw	Minisim POMS/Synw	Minisim POMS/Synw	
12-13		POMS/Synw Lunch	Lunch	Lunch	Lunch	Lunch	
13-14		Simulator	Simulator	Simulator	Simulator	Simulator	
14-15		EEG	EEG	EEG	EEG	EEG	
15-16		Minisim	Minisim POMS/Synw	Minisim POMS/Synw	Minisim POMS/Synw	Minisim POMS/Synw	
16-17		POMS/Synw					
17-18		Simulator	Simulator	Simulator	Simulator	Simulator	
18-19	Arrive Med exam	EEG	EEG	EEG	EEG	EEG	
19-20	EEG Hook up	Minisim POMS	Minisim POMS Synwork	Minisim POMS Synwork	Minisim POMS Synwork	Minisim POMS Synwork	
20-21		Synwork Dinner	Dinner	Dinner	Dinner	Dinner	
21-22		pt	pt	pt	pt	pt	
22-23		Shower POMS	Shower	Shower POMS	Shower	Shower POMS	
23-24			POMS		POMS		

Note: DEX = Dexedrine dose (10 mg), PBO = Placebo

FIG. 10.1. Test schedule for the simulator dextroamphetamine study.

(straight and levels, right and left standard turns, climbs and descents, and a left descending turn), and formation flight. Following the flights, EEG data were collected, with eyes open and eyes closed, from electrodes placed at Fz, Cz, Pz, and Oz (referenced to linked mastoids A1 and A2). Next, POMS questionnaires (for ratings of tension, depression, anger, vigor, fatigue, and confusion) and the SYNWORK (a concurrent multitasking auditory monitoring, visual monitoring, mathematics, and memory test) were administered.

Flight Performance

The flight data revealed there were overall drug-related performance effects on every maneuver with the exception of the hovers and the formation flight. The low-level navigation, climbs, descents, left-descending turns, left and right turns, and straight-and-levels were improved by dextroamphetamine in comparison to placebo. In addition, there were interactions between drug and time of day on the straight and levels, climbs, descents, and left standard-rate turns. In each case, there was no difference between performance as a function of dextroamphetamine versus placebo at the 01:00 flight when subjects had not yet begun to experience the full effects of sleep deprivation (2 hours past their normal bedtimes). However, dextroamphetamine consistently produced better performance than placebo at 05:00 and 09:00 hours when the combined fatigue from sleep loss and the circadian trough tended to be greatest. The 13:00 and 17:00 flights in these maneuvers were variably affected by the drug conditions. In the straight-and-levels and descents, dextroamphetamine continued to markedly improve afternoon performance as it had done earlier in the day. However, during the climbs and left standard-rate turns, there were no drug-related differences at 13:00 and 17:00 hours. These effects are depicted in Fig. 10.2. A close examination of each of the drug by time-of-day effects suggests that dextroamphetamine essentially maintained performance at predeprivation levels throughout the sleep-loss period. However, under placebo, marked decrements occurred, particularly at 05:00 and 09:00 hours, followed by a tendency toward a circadian-influenced recovery at 13:00 and 17:00 hours. Overall, performance under dextroamphetamine was significantly better than performance under placebo.

EEG

Drug effects consistent with flight performance changes were found in the EEG. Under placebo, there were overall increases in slow-wave (delta and theta) activity, normally associated with decreased alertness and/or sleep deprivation (Comperatore et al., 1993; Lorenzo et al., 1995; Pigeau, Heslegrave, & Angus, 1987). Dextroamphetamine mitigated this effect as it did the decrements in flight performance. There also were several drug-by-eyes interactions within the delta, theta, and alpha bands. The delta and theta effects were clearer under eyes closed

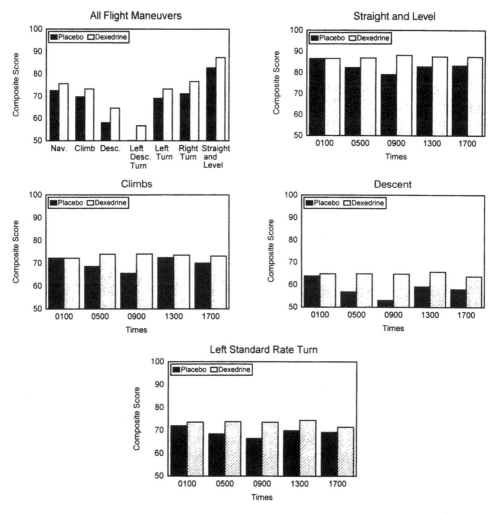

FIG. 10.2. Drug-related performance effects overall and across sessions on simulator maneuvers.

than eyes opened. In fact, drug-related changes in delta at Fz, Cz, and Pz were observed only during eyes closed. Similar changes occurred in theta at Cz and Pz, where, although the differences were significant both during eyes open and eyes closed, they were most pronounced under eyes closed. Reductions in alpha activity under placebo versus dextro-amphetamine at Fz and Cz (indicating subjects were more alert under dextroamphetamine) were similarly due to more pronounced differences during eyes closed rather than eyes open (see Fig. 10.3). Taken together, these condition-by-eyes interactions show that shifts in

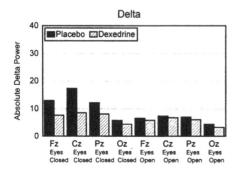

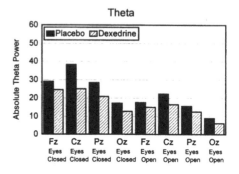

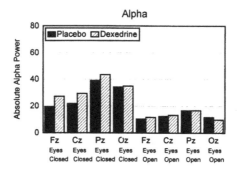

FIG. 10.3. Drug-related EEG effects under eyes open and eyes closed by electrode site—dextroamphetamine simulator study.

alertness were more pronounced while subjects were sitting quietly with their eyes closed because the increased temptation to sleep was greater under these circumstances.

POMS

Subjective ratings of vigor, fatigue, confusion, and anger agreed with objectively measured performance/alertness. The first POMS of the day (administered 1 hour, 20 minutes prior to the 05:00 flight) indicated greater feelings of vigor-activity under dextroamphetamine than placebo, while there were greater feelings of fatigue-inertia and confusion-bewilderment under placebo than dextroamphetamine at this time. The drug effects were evident 1 hour, 40 minutes after the second and third flights as well, showing that subjects were feeling the effects of sleep deprivation under placebo and deriving the most beinefit from dextroamphetamine, especially during the morning. Later in the day, subjects continued to feel more vigorous and less fatigued under dextroamphetamine, although differences in confusion scores between the two conditions disappeared (see Fig. 10.4). Regardless of the testing session, overall drug effects on the anger-hostility scale indicated greater anger under placebo than under dextroamphetamine.

The most uniform dextroamphetamine-related improvements in flight performance occurred at 05:00 and 09:00 hours, and these were bracketed by the 03:40 to 11:40 POMS. There also were overall improvements in both flight performance and CNS indicators of alertness throughout the test days, and this finding is consistent with overall improvements in vigor and fatigue. The fact that there was greater vigor and less fatigue under dextroamphetamine than placebo as late as 22:25 hours showed dextroamphetamine continued to exert its effects long after the last dose was administered.

SYNWORK

There were virtually no effects of either sleep deprivation or dextroamphetamine on performance of the Sternberg memory, arithmetic, visual monitoring, or auditory monitoring tasks. The only exception was a tendency toward decrements in auditory vigilance under dextroamphetamine at the last session of the day; however, this was seen only in the females and was not statistically significant.

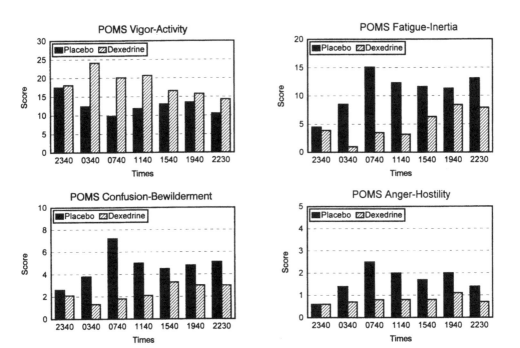

FIG. 10.4. Drug-related effects on mood across sessions—dextroamphetamine simulator study.

The fact that there were no substantial effects on SYNWORK despite striking changes in flight performance, mood, and EEG highlights a methodological problem in the cognitive testing approach. Specifically, a test session longer than 10 minutes should have been used because past research has shown the effects of sleep loss can often be overcome during short tests (Wilkinson, 1969).

Polysomnography

The data indicated recovery sleep following dextroamphetamine was not as restful as sleep following placebo, but was better than sleep recorded on the baseline night. Dextroamphetamine degraded sleep quality (increased time awake after sleep onset, Stages 1 and 2 sleep, movement time, and rapid eye movement [REM] latency, and reduced the percentage of REM sleep) compared to placebo. However, sleep after dextroamphetamine was more restful than baseline sleep, with a shorter sleep latency, less awake time after sleep onset, and more Stage 4 sleep (see Fig. 10.5). These results indicate the effects of dextroamphetamine (15 hours after the last dose) were not as strong as the need for sleep following 40 hours of continuous wakefulness.

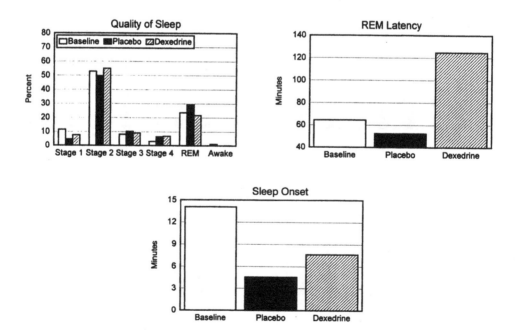

FIG. 10.5. Drug effects on sleep parameters—dextroamphetamine simulator study

Summary

In summary, properly administered 10-mg doses of dextroamphetamine effectively prevented most of the performance losses attributable to moderate sleep deprivation, without creating unwanted side effects. Also, because there were physiological changes (EEG) accompanying the performance improvements, it can be concluded that the positive effects of dextroamphetamine are based on changes in physiological arousal and not practice, motivation, or some other extraneous factor. However, it remained to be determined whether the results of this simulator study would generalize to the actual in-flight environment.

AN IN-FLIGHT EVALUATION
OF DEXTROAMPHETAMINE

The in-flight evaluation of dextroamphetamine in sustained operations was a systematic replication of the earlier simulator study. Once again, the primary objective was to evaluate the usefulness of dextroamphetamine for situations in which personnel are offered no opportunities for sleep. The replication was performed because, although modern simulators closely emulate actual flight conditions, pilots are always aware of the difference between flying a platform that is aerodynamically suspended in the air 2,000 feet above the ground and one that is anchored via the pistons and cylinders of a steel hydraulics system 15 feet above the floor. Would this knowledge affect the motivation of pilots sufficiently to eliminate the deprivation-related performance decrements found under the placebo condition in the simulator or would the actual flight environment (with its noise and vibration) worsen the effects of sleep loss? Assuming there were no other differences between the simulator and flight environments, the motivational difference could conceivably raise the level of performance under placebo to the extent that the dextroamphetamine and placebo scores did not differ. Conversely, if the noise and vibration of the flight environment exerted some hypnotic effect, the difference between dextroamphetamine and placebo might be even larger than predicted. Besides these two possibilities, it was also possible that wind turbulence differences in the two situations (zero turbulence in the simulator vs. a full range of unpredictable turbulence in the aircraft) would create so much error variance that no statistically significant advantage to dextroamphetamine could be found in the real world.

To address these concerns, 10 UH-60 pilots were exposed to 1 week of testing that included two separate 40-hour periods of continuous wakefulness very similar to those used in the earlier simulator study. Subjects completed flights in a specially instrumented UH-60 helicopter, as well as electroencephalographic, cognitive, and mood tests. During one deprivation period, subjects received three separate 10-mg doses of dextroamphetamine 1 hour prior to each of the first three sessions, which began at 01:00, 05:00, and 09:00 hours. During the other period, subjects received placebos. As in the simulator study, subjects were not administered medication prior to the 13:00 and 17:00 sessions. A double-blind, counterbalanced administration scheme was used.

As shown in Fig. 10.6, each session began with a 1-hour flight in the UH-60 aircraft consisting of upper-altitude airwork (straight and levels, right and left

Time	Sunday	Monday	Tuesday	Wednesday	Thursday	Friday	Saturday
00-01				DEX/PBO		DEX/PBO	
01-02				Aircraft Flight		Aircraft Flight	
02-03				EEG Dsktp sim		EEG Dsktp sim	
03-04				POMS/CFF MATB		POMS/CFF MATB	
04-05				DEX/PBO		DEX/PBO	
05-06				Aircraft Flight		Aircraft Flight	
06-07				EEG Dsktp sim		EEG Dsktp sim	
07-08		Wake up	Wake up	POMS/CFF MATB	Wake up	POMS/CFF MATB	Wake up Breakfast
08-09		Testdose Breakfast	Breakfast	DEX/PBO Breakfast	Breakfast	DEX/PBO Breakfast	Release
09-10		Aircraft Flight	Aircraft Flight	Aircraft Flight	Aircraft Flight	Aircraft Flight	
10-11		EEG Dsktp sim	EEG Dsktp sim	EEG Dsktp sim	EEG Dsktp sim	EEG Dsktp sim	
11-12		POMS/CFF MATB	POMS/CFF MATB	POMS/CFF MATB	POMS/CFF MATB	POMS/CFF MATB	
12-13		Lunch	Lunch	Lunch	Lunch	Lunch	
13-14		Aircraft Flight	Aircraft Flight	Aircraft Flight	Aircraft Flight	Aircraft Flight	
14-15		EEG Dsktp sim	EEG Dsktp sim	EEG Dsktp sim	EEG Dsktp sim	EEG Dsktp sim	
15-16		POMS/CFF MATB	POMS/CFF MATB	POMS/CFF MATB	POMS/CFF MATB	POMS/CFF MATB	
16-17							
17-18		Aircraft Flight	Aircraft Flight	Aircraft Flight	Aircraft Flight	Aircraft Flight	
18-19	Arrive Med exam	EEG Dsktp sim	EEG Dsktp sim	EEG Dsktp sim	EEG Dsktp sim	EEG Dsktp sim	
19-20	EEG Hook up	POMS/CFF MATB	POMS/CFF MATB	POMS/CFF MATB	POMS/CFF MATB	POMS/CFF MATB	
20-21		Dinner	Dinner	Dinner	Dinner	Dinner	
21-22		pt	pt	pt	pt	pt	
22-23		Shower POMS	Shower	Shower POMS	Shower	Shower POMS	
23-24			POMS		POMS		

Note: DEX = Dexedrine dose (10 mg), PBO = Placebo

FIG. 10.6. Test schedule for the in-flight dextroamphetamine study.

standard turns, climbs and descents, a left-descending turn, and an instrument landing system [ILS] approach). The hovers, low-level navigation, and formation-flight portions used in the simulator were not performed for safety reasons. Following the flights, subjects were returned to the laboratory where EEG data were collected, with eyes open and eyes closed, from electrodes placed at Fz, Cz, Pz, and Oz (referenced to linked mastoids A1 and A2). Next, POMS questionnaires and the MATB (a concurrent multitasking aviation simulation consisting of monitoring lights and dials, managing fuel and radios, and concurrently performing an unstable tracking task) were administered.

Flight Performance

Dextroamphetamine significantly improved the ability of aviators to fly a UH-60 helicopter during the final 23 hours of the 40-hour periods of continuous wakefulness. Heading and airspeed control were more precisely maintained (lower root mean square [RMS] errors) after dextroamphetamine than after placebo in the straight and levels; roll control was better in the left turns; heading and slip control were improved during the climbs; heading, airspeed, roll, and vertical speed control were more accurately maintained in the descent; vertical speed control was better in the left descending turn; and localizer tracking was more precise during the ILS. Although there were no overall drug effects on the three right turns, roll control was improved under dextroamphetamine in comparison to placebo on the last turn in which the UH-60's computerized flight-path stabilization system was disengaged. Dextroamphetamine was particularly helpful for maintaining both roll and vertical-speed control accuracy at 05:00 and 09:00 hours during the left descending turn (note the smaller RMS errors at these times in Fig. 10.7). There were 2 instances out of 19 in which a reversal of the expected drug effects occurred (vertical speed control was better under placebo on the first climb and slip control was better under placebo at 05:00 hours on the ILS), but these were exceptions and not the rule. Although these results are consistent with those from the simulator study, there were fewer significant drug effects than expected. This probably resulted because turbulence, traffic delays, radio distractions, and environmental changes absent in the simulator obscured drug-related effects in the actual aircraft. Also, the in-flight study participants benefitted from the alerting effects of frequent sunlight exposures, periodic walks outside of the laboratory, and changes in scenery associated with traveling from the Laboratory to the airfield every 4 hours. The subjects who flew the simulator had fewer opportunities to move about because the simulator room was only a short walk down the hall.

EEG

The fact that dextroamphetamine improved flight performance despite sleep deprivation is consistent with changes in the EEG. In particular, EEG theta

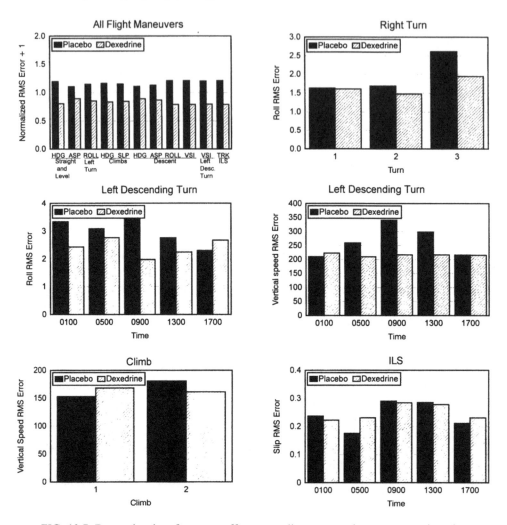

FIG. 10.7. Drug-related performance effects overall, among various maneuver iterations, and across sessions on flight maneuvers.

activity, which is known to increase as a function of sleep deprivation, was reduced by dextroamphetamine in comparison to placebo, especially under eyes closed. Because elevations in theta are associated with cognitive impairments (Lorenzo et al., 1995), reductions in theta suggest improvements in performance under dextroamphetamine were probably due to enhancements in CNS arousal. Further evidence for increased CNS arousal under dextroamphetamine was seen in the fact that EEG alpha activity was higher under the drug relative to placebo when subjects closed their eyes (see Fig. 10.8). Because alpha suppression and

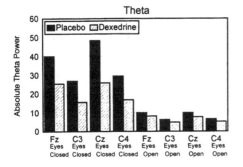 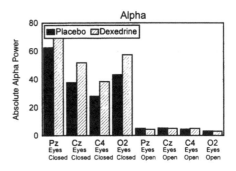

FIG. 10.8. Drug-related EEG effects under eyes open and eyes closed by electrode site—dextroamphetamine in-flight study.

increased theta are primary indicators of the onset of sleep, these changes show that subjects suffered less from drowsiness and sleepiness-related performance errors after dextroamphetamine than placebo, findings consistent with those from the simulator study.

POMS

Subjective mood states improved after dextroamphetamine, especially in the morning and the middle of the day as was the case in the simulator study. In comparison to placebo, depression-dejection ratings were lower under dextroamphetamine at 07:25 hours (1.5 hours prior to the third flight); fatigue-inertia and vigor-activity scores were improved by dextroamphetamine 1.5 hours before the second, third, and fourth, as well as after the fifth flight of the day. Confusion-bewilderment scores were lower under dextroamphetamine before the second, third, and fourth flights and at the end of the day (see Fig. 10.9). Thus, in addition to improved performance and physiological arousal under dextroamphetamine, feelings of alertness were sustained by the drug throughout sleep-deprivation.

MATB

Based on simulator studies and other literature (Caldwell et al., 1994; Caldwell et al., 1995; Wilkinson, 1969), it was determined that a cognitive testing session of at least 20 minutes would be necessary to detect cognitive and vigilance decrements on a computerized task. Thus, the MATB was set to run for 20 minutes each time. There were a variety of effects that indicated dextroamphetamine was effective in sustaining aviator performance. Subjects were faster at responding to the onset of warning lights and the presence of dial deviations, slightly better at completing the communications task, and more accurate in tracking a target

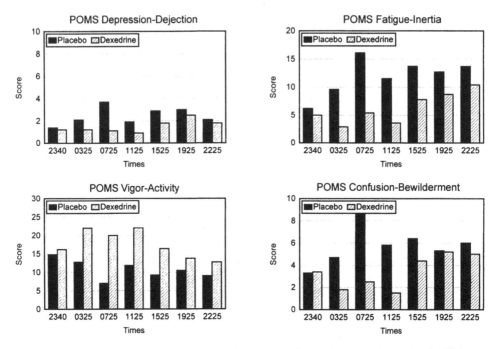

FIG. 10.9. Drug-related effects on mood across sessions—dextroamphetamine in-flight study.

under dextroamphetamine than under placebo. In addition, drug-by-session interactions on time-out errors for dials, total errors for communications, and RMS errors on the tracking task indicated many of the largest differences between dextroamphetamine and placebo occurred in the morning hours around the times of the greatest differences in some aspects of flight performance and mood. There were more time-out errors for dials under placebo than dextroamphetamine at 1.5 hours prior to the third, fourth, and fifth flights, and more total communications errors and tracking errors under placebo than dextroamphetamine before the third flight (see Fig. 10.10). These data agreed with findings from other measures that indicated the greatest benefit from dextroamphetamine occurred during the times of day when fatigue levels were most severe (from about 04:00 hours until 12:00 hours). It is noteworthy that the MATB tracking task that required the most vigilance was one of the tasks most affected by fatigue, consistent with the notion that sleep deprivation affects vigilance-based tasks more than others (especially longer duration ones; Wilkinson, 1964).

Polysomnography

Sleep was slightly more disrupted on the recovery night after dextroamphetamine than after placebo. Stages 1, 2, and REM, movement time, REM latency, and

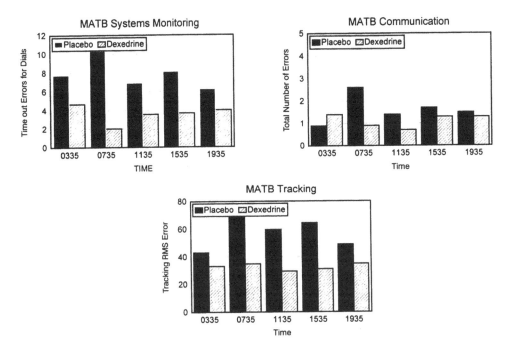

FIG. 10.10. Drug-related effects on the MATB by session—dextroamphetamine in-flight study.

sleep latency increased on the night after dextroamphetamine versus placebo. Stage 4 was unaffected. Comparisons to baseline revealed more Stage 1 during baseline than during either postdeprivation sleep period and more Stage 2 during baseline than during sleep following placebo. Stage 4 sleep was lower during baseline than after either dose, and REM sleep was lower during baseline and after dextroamphetamine than after placebo. Sleep onset was slowest on the baseline night (see Fig. 10.11).

Summary

In summary, the results of the in-flight evaluation were consistent with the earlier simulator study in that dextroamphetamine effectively maintained several aspects of flight performance, cognitive performance, psychomotor skill, positive psychological mood, and CNS arousal of sleep deprived pilots better than placebo. Although dextroamphetamine did not improve every variable investigated, there were sufficient positive effects consistent with those found in the earlier simulator study to conclude that dextroamphetamine is efficacious for the short-term maintenance of aviator performance in sustained operations. Prophylactic administration of this medication is particularly beneficial for

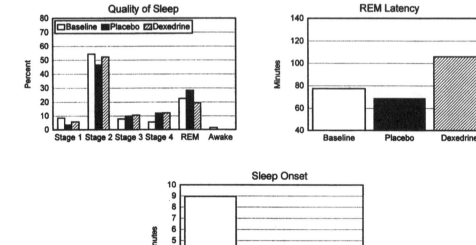

FIG. 10.11. Drug effects on sleep parameters—dextroamphetamine in-flight study.

preventing the dangerous reductions in aviator performance and alertness that are most evident between 03:00 and 10:00 in the morning. However, the issue of whether or not longer term (3 to 4 day) use of dextroamphetamine is a viable option remains to be addressed. It may be that the short-term benefits disappear after 1 to 2 days because of sleep pressure, drug tolerance, or physiological stresses. Multidisciplinary studies involving job-related performance, tests of basic cognitive skills, and physiological indices of CNS arousal should be used to fully characterize the positive as well as negative aspects of stimulant use before new policies are drafted for the operational environment.

A LABORATORY STUDY OF HYPNOTIC-INDUCED PROPHYLACTIC NAPPING

Up to this point, the focus has been on using stimulants to sustain alertness in situations where sleep is impossible. However, sustained operations do not necessarily entail total sleep deprivation. Frequently, there are brief opportunities for sleep even though the total amount is less than normal. If properly scheduled, these opportunities can significantly sustain alertness and performance. Naps ranging from 15 minutes to several hours in duration can either restore the performance of already sleep deprived personnel or protect personnel who are about to experience sleep deprivation from the full consequences of inadequate sleep (Dinges & Broughton, 1989).

A nap prior to sleep loss (a *prophylactic nap*) improves performance compared to no nap. Schweitzer, Muehlback, and Walsh (1992) assessed subjects who received a 2- to 3-hour nap before a night work shift (with concurrent sleep loss), and found that, although the morning circadian trough was still evident, the nap attenuated performance declines. Bonnet (1991) found that subjects who napped before 52-hours of continuous wakefulness performed better and were more alert for up to 24 hours of sleep loss compared to those who did not nap.

Thus, naps are beneficial for reducing performance decrements normally associated with total sleep deprivation. However, proper implementation of strategic naps can be difficult. It is often impossible to schedule naps for everyone during times when sleep onset and maintenance are easiest (during the circadian trough). Instead, some will need to sleep during difficult times (i.e., midmorning or a few hours prior to the normal bedtime). Also, environmental factors (heat, light, noise, etc.) may hinder napping. Thus, it is necessary to provide a way for personnel to obtain sleep whenever opportunities arise. One possibility is to use a short-acting hypnotic such as zolpidem tartrate (Ambein®). However, before proposing this solution, the utility of a short nap, placed at the wrong time in the circadian phase (worst case scenario) and initiated with and without zolpidem tartrate must be established.

Eighteen aviators were used to evaluate the efficacy of prophylactic naps induced with 10 mg zolpidem tartrate in comparison to placebo naps and a forced rest period for sustaining alertness and performance. Orders were counterbalanced and drug administrations were double blind. Each subject received all three interventions (zolpidem nap, placebo nap, or no nap) from 21:00 to 23:00 hours on the evening prior to sleep loss. Testing was conducted (beginning 2 hours after the intervention) from 01:00 until 22:00 hours. Sessions began with a VAS measuring alertness, anxiety, energy, confidence, irritability, nervousness, sleepiness, and talkativeness (Penetar et al., 1993), and a POMS. Next, subjects completed a 30-minute episode of the first cognitive test battery (the MATB). Afterward, subjects performed the RTSW in which they layed in bed for up to 20 minutes with instructions to remain awake while polysomnographic data were continuously recorded from EEG sites C3, C4, O1, and O2 (referenced to contralateral mastoids) to determine objectively when and if they fell asleep. A resting eyes open/eyes closed EEG was then performed (data were recorded from Fz, Cz, Pz, C3, C4, O1, and O2 referenced to linked mastoids). Next, the VAS and POMS were repeated prior to a second cognitive battery (SYNWORK). Finally, the VAS, POMS, and RTSW were performed once more (see Fig. 10.12). During each of the naps, sleep quality was measured via polysomnography.

Time	Sunday	Monday	Tuesday	Wednesday	Thursday	Friday	Saturday	Sunday	Monday	Tuesday
0100				VAS/POMS		VAS/POMS		VAS/POMS		
0110				MATB		MATB		MATB		
0200				VAS/Vitals		VAS/Vitals		VAS/Vitals		
0210				RTSW		RTSW		RTSW		
0235				EEG/EP		EEG/EP		EEG/EP		
0300 / 0310				VAS/POMS Synwork		VAS/POMS Synwork		VAS/POMS Synwork		
0330				MiniSim		MiniSim		MiniSim		
0400				VAS/Vitals Elec Rep		VAS/Vitals Elec Rep		VAS/Vitals Elec Rep		
0410				RTSW		RTSW		RTSW		
0500				VAS/POMS MATB		VAS/POMS MATB		VAS/POMS MATB		
0600 / 0605				VAS/Vitals Elec Rep		VAS/Vitals Elec Rep		VAS/Vitals Elec Rep		
0610				RTSW		RTSW		RTSW		
0635				EEG/EP		EEG/EP		EEG/EP		
0700				VAS/POMS Synwork		VAS/POMS Synwork		VAS/POMS Synwork		
0730				MiniSim		MiniSim		MiniSim		
0800		Wakeup	Wakeup	VAS/Vitals	Wakeup	VAS/Vitals	Wakeup	VAS/Vitals	Wakeup	Wakeup
0810				RTSW		RTSW		RTSW		
0830		Breakfast	Breakfast	Breakfast	Breakfast	Breakfast	Breakfast	Breakfast	Breakfast	Breakfast
0900 / 0910		VAS/POMS MATB	VAS/POMS MATB	VAS/POMS MATB	VAS/POMS MATB	VAS/POMS MATB	VAS/POMS MATB	VAS/POMS MATB	VAS/POMS MATB	Debrief
1000		VAS/Vitals	VAS/Vitals	VAS/Vitals	VAS/Vitals	VAS/Vitals	VAS/Vitals	VAS/Vitals	VAS/Vitals	Release
1010		RTSW	RTSW	RTSW	RTSW	RTSW	RTSW	RTSW	RTSW	
1035		EEG/EP	EEG/EP	EEG/EP	EEG/EP	EEG/EP	EEG/EP	EEG/EP	EEG/EP	
1100 / 1110		VAS/POMS Synwork	VAS/POMS Synwork	VAS/POMS Synwork	VAS/POMS Synwork	VAS/POMS Synwork	VAS/POMS Synwork	VAS/POMS Synwork	VAS/POMS Synwork	
1130		MiniSim	MiniSim	MiniSim	MiniSim	MiniSim	MiniSim	MiniSim	MiniSim	
1200		VAS/Vitals	VAS/Vitals	VAS/Vitals	VAS/Vitals	VAS/Vitals	VAS/Vitals	VAS/Vitals	VAS/Vitals	
1210		RTSW	RTSW	RTSW	RTSW	RTSW	RTSW	RTSW	RTSW	
1235		Lunch	Lunch	Lunch	Lunch	Lunch	Lunch	Lunch	Lunch	
1300		VAS/POMS	VAS/POMS	VAS/POMS	VAS/POMS	VAS/POMS	VAS/POMS	VAS/POMS	VAS/POMS	
1310		MATB	MATB	MATB	MATB	MATB	MATB	MATB	MATB	
1400		VAS/Vitals	VAS/Vitals	VAS/Vitals	VAS/Vitals	VAS/Vitals	VAS/Vitals	VAS/Vitals	VAS/Vitals	
1410		RTSW	RTSW	RTSW	RTSW	RTSW	RTSW	RTSW	RTSW	
1435		EEG/EP	EEG/EP	EEG/EP	EEG/EP	EEG/EP	EEG/EP	EEG/EP	EEG/EP	
1500 / 1510	Arrive/ Inservice	VAS/POMS Synwork	VAS/POMS Synwork	VAS/POMS Synwork	VAS/POMS Synwork	VAS/POMS Synwork	VAS/POMS Synwork	VAS/POMS Synwork	VAS/POMS Synwork	
1530	Inservice	MiniSim	MiniSim	MiniSim	MiniSim	MiniSim	MiniSim	MiniSim	MiniSim	
1600	Medical	VAS/Vitals	VAS/Vitals	VAS/Vitals	VAS/Vitals	VAS/Vitals	VAS/Vitals	VAS/Vitals	VAS/Vitals	
1610		RTSW	RTSW	RTSW	RTSW	RTSW	RTSW	RTSW	RTSW	
1700 / 1710	Electrode Hook-up	VAS/POMS MATB	VAS/POMS MATB	VAS/POMS MATB	VAS/POMS MATB	VAS/POMS MATB	VAS/POMS MATB	VAS/POMS MATB	VAS/POMS MATB	
1800	Vitals/	VAS/Vitals	VAS/Vitals	VAS/Vitals	VAS/Vitals	VAS/Vitals	VAS/Vitals	VAS/Vitals	VAS/Vitals	
1810	Dinner	RTSW	RTSW	RTSW	RTSW	RTSW	RTSW	RTSW	RTSW	
1835		EEG/EP	EEG/EP	EEG/EP	EEG/EP	EEG/EP	EEG/EP	EEG/EP	EEG/EP	
1900 / 1910		VAS/POMS Synwork	VAS/POMS Synwork	VAS/POMS Synwork	VAS/POMS Synwork	VAS/POMS Synwork	VAS/POMS Synwork	VAS/POMS Synwork	VAS/POMS Synwork	
1930		MiniSim	MiniSim	MiniSim	MiniSim	MiniSim	MiniSim	MiniSim	MiniSim	
2000		VAS/Vitals	VAS/Vitals	VAS/Vitals	VAS/Vitals	VAS/Vitals	VAS/Vitals	VAS/Vitals	VAS/Vitals	
2010		RTSW	RTSW	RTSW	RTSW	RTSW	RTSW	RTSW	RTSW	
2035		Dinner	Dinner	Dinner	Dinner	Dinner	Dinner	Dinner	Dinner	
2100	Vitals	PT	Drug/nap	PT	Placebo/nap	PT	Rest-NoDrug	PT	PT	
2130			PT/Shower			PT/Shower		PT/Shower	PT/Shower	
2200										
2300 (2330)			Wakeup/VAS Vitals(PT)		Wakeup/VAS Vitals (PT)		VAS/Vitals Vitals (PT)			
2400			Shower		Shower		Shower			

FIG. 10.12. Test schedule for the zolpidem napping study.

POMS

The zolpidem nap was better at sustaining vigor than forced rest and slightly better than the placebo nap, especially from 05:00 to 09:00 hours (when alertness suffered most with no sleep). Fatigue was lower after the zolpidem nap than after forced rest, whereas the placebo nap produced only marginally less fatigue

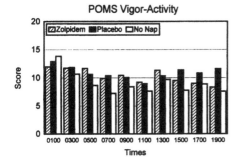
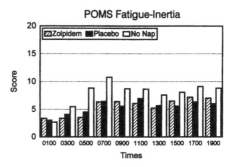

FIG. 10.13. Mood (POMS) effects across conditions and sessions—napping study.

(p=.07). The benefit of naps on mood were most apparent when the nap was facilitated by zolpidem tartrate (see Fig. 10.13).

VAS and RTSW

Alertness, energy, and sleepiness were measured subjectively with the VAS and objectively with the RTSW Subjects were less alert and energetic and more irritable and sleepy after forced rest than after one or both napping conditions. After rest only, the loss of alertness persisted from 04:00 to 11:00 hours, and the low energy, increased sleepiness, and elevated irritability lasted from 04:00 to 08:00 hours. Feelings of talkativeness dropped at 05:00 and 07:00 hours when napping was not used. Sleepiness later in the day was greater after forced rest than after napping, whereas confidence early in the day was greater after naps than forced rest. The benefits from napping were greatest after zolpidem (see Fig. 10.14). RTSW data showed subjects remained awake longer after the zolpidem nap than after the placebo nap or rest only (see Fig. 10.15). Prior to sleep loss, subjects remained awake for 17.7 minutes, whereas during sleep deprivation subjects remained awake for an average of 11.7 minutes after the zolpidem nap, 9.4 minutes after the placebo nap, and only 6.3 minutes after forced rest.

The benefits from napping were not apparent immediately after subjects were awakened. VAS data collected about 5 minutes after the 2-hour naps revealed that alertness, energy, confidence, and talkativeness were lower and irritability and sleepiness were higher after the zolpidem and placebo naps than after forced rest. This is consistent with literature that describes transient impairments immediately upon awakening from sleep (Dinges, 1989). These sleep inertia problems disappeared before 01:00 hours, with the exception of the alertness decrement, which persisted until 01:00 hours. It is unclear whether performance would have suffered along with mood during these times because the first test was not given until 01:10 hours; however, it has been suggested that the mood disruptions caused by sleep inertia outlast the performance decrements (Naitoh & Angus, 1989).

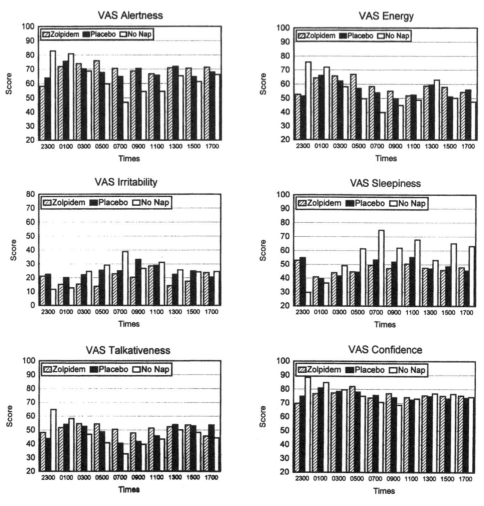

FIG. 10.14. Mood (VAS) effects across conditions and sessions—napping study.

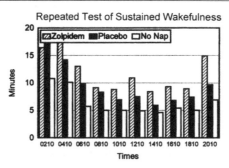

FIG. 10.15. Minutes asleep on the RTSW across conditions and sessions—napping study.

MATB and SYNWORK

The effects of sleep deprivation on the MATB were most severe at 09:10 hours, especially after rest only. At this time, subjects were slower at responding to warning lights after both the placebo nap and the rest-only condition than after the zolpidem nap and slower at responding to dial deviations after rest only than after either napping intervention. Tracking errors at 09:10 hours were larger after forced rest than after the zolpidem nap, and tracking skill was poorer after the placebo nap than after the zolpidem nap. On the radio management task, reaction time was affected as well. Although there was a tendency for this to have occurred at 09:10 hours, the significant differences were found only at 17:10 hours. At this point, reaction times were faster under both the zolpidem and rest-only conditions than under the placebo condition (see Fig. 10.16). Analysis of the SYNWORK was generally supportive of what was found with the MATB (see Fig. 10.17). Overall, performance suffered the most between about 07:00 and 11:00 hours after a single night of sleep deprivation, consistent with the mood and alertness decrements, which were most severe at these times. Napping prior to sleep loss attenuated many of these problems, especially on tasks that required

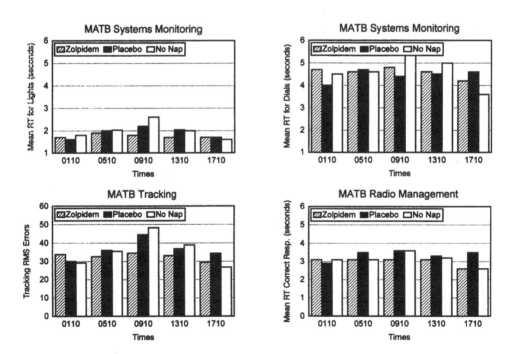

FIG. 10.16. Performance on the MATB across conditions and sessions—napping study.

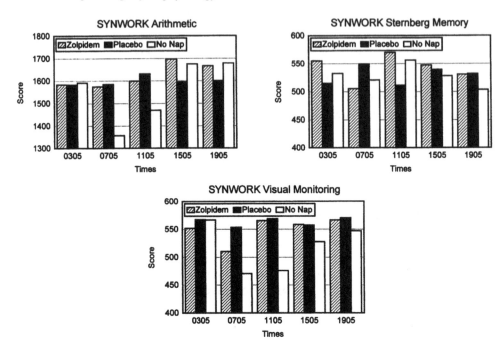

FIG. 10.17. Performance on the SYNWORK across conditions and sessions—napping study.

vigilance and rapid responding. The zolpidem nap tended to be superior to a "natural" nap.

EEG

Eyes open/eyes closed EEGs provided further insight into physiological arousal. Although increases in theta and decreases in alpha activity occurred as a function of time of day (independent of condition), napping did not attenuate these effects as much as expected. Delta activity (at Cz) increased from the early morning to later in the day (especially at 10:30 hours) after rest only in comparison to napping, but theta was unaffected. Both naps attenuated the deprivation-related reductions in occipital alpha in comparison to forced rest. Also, the overall amount of occipital alpha was lower after forced rest than after the placebo nap, with a similar tendency after the zolpidem nap. These changes in alpha activity probably resulted from an increased sleep tendency after the forced-rest period since alpha attenuation is associated with increased sleepiness. This is consistent with the RTSW and VAS, which showed a more rapid decrease in alertness following rest only. There were also increases in beta activity that indicated the superiority of napping. More beta was recorded from C4, Pz, and Ol after the zolpidenvinduced nap than after forced rest, and more was recorded from Pz and Ol after the placebo nap than after rest only (see Fig. 10.18). Thus, it appears that CNS activation was improved by napping in that beta activity was more pronounced and attenuations in alpha were less severe throughout sleep deprivation. The fact that these differences were not more robust may have been due to large amounts of intra-subject variability or the short duration of the EEG evaluations (about 5 minutes each).

Polysomnography

Sleep onset was more rapid, the amount of sleep was longer, and sleep was more restful (less Stage 1 and more Stage 4) after zolpidem than placebo. Subjects fell asleep almost twice as fast after zolpidem than after placebo, which provided them with significantly more sleep during the 2-hour period (see Fig. 10.19). Longer sleep duration coupled with more restful sleep under zolpidem no doubt contributed to the superiority of the zolpidem nap for sustaining mood, alertness, and performance.

Summary

Prophylactic naps were beneficial in sustaining the mood, alertness, and performance of personnel in continuous wakefulness despite the fact that the

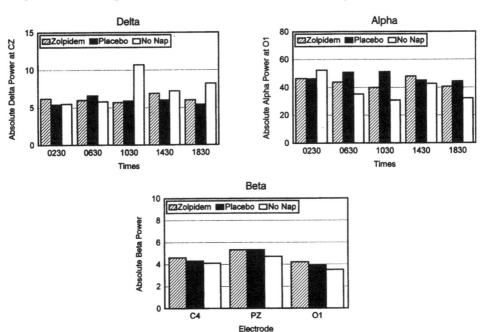

FIG. 10.18. Effects of conditions and sessions on EEG activity—napping study.

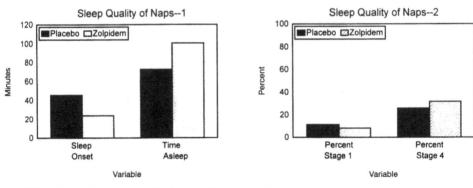

FIG. 10.19. Nap sleep quality by condition—napping study.

naps were placed during the time of day at which sleep is difficult to obtain. With the exception of sleep inertia after awakening, mood/alertness ratings were more positive after one or both of the naps than after forced rest. The mood findings were reinforced by objective, physiological indices that naps increased the subjects' abilities to maintain wakefulness. Mood and alertness decrements were most severe between 04:00 and 11:00 hours. Cognitive test results were often consistent with the self-report and physiological data, although many aspects of cognitive functioning were unaffected by sleep deprivation. Subjects performed fuel management and auditory monitoring about as well after total sleep deprivation as after naps, but subjects were better able to monitor systems, respond to warning lights, manage radios, track targets, and perform mental calculations after napping than after rest only, especially between 07:00 and 11:00 hours. Sleep quality in the zolpidem nap was better than sleep after placebo, which provides an explanation for the mild superiority of zolpidem-induced naps. Thus, in situations where personnel are provided with only limited opportunities to sleep, which may occur at less than optimal times in the circadian phase, zolpidem should be considered. However, zolpidem should only be administered immediately prior to the sleep period because of its rapid onset of action. A test dose of the medication is recommended prior to operational use. Care should be taken to allow sufficient time from awakening until returning to work to ensure sleep-inertia has subsided.

THE EFFICACY OF TRIAZOLAM FOR PROMOTING SLEEP IN COMBAT

Thus far, two strategies have been discussed for possible application in situations where sleep deprivation is imposed due to workload requirements that make normal sleep impossible. Depending on the exact nature of the situation, either stimulants or naps can attenuate the deprivation-related declines in performance, mood, and alertness. However, there also are circumstances under which 8 hours of sleep are possible, but sleep loss occurs because of discomfort, anxiety, circadian disruptions, or other factors. In fact, during prolonged military conflicts, this last scenario may be responsible for more day-to-day declines in aviator effectiveness than the less frequent episodes of sustained operations discussed earlier. Thus, it is necessary to provide some reliable method for promoting sleep in the operational environment. A fast-acting hypnotic with a short half-life and few side effects would be the best choice since hypnotics are capable of rapidly promoting and maintaining sleep on demand.

During Operation Desert Shield/Storm, triazolam (Halcion®) was the sleeping aid of choice for aviation personnel who were unable to sleep in the field. This drug was selected due to its availability, effectiveness, and short half-life. Both laboratory and clinical trials have shown triazolam to be an effective hypnotic (Goetzke et al., 1983; Ogura et al., 1980; Vogel et al., 1975) in that it reduces sleep latency and the number of awakenings after sleep onset, increases total sleep time, increases the amount of Stage 2 sleep, increases the latency of the first REM period without altering the overall percentage of REM sleep during the night, and does not significantly reduce slow wave sleep. A flight surgeon could prescribe triazolam (0.25 mg) to an aviator if natural means of promoting sleep were ineffective, provided a 6-hour grounding period was imposed.

Since the end of Operation Desert Storm, the use of triazolam has been questioned by the aviation community because no controlled studies were conducted to determine the effects of this sleeping aid on the performance of aviators. There were concerns that aviators administered a sleeping aid could experience problems (grogginess, confusion, difficulty awakening, etc.) if called upon to awaken and fly in an emergency situation shortly after drug administration. Such concerns were realistic given that nonaviation studies suggested triazolam may exert performance effects for 8 to 10 hours postdose (Balkin et al., 1988; Bornstein, Watson, & Kaplan, 1985; Ogura et al., 1980; Roache & Griffiths, 1985; Spinweber & Johnson, 1982). Gorenstein and Gentil (1983) found that triazolam (0.5 mg) produced hangover effects in the morning as well as reduced motor performance, and Roth et al. (1980) found decrements in morning recall of a memory set after triazolam (0.5 mg); however, Spinweber and Johnson (1982) indicated performance was affected only up to 5 hours postdose. With regard to possible difficulties awakening from a triazolam-induced sleep, it had been determined that awakening to a smoke detector alarm was slower during nights when subjects were administered triazolam (0.25 or 0.5 mg) than after receiving placebo (Johnson et al., 1987). Some subjects completely failed to awaken after 0.5 mg triazolam when they were in slow wave sleep, despite the fact that, by morning, all subjects were easily awakened, regardless of the dose. Unfortunately, none of these studies involved aviation-oriented performance tasks, few used the currently recommended clinical dose of triazolam (0.25 mg), and none simulated the problem of awakening an aviator and requiring complex performance shortly after a triazolam-induced sleep. For these reasons, it was necessary to conduct an aviator performance study.

Ten Army aviators were administered triazolam (0.25 mg) on 2 nights and placebo on 2 nights. On 1 of the nights under each condition, subjects were allowed 8 hours of uninterrupted sleep, and on the other night, subjects were awakened 2 hours postdrug and required to fly a mission in a UH-60 helicopter simulator prior to returning to bed. Twice during the daytime following both uninterrupted or interrupted sleep, subjects completed eyes open/eyes closed EEG assessments (recorded from Fz, C3, Cz, C4, Pz, Ol, and O2 referenced to linked mastoids); the 7-point Stanford Sleepiness Scale (SSS); and the memory search, logical reasoning, and arithmetic subtests of the PAB. The first session was at 09:00 hours and the second was at 16:00 hours. In addition, subjects completed two flight missions consisting of hovers, low-level navigation, upper-altitude air work (straight and levels, right and left standard turns, climbs and descents, and a left-descending turn), and formation flight each day (the same flight profile used in the simulator studies of dextroamphetamine). On the days following interrupted sleep, flights were at midnight and 13:00 hours, and on the days after full nights of sleep the flights were at 06:00 and 13:00 hours. Each flight was preceded by an additional SSS. As was the case with all the previous studies, sleep quality was assessed with polysomnography each night. The schedule is shown in Fig. 10.20.

Flight Performance

Even though aviators were able to fly the profile without immanently dangerous decrements under triazolam, differences were evident on most maneuvers. Of the nine maneuvers flown, performance on five deteriorated following a full night of sleep with triazolam compared to a placebo, regardless of the session. Degradation was apparent in roll control during the low-level navigation, roll during the straight and levels, turn rate during the left turns, airspeed and vertical speed control during the climbs, and heading and roll control during the descents (composite flight scores are depicted in Fig. 10.21). In the 10-foot hovering turn, the first left turn, and the third right turn, triazolam was associated with poorer performance than with placebo during the 06:00 flight (8-hours postdose), but not the 13:00 flight (15-hours postdose). In the second left turn, triazolam produced poorer performance than placebo in the afternoon. These effects are depicted in Fig. 10.21. On the interrupt night, at the 00:30 flight (2-hours postdose), there were performance decrements on slip control in the left turns, and turn rate, altitude, and airspeed control in the right turns, after triazolam relative to placebo (see Fig. 10.22), but no other drug-related effects were significant. Taken together, these results suggest some hangover effects following triazolam that are cause for concern when using this medication in the operational aviation environment.

Time	Monday	Tuesday	Wednesday	Thursday	Friday	Saturday	Sunday	Monday	Tuesday	Wednesday	Thursday
00-01								Wake-up SSS/Sim		Wake-up SSS/Sim	
01-02								Sim		Sim	
02-03											
03-04											
04-05											
0505											
0530	Wake-up	Wake-up	Wake-up	Wake-up	Wake-up	Wake-up			Wake-up		Wake-up
06-07	SSS/Sim	SSS/Sim	SSS/Sim	SSS/Sim	SSS/Sim	SSS/Sim			SSS/Sim		SSS/Sim
07-08											
0730	Breakfast	Breakfast	Breakfast	Breakfast	Breakfast	Breakfast	Wake-up	Breakfast	Wake-up	Breakfast	
08-09	Free	Free	Free	Free	Free	Free	SSS Brkft	Free	SSS Brkft	Free	
09-10	EEG/EP SSS/PAB	EEG/EP SSS/PAB	EEG/EP SSS/PAB	EEG/EP SSS/PAB	EEG/EP SSS/PAB	EEG/EP SSS/PAB	EEG/EP SSS/PAB	EEG/EP SSS/PAB	EEG/EP SSS/PAB	EEG/EP SSS/PAB	
10-11	Free	Free	Free	Free	Free	Free	Free	Free	Free	Free	
11-12	PAB	PAB	Free	Free	Free	Free	Free	Free	Free	Free	
12-13	Lunch	Lunch	Lunch	Lunch	Lunch	Lunch	Lunch	Lunch	Lunch	Lunch	
13-14	SSS/Sim	SSS/Sim	SSS/Sim	SSS/Sim	SSS/Sim	SSS/Sim	SSS/Sim	SSS/Sim	SSS/Sim	SSS/Sim	
14-15	Arrive	Free	Free	Free	Free	Free	Free	Free	Freee	Free	Free
15-16		Free	Free	Free	Free	Free	Free	Free	Free	Free	Free
16-17	Consent	EEG/EP SSS/PAB	EEG/EP SSS/PAB	EEG/EP SSS/PAB	EEG/EP SSS/PAB	EEG/EP SSS/PAB	EEG/EP SSS/PAB	EEG/EP SSS/PAB	EEG/EP SSS/PAB	EEG/EP SSS/PAB	EEG/EP SSS/PAB
17-18	Med	Free	Free	Free	Free	Free	Free	Free	Free	Free	Debrief Release
18-19	Dinner	Dinner	Dinner	Dinner	Dinner	Dinner	Dinner	Dinner	Dinner	Dinner	
19-20	Free	Free	Free	Free	Free	Free	Free	Free	Free	Free	
20-21	Electrode Hook-up	Electrode Repair	Electrode Repair	Electrode Repair	Electrode Repair	Electrode Repair	Electrode Repair	Electrode Repair	Electrode Repair	Electrode Repair	
21-22		SSS	SSS Drug	SSS	SSS Drug	SSS	SSS Placebo	SSS	SSS Placebo	SSS	
22-23											
23-24											

FIG. 10.20. Test scedule for the triazolam study.

SSS

Although there was a tendency toward greater sleepiness 20 minutes after awakening than later in the day and an increase in subjective sleepiness 15 minutes before bedtime as compared to the late morning and afternoon ratings, SSS ratings were unaffected by triazolam relative to placebo. This may have been

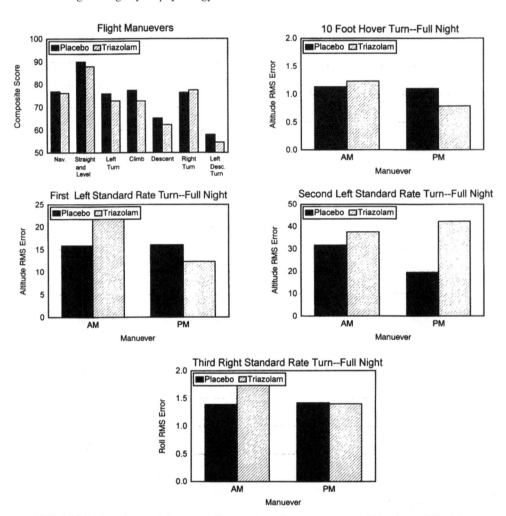

FIG. 10.21. Next-day performance effects on simulator maneuvers following a full night of sleep after nighttime triazolam administration.

due to large variability among the subjects' responses, an unwillingness on the part of aviators to express sleepiness during duty periods, or a lack of sensitivity from an instrument with only a 7-point rating scheme.

PAB

There were no differences between drug conditions in cognitive performance tests. The MAST-6, a scanning and memory test, did show that morning performance was slower than afternoon performance, as did the logical reasoning

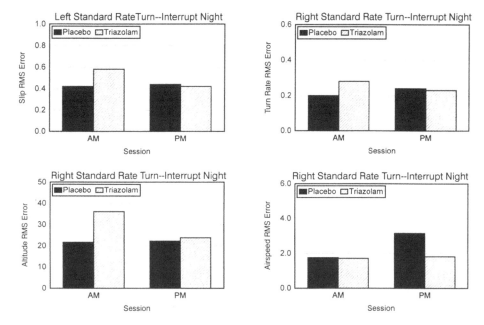

FIG. 10.22. Performance effects on simulator maneuvers 2 hours and 15 hours post-dose—triazolam study.

test; however, this effect was not influenced by whether the subjects had received placebo or triazolam. Because the first PAB was 12 hours postdose, it is probable that any drug effects had already dissipated.

EEG

The resting EEG conducted to determine the physiological state of the subjects following placebo and triazolam indicated more slow activity (alpha, theta, and delta) following the placebo night than following the triazolam night (see Fig. 10.23). Because slow activity (associated with relaxation or sedation) appeared after placebo rather than triazolam, the effects of the active hypnotic apparently had dissipated by the time of the first EEG session, about 12 hours postdose. The slower activity that occurred after the placebo night may have been due to less restful sleep since sleep deprivation is known to increase slow-wave EEG. Even though there was more slow-wave sleep on the placebo nights than on the triazolam nights, the sleep during the placebo night showed more light sleep (Stage 1) and more awake time, which probably led to lower daytime alertness.

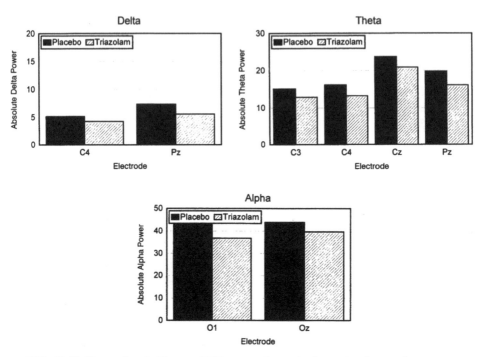

FIG. 10.23. Drug related effects on EEG across electrode site—triazolam study.

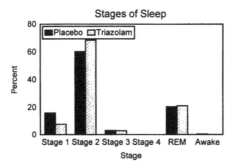

FIG. 10.24. Effects of triazolam on sleep architecture.

Polysomnography

There was more slow wave sleep after placebo than after triazolam, an effect inconsistent with some of the previous literature that indicates that triazolam has no effect on slow-wave sleep. However, there was more Stage 2 sleep following triazolam than placebo, as has been shown in other studies. In addition, there was less Stage 1 and awake time after sleep onset following triazolam than following placebo (see Fig. 10.24). This would indicate that sleep with triazolam was more restful than sleep with placebo. On the sleep interrupt night, triazolam was especially helpful in returning to sleep after the 00:30 flight. Of the 10 subjects tested, 8 were asleep in less than 10 minutes under triazolam, whereas 6 were still awake after 30 minutes under placebo. A point of operational significance is that the amount of time it took for subjects to awaken in the morning tended to be longer following triazolam than following placebo (7.7 vs. 5.9 seconds, respectively). Although this effect was not statistically significant due to the high variability among the subjects, it is a point of concern. Individual reactions to a sleeping aid should be assessed before a soldier is given a hypnotic and then required to awaken in the morning with only the usual wake-up cue. Of greater concern is the fact that some subjects responded very slowly to the midnight wake-up call after triazolam. It was expected that most people would react slowly to the early wake-up call because the drug should be near its peak approximately 2 hours postadministration. However, a few subjects were particularly slow, and one had to be awakened by the technician after the wake-up signal was sounded for 6 minutes. Because of this type of individual variability, test dosing for aviators is essential.

Side Effects

Two subjects were unable to recall the details of the mission flown at 00:30 hours following triazolam administration. Although they remembered having flown the simulator, they were not able to recall (on the next day) some of the portions of the 2 hours they were awake and performing tasks. This is a concern for the aviation community because the details of flight missions often must be reported to other unit members. Although every person who consumes triazolam does not experience this problem, it does occur with some people, particularly when awakened early after administration of triazolam. Once again, a test dose of the drug is imperative prior to relying on this or other medications in the operational environment.

Summary

Performance is degraded somewhat by triazolam administered 8 hours before a flight, although, curiously, it has almost no effects when administered 2 hours before a flight. It appears that aviators should be capable of flying missions after triazolam, but it should be noted that slight decrements observed in this well-practiced scenario may underestimate performance difficulties in a more unpredictable situation. Clearly, it is important to administer a test dose under controlled conditions, including a wake-up shortly following administration, to test for unusual effects from the drug such as amnesia or unusually long responses to a wake-up call. Further research is needed to determine responses to emergency situations.

CONCLUSIONS

As discussed earlier, the combination of behavioral and cognitive measures, physiological data, and questionnaires is beneficial in studying pharmacological interventions to enhance performance (as well as a whole range of other topics). Without this wide variety of measures, some important effects may have been missed altogether, and it would not have been possible to know whether observed performance and mood changes were attributable to physiological shifts in arousal, changes in motivation, or possibly even some form of experimenter or subject bias. The ability to collect EEG data in order to assess physiological arousal, questionnaire data to monitor subjective impressions of mood and alertness, computerized test data to evaluate basic cognitive skills, and simulator and/ or aircraft flight data to assess actual operational performance all contribute to answering the research questions. For example, analysis of the SYNWORK data from the simulator dextroamphetamine studies indicated a 10-minute test was too short to show effects of sleep deprivation, however by having physiological data as well as an actual flight-performance measure, the deleterious effects of sleep deprivation were captured. Had a full array of tests not been implemented, it may have been erroneously concluded that sleep loss would not cause operationally significant problems or that dextroamphetamine would not be an effective countermeasure. Another example is the SSS used in the triazolam study. For aviators who are reluctant to express that they are less than fully capable when on flight duty, this test was not sensitive to sleepiness encountered throughout the day; however, the EEG measures that were collected in conjunction with the SSS were not susceptible to subjective factors. These data revealed that although subjects were not expressing alertness problems, decrements in CNS arousal were in fact present.

Another advantage of utilizing electrophysiological measures in conjunction with more standard performance tests is that physiological measures can provide the explanation for observed performance decrements or suggest that performance may begin to deteriorate at some future time that has yet to be established. An example of the explanatory utility of electrophysiological techniques can be seen in the zolpidem-napping study in which changes in both mood and performance occurred concurrently with objectively measured reductions in physiological arousal In addition, the fact that zolpidem-induced naps attenuated the negative aspects of sleep loss better than placebo naps was at least partially attributable to the better sleep quality (measured electrophysiologically) after zolpidem. An example of the utility of EEG measures for suggesting a future performance problem could be seen in the dextroamphetamine studies. Note that although dextroamphetamine was effective for attenuating the impact of sleep deprivation, the medication did create modest reductions in the quality of recovery sleep. Although these reductions did not create problems in the short term, it is quite possible that chronic use of this medication may produce cumulative difficulties in recovery from sleep deprivation that could become operationally significant over time.

In addition to using sensitive measures to cover a wide range of factors related to adequate performance assessment, it is important that these measures be collected at times that will answer the question being asked. When surveying the problems encountered in sustained operations, it is imperative that data measuring behavior, mood, performance, and physiological status be collected throughout the entire period of interest. As the data in these studies indicate, performance fluctuates with time of day. Without frequent sampling intervals, erroneous conclusions may be made. All the sleep deprivation studies discussed here show a circadian fluctuation, and because of the repeated testing scheme, these fluctuations were accurately captured so that decisions could be made concerning which times of day were most susceptible to the effects of sleep deprivation. In addition, side effects from the pharmacological interventions were recorded throughout the intervention period so that idiosyncratic effects from these medications could be sampled and information concerning these effects could be evaluated.

Besides using a wide array of assessment techniques, it is important to maintain some continuity of methods across studies when possible. Note that in the studies presented earlier, many of the same core tests were used despite modifications in the overall approach. Resting EEGs were employed in all four investigations, and the use of this measure made it possible to comparably evaluate drug and/or sleep deprivation effects across each one. By using the POMS in both dextroamphetamine studies and in the zolpidem study, it became possible to assess the comparability of sleep-deprivation effects across investigations and to evaluate the relative sensitivity of a newer technique (the VAS) in the zolpidem study. Because both the EEG and POMS assessments were completed at similar times in three of the investigations, data could be combined into a larger pool to further establish the effects of sleep loss using more powerful statistical approaches than the ones possible in each individual study. In addition, reliance on the same or similar cognitive-performance metrics across three investigations permitted a determination of the effects of test duration on the sensitivity of the test to the impact of sleep loss. By using 10 minutes of SYNWORK in one of the dextroamphetamine studies and 20 minutes of SYNWORK in the zolpidem study, it was possible to determine that at least a 20-minute test is needed to begin to accurately assess the effects of sleep deprivation on basic cognitive skills (roughly the same amount of sleep loss was encountered in both studies). In the meantime, it was possible to evaluate the potential utility of new tests (such as the MATB) by comparing the results with a more familiar instrument (SYNWORK).

One other methodological point is that in order to obtain accurate indications of performance, which can be accurately generalized to the actual situation of interest, the subject pool should reflect the population to which the results will be applied and, if possible, at least one measure should sample some aspects of actual job performance. By using a representative sample, factors such as motivation, training, personality, and so on, are less likely to confound the ultimate conclusions of the investigation. By using actual job-based performance measures, factors such as task complexity, practice effects, and boredom are minimized. Furthermore, particularly in applied research, the target population (to which the research findings will be applied) is more likely to accept conclusions and recommendations when they are based on a study sample that shares as many features as possible with the target group and when they are based on the sample's completion of tests that possess fairly high levels of face validity. Not only are the results more accurately generalized, but they are also more likely to be utilized by the population of individuals for which the research was originally formulated.

REFERENCES

Åkerstedt, T. (1995). Work hours, sleepiness, and the underlying mechanisms. *Journal of Sleep Research, 4*(Suppl. 2), 15–22.

Angus, R.G., Pigeau, R.A., & Heslegrave, R.J. (1992). Sustained-operations studies: From the field to the laboratory. In C.Stampi (Ed.), *Why we nap* (pp. 217–244). Boston: Birkhauser.

Balkin, T.J., O'Donnell, V.M., Kamimori, G.H., Simon, L.M., Andrade, J.R., Redmond, D. P., & Belenky, G. (1988). Efficacy and safety of triazolam for deployment operations. In *Army science conference proceedings, Vol 1*, 69–80.

Bonnet, M.H. (1991). The effect of varying prophylactic naps on performance, alertness and mood throughout a 52-hour continuous operation. *Sleep, 14*(4), 307–315.

Bornstein, R.A., Watson, G.D., & Kaplan, M.J. (1985). Effects of flurazepam and triazolam on neuropsychological performance. *Perceptual and Motor Skills, 60,* 47–52.

Caldwell, J.A. (1995). Assessing the impact of stressors on performance: Observations on levels of analyses. *Biological Psychology, 40*(1, 2), 197–208.

Caldwell, J.A., Caldwell, J.L., Crowley, J. S., Jones, H.D., Darling, S.R., Wallace, S.L., Woodrum, L.C., & Colon, J.A. (1994). *Effects of dextroamphetamine on helicopter pilot performance: A UH-60 simulator study* (USAARL Tech. Rep. No. 94–43). Fort Rucker, AL: US Army Aeromedical Research Laboratory.

Caldwell, J.A., Crowley, J.S., Caldwell, J.L., Jones, H.D., Darling, S., Wallace, S., & Pegues, A. (1995). *Sustaining female UH-60 helicopter pilot performance with Dexedrine during sustained operations: A simulator study* (USAARL Tech. Rep. No. 95–27). Fort Rucker, AL: US Army Aeromedical Research Laboratory.

Comperatore, C.A., Caldwell, J.A., Stephens, R.L., Mattingly, A., Chiaramonte, J., & Trast, S.T. (1993). *The use of electrophysiological and cognitive variables in the assessment of degradation during periods of sustained wakefulness* (USAARL Rep. No. 93–5). Fort Rucker, AL: US Army Aeromedical Research Laboratory.

Cornum, K.G. (1992, November). *Sustained operations: A F-15 squadron in the Gulf War*. Minutes of the Department of Defense Human Factors Engineering Technical Group 29th meeting. Huntsville, AL.

Dinges, D.F. (1989). Napping patterns and effects in human adults. In D.F.Dinges & R.J. Broughton (Eds.), *Sleep and alertness: Chronobiological, behavioral, and medical aspects of napping* (pp. 171–204). New York: Raven Press.

Dinges, D.F., & Broughton, R.J. (Eds.). (1989). *Sleep and alertness: Chronobiological, behavioral, and medical aspects of napping*. New York: Raven Press.

Emonson, D.L., & Vanderbeek, R.D. (1995). The use of dextroamphetamine in support of tactical air operations during Operation Desert Shield/Storm. *Aviation, Space, and Environmental Medicine, 66,* 260–263.

Goetzke, E., Findeisen, P., & Welbers, I.B. (1983). Comparative study on the efficacy of and the tolerance to the triazolodiazepines, triazolam, and brotizolam. *British Journal of Clinical Pharmacology, 16,* 407S–412S.

Gorenstein, C., & Gentil, V. (1983). Residual and acute effects of flurazepam and triazolam in normal subjects. *Psychophysiology, 10*(4), 431–436.

Horne, J.A. (1978). A review of the biological effects of total sleep deprivation in man. *Biological Psychology, 7,* 55–102.

Johnson, L.C., Spinweber, C.L., Webb, S.C., & Muzet, A.G. (1987). Dose level effects of triazolam on sleep and response to a smoke detector alarm. *Psychopharmacology, 91,* 397–402.

Krueger, G.P. (1989). Sustained work, fatigue, sleep loss, and performance: A review of the issues. *Work & Stress, 3*(2), 129–141.

Lorenzo, I., Ramos, C.A., Guevara, M.A., & Corsi-Cabrera, M. (1995). Effect of total sleep deprivation on reaction time and waking EEG activity in man. *Sleep, 18*(5), 346–354.

Naitoh, P., & Angus, R.G. (1989). Napping and human functioning during prolonged work. In D.F.Dinges & R.J.Broughton (Eds.), *Sleep and alertness: Chronobiological, behavioral, and medical aspects of napping* (pp. 221–246). New York: Raven Press.

Ogura, C., Nakazawa, K., Majima, K., Nakamura, K., Ueda, H., Umezawa, Y., & Wardell, W. M. (1980). Residual effects of hypnotics: Triazolam, flurazepam, and nitrazepam. *Psychopharmacology, 68,* 61–65.

Penetar, D., McCann, U., Thorne, D., Kamimori, G., Galinski, C., Sing, H., Thomas, M., & Belenky, G. (1993). Caffeine reversal of sleep deprivation effects on alertness and mood. *Pharmacology, 112,* 359–365.

Pigeau, R.A., Heslegrave, R.J., & Angus, R.G. (1987). Psychophysiological measures of drowsiness as estimators of mental fatigue and performance degradation during sleep deprivation. In *Electric and magnetic activity of the central nervous system: Research and clinical applications in aerospace medicine* (AGARD Rep. No. CP-432, 21–1/21–16). Neuilly sur Seine, France: Advisory Group for Aerospace Research and Development.

Roache, J.D., & Griffiths, R.R. (1985). Comparison of triazolam and pentobarbital: Performance impairment, subjective effects and abuse liability. *The Journal of Pharmacology and Experimental Therapeutics, 234,* 120–133.

Roth, T., Hartse, K.M., Saab, P.G., Piccione, P.M., & Kramer, M. (1980). The effects of flurazepam, lorazepam, and triazolam on sleep and memory. *Psychopharmacology, 70,* 231–237.

Schweitzer, P.K., Muehlback, M.J., & Walsh, J.K. (1992). Countermeasures for night work performance deficits: The effect of napping or caffeine on continuous performance at night. *Work & Stress, 6*(4), 355–365.

Senechal, P.K. (1988). Flight surgeon support of combat operations at RAF Upper Heyford. *Aviation, Space, and Environmental Medicine, 59,* 776–777.

Spinweber, C.L., & Johnson, L.C. (1982). Effects of triazolam (0.5 mg) on sleep, performance, memory, and arousal threshold. *Psychopharmacology, 76,* 5–12.

Vogel, G., Thurmond, A., Gibbons, P., Edwards, K., Sloan, K.B., & Sexton, K. (1975). The effect of triazolam on the sleep of insomniacs. *Psychopharmacologia, 41,* 65–69.

Wilkinson, R.T. (1969). Some factors influencing the effect of environmental stressors upon performance. *Psychological Bulletin, 72*(4), 260–272.

Wilkinson, R.T. (1964). Effects of up to 60 hours sleep deprivation on different types of work. *Ergonomics, 17,* 175–186.

Chapter 11

A Biocybernetic System
for Adaptive Automation

Mark W.Scerbo
Frederick G.Freeman
Peter J.Mikulka
Old Dominion University

Adaptive automation refers to systems that can adjust their mode or level of operation dynamically. Unlike traditional forms of automation where the operator is responsible for initiating changes in the state of the system, in adaptive technology both the operator and the system can initiate changes (Hancock & Chignell, 1987; Rouse, 1976; Scerbo, 1996). Parasuraman, Bahri, Deaton, Morrison, and Barnes (1992) argue that adaptive automation can create a tighter coupling between an operator's level of workload and the degree or mode of automation in the system.

Interest in adaptive automation is fueled by concerns over the difficulties operators have when working with complex systems with multiple modes of automation (Woods, 1996). Wiener and Curry (1980) indicate that it is not uncommon for pilots to become confused trying to keep track of the numerous modes of operation in modern flight-management systems. Further, Wiener (1989) argues that automated systems may actually become a burden at the times when a reduction in workload is needed the most. On the other hand, Parasuraman and his colleagues have shown that continued reliance on automation can impair the ability to detect system failures (see Parasuraman, Mouloua, Molloy, & Hilburn, 1996). An adaptive system would circumvent many of these problems by modifying its level or mode of operation in response to changes in the demands placed on the operator.

One of the critical problems facing designers of adaptive systems centers around the mechanisms for monitoring changes in workload and switching among modes of automation. Morrison and Gluckman (1994) offered three possibilities. The first method would be to monitor the operator's performance and implement changes according to some criterion. The utility of this procedure, however, is limited in that it requires continuous performance on a task in order to be effective. The second method improves upon this notion by appealing to operator performance models (Rouse, Geddes, & Curry, 1987–1988). Under this method, the system orchestrates changes in automation by making predictions about the future state of the system based upon information about current system status, external events, and operator performance. Unfortunately, this method also relies upon measures of operator performance. The third technique involves the use of psychophysiological measures for triggering changes in the modes of operation.

There are several advantages to this method. First, psychophysiological measures can be obtained in the absence of any overt behavior on the part of the operator. Second, they can be obtained continuously and provide a rapid response to changes in the state of the operator. Third, they can provide an index of workload without causing interference in the task at hand. Despite these advantages, psychophysiological indices are not without their drawbacks. They can be expensive to obtain and are sometimes difficult to interpret. Also, as Kramer (1991) indicates, they are often confounded with other sources of noise. Thus, although psychophysiological measures may be the most promising candidates for implementing adaptive automation, at present they continue to be restricted to primarily laboratory applications.

PHYSIOLOGICAL MEASURES

There are many physiological indices that have been shown to reflect differences in workload (Kramer, 1991; Wilson & O'Donnell, 1988). Some of these include pupil diameter, eye blinks, electrodermal activity, heart rate, heart rate variability, respiration, and the event-related potential (ERP). However, not all are suited to drive changes among modes of automation in real time and some are more invasive than others. For example, measurement of respiratory or pupillary responses requires equipment configurations that are largely impractical outside of the laboratory. Eye blinks may be insensitive to brief changes in processing demands (Kramer, 1991). Moreover, ERPs are time-locked to specific events and must typically be averaged across several stimulus presentations, off-line, to be made suitable for analysis.

One promising psychophysiological measure is electrical brain activity as measured by electroencephalogram (EEG). The EEG represents fluctuations in voltage recorded from the scalp that fall between 1 and 40 Hz. This frequency range is often subdivided into narrower bands named delta (1–3 Hz), theta (4–7 Hz), alpha (8–13 Hz), and beta (13–20 Hz).

Many investigators have examined the relationship between these EEG bands and states of arousal, attention, and workload (Davidson, 1988; O'Hanlon & Beatty, 1977; Parasuraman, 1983). For instance, activity in the delta band is associated with sleep and is not typically used in assessing workload. Theta activity, on the other hand, usually reflects lapses in attention. Beatty and O'Hanlon (1979) found that observers who could suppress their theta activity improved their vigilance performance. Alpha activity is associated with an awake but relaxed state and appears to be particularly sensitive to changes in workload (Gale & Edwards, 1983). Lastly, alert states are reflected by a preponderance of beta activity. In a simulated radar monitoring task, O'Hanlon and Beatty (1977) found that better detection performance coincided with higher levels of beta activity as well as lower levels of alpha and theta activity.

Like other psychophysiological measures, EEG has its drawbacks, the most important of which may be its susceptibility to noise. A good deal of care must be exercised to obtain signals that are relatively free of ocular or other muscular activity. Further, in some instances, EEG may not distinguish between mental workload and emotional or physical load. Despite these disadvantages, EEG is one of the more reliable and well-investigated psychophysiological measures (Kramer, 1991). Moreover, useful data have been obtained outside of the laboratory in more adverse environments such as flight simulators and genuine aircraft (Natani & Gomer, 1981; Sterman, Schummer, Dushenko, & Smith, 1987).

POWER BAND RATIOS

Recently, several investigators have discussed the utility of power band ratios as an index of engagement. For instance, Streitberg, Röhmel, Herrmann, and Kubicki (1987) showed that incorporating the collective activity among multiple power bands was useful in distinguishing among stages of vigilance and wakefulness. Also, Lubar and his associates (Lubar, 1991; Lubar, Swartwood, Swartwood, & O'Donnell, 1995) observed higher theta to beta ratios, particularly over the frontal cortex, for individuals with Attention Deficit/Hyperactivity Disorder compared to controls. Lubar recommended using indices of theta/beta and alpha/beta to discriminate between children with ADD and those without ADD.

In a recent experiment, Cunningham, Scerbo, and Freeman (in press) examined the relationship between various EEG ratio band indices and daydreaming. They had participants perform a 40-minute target detection vigilance task. In addition, the participants were asked to indicate instances where they realized they had just experienced a task-unrelated image or thought (TUIT; Giambra, 1993) by pressing a key on the computer keyboard. EEG activity occurring 30 seconds before and after the TUIT was examined. Observers in the midst of a daydream were expected to exhibit lower electrocortical activity immediately before the report of a TUIT and higher activity after the TUIT once the observers realized they had been daydreaming and then needed to redirect their attention to the task at hand.

Cunningham et al. (in press) recorded the EEG activity from sites F3, F4, Fz, Cz, P3, P4, and Pz and determined the absolute power in the alpha, beta, and theta bands at each site. In addition, the EEG activity from these seven sites was converted into three different band ratios: beta/alpha, beta/theta, and beta/(alpha+theta).

The results showed that the participants became less vigilant and reported more TUITs over the course of the session. The measures of absolute power in the alpha, beta, and theta frequency bands did not show any differences across the pre-TUIT and post-TUIT intervals. However, significant increases from pre- to post-TUIT in the value of two of the power band ratios, beta/(alpha+theta) and beta/alpha, were observed for all parietal lobe sites (Pz, P3, and P4). Cunningham et al. (in press) concluded power band ratios may provide a more sensitive measure of fluctuations in attention than more traditional measures of absolute power within any individual frequency band.

PHYSIOLOGICALLY BASED ADAPTIVE SYSTEMS

In the early 1980s, Gomer (1981) proposed adaptive computer systems in which the distribution of responsibilities between operator and computer could be modified in real time. He argued that psychophysiological indices of workload (e.g., EEG or ERP) could serve to trigger such changes.

Since that time, little progress has been made in the area of psychophysiologically based adaptive systems. To date, there are only a few examples of such systems. Yamamoto and Isshiki (1992) described one system that uses skin conductance responses (SCR) to maintain alertness. In their system, skin conductance is measured continuously for spontaneous changes. If 3 minutes elapse without a change, an auditory alarm sounds. The subjects in their study were expected to expend effort at the sound of the alarm and therefore change their SCR, which in turn would shut off the alarm. The system described by Yamamoto and Isshiki functions as a simple alertness monitor and therefore does not truly represent an application of adaptive automation. Recently, however, a prototype of an adaptive system for moderating operator workload that is driven by changes in EEG has been described.

A BIOCYBERNETIC SYSTEM

Pope, Bogart, and Bartolome (1995) developed a closed-loop system to regulate task demands in response to changes in mental engagement. The system was

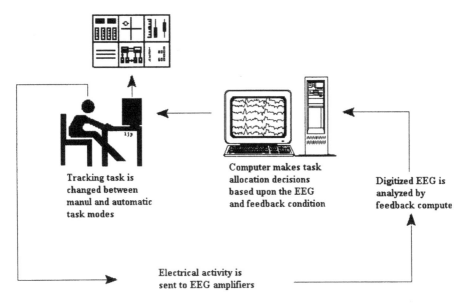

Tracking task is
changed between
manul and automatic
task modes

Computer makes task
allocation decisions
based upon the EEG
and feedback condition

Digitized EEG is
analyzed by
feedback compute

Electrical activity is
sent to EEG amplifiers

FIG. 11.1. Hardware configuration for the biocybernetic closed loop system.

designed to determine the optimal task allocation between the human and the system based upon a real-time measure of EEG activity.

Pope et al. (1995) employed several power band ratios or engagement indices to drive the biocybernetic system based on the work of Lubar (1991) and other EEG researchers (e.g., Offenloch & Zahner, 1990; Streitberg, Röhmel, Herrmann, & Kubicki, 1987). Recording from parietal and central sites, they found that the index, beta/(alpha+theta), was more sensitive to experimental manipulations than two other indices, beta/alpha or 1/alpha index.

The configuration of the closed-loop system is shown in Fig. 11.1. EEG signals are recorded from the scalp, amplified, and then digitized and stored to disk. The digitized signals are then converted back to analog and sent to a LabView Virtual Instrument (VI). The VI determines the power in the alpha, beta, and theta bands for all recording sites and also calculates the engagement index. The VI also orchestrates the changes between task modes and sends this information to the PC running the operator's task.

The system operates with a moving window procedure. EEG is recorded for 40 seconds to determine the initial value of the engagement index. The window is advanced 2 seconds and the index is recomputed. Pope et al. (1995) argue that an increase in the slope of two successive values of the index reflects an increase in engagement. Likewise, a decreasing slope reflects a decrease in engagement.

Pope et al. (1995) had their participants perform the compensatory tracking task from the Multiple-Attribute Task (MAT) Battery (Comstock & Arnegard, 1992). The MAT is a PC-based suite of tasks that draws upon the kinds of information processing skills used by pilots in typical flight-management tasks. The screen is divided into four windows that contain a compensatory tracking task, a series of lights and gauges that must be monitored for critical events, a communications task, and a resource management task requiring the operator to maintain an optimal level of fuel in two tanks by directing the flow of fuel though a series of pumps. In their study, all of the tasks remained in automatic mode except the tracking task which shifted between automatic and manual modes. The current mode of operation was always displayed on the screen.

The participants in their study performed the task under four different engagement indices: beta/alpha and beta/(alpha+theta) combined from sites Cz, Pz, P3, and P4; alpha at T5 and P3/alpha at Cz and Pz; and alpha at O1/alpha at O2. The duration of each trial under each index was 16 minutes. Within each trial, the participants worked under alternating 4-minute blocks of positive and negative feedback. In the negative feedback condition, the tracking task was switched to or maintained in automatic mode when the slope of the index was increasing. Likewise, the task was switched to or maintained in manual mode when the value of the slope decreased. The system operated in the opposite manner under the positive feedback condition. Pope et al. (1995) expected that the system would switch back and forth between automatic and manual modes more frequently under negative feedback in order to maintain a stable level of engagement. On the other hand, they expected that the positive feedback condition would produce longer episodes at extreme values of the index and, therefore, result in fewer changes between modes. Moreover, the engagement indices were evaluated according to their ability to generate the expected differences in system performance between the positive and negative feedback conditions.

The results showed that three indices, beta/alpha, beta/(alpha+theta), and alpha/alpha, at the temporal, parietal, and central sites distinguished between positive and negative conditions, but that the best discriminator was the index beta/(alpha+theta). Moreover, a second experiment in which EMG (42–100 Hz) or high beta (38–42 Hz) was substituted for the numerator in this index led Pope et al. (1995) to conclude the presence of any of these frequencies in the beta/(alpha+theta) index would weaken, but not eliminate the ability of this index to discriminate between feedback conditions.

The results of the study by Pope et al. (1995) are significant in two respects. First, they demonstrate that it is feasible to allocate tasks in real time in a biocybernetic closed-loop system triggered by psychophysiological signals, i.e., changes in EEG indices of engagement. Second, of the several candidate indices examined, beta/(alpha+theta) appears to be the most sensitive and robust index for reliable system operation under both positive and negative feedback.

AN EXAMINATION OF THE ENGAGEMENT
INDEX AND PERFORMANCE

Although the system described by Pope et al. (1995) represents a significant step forward in creating adaptive technology triggered by physiological indices, Prinzel, Scerbo, Freeman, and Mikulka (1997) offered two criticisms of the original study. First, Pope et al. assessed the performance of the system by measuring the number of task allocations between modes of operation. Prinzel et al. suggested that a more valid measure might be the value of the engagement index itself. Second, although participants were asked to perform a tracking task, these data were not analyzed. Prinzel et al. argued that the benefits of any adaptive system, ultimately, should be reflected in operator performance.

To address these limitations, Prinzel et al. (1997) conducted a second study that essentially replicated the experimental conditions described by Pope et al. (1995) with a few differences. First, they examined system operation with the indices 1/alpha, beta/alpha, and beta/(alpha+theta). Also, the system configuration used by Prinzel et al. was similar to the one used by Pope et al. except that EEG was recorded from only four sites: Cz, Pz, P3, and P4. In addition, they analyzed the root mean square error (RMSE) on the tracking task.

Like Pope et al. (1995), these investigators found that the system made more switches between automatic and manual modes in the negative as opposed to positive feedback condition. This effect, however, was moderated by index. Specifically, the difference in task allocations between the positive and negative feedback conditions was only significant for the beta/(alpha+theta) index. This result provides further evidence for the superiority of beta/(alpha+theta) as an index of engagement. The other two indices did not distinguish between positive and negative feedback conditions.

More important, however, Prinzel et al. (1997) observed an interaction between feedback condition and task mode when analyzing the value of the engagement index, beta/(alpha+theta). The nature of this interaction is shown in Table 11.1. Under negative feedback, the value of the engagement index was higher in the automatic mode than the in the manual mode. This pattern was reversed under

TABLE 11.1
Mean Values of EEG Engagement Index (beta/(alpha+theta)) for Task Mode and Feedback Conditions

	Task Mode	
	Automatic	Manual
Negative Feedback	31	21
Positive Feedback	20	30

positive feedback. In addition, when examining the tracking performance under the manual mode, Prinzel et al. found that error scores were lower under negative as opposed to positive feedback.

Collectively, the findings of Prinzel et al. (1997) help to validate the operation of the biocybernetic system, particularly in the negative feedback condition. Specifically, the value of index beta/(alpha+theta) is lower when individuals are actively engaged in the tracking task than when they are merely observing it under automatic mode. Further, examination of the RMSE scores showed that the system produced better performance under negative feedback.

WORKLOAD

In a second study, Prinzel, Freeman, Scerbo, Mikulka, and Pope (in press) addressed another limitation of the study by Pope et al. (1995). Although one of the original goals of the system was to moderate workload and maintain a stable level of engagement, no measures of workload were reported.

To address some of these limitations, Prinzel et al. (in press) manipulated the workload level in their study. In the low workload condition, participants performed only the tracking portion of the MAT task. The monitoring and resource management tasks remained in automatic mode. In the high workload condition, however, participants performed all three tasks simultaneously (i.e., the monitoring and resource management tasks remained in manual mode). In both workload conditions only the tracking task switched between automatic and manual modes. Prinzel et al. hypothesized that the system would make more task allocations under the high workload condition because of the operator's need to address the unpredictable demands of three different tasks.

Second, to check the workload manipulation the participants were asked to complete the NASA-Task Load Index (TLX; Hart & Staveland, 1988). The TLX is a subjective measure of workload composed of six rating scales that address the contributions of various sources of workload. Prinzel et al. (in press) had their participants complete the TLX after each 16-minute period.

The experimental conditions were essentially the same as in the previous study by Prinzel et al. (1997) with one primary difference. Because Prinzel, Pope and their colleagues found the index beta/(alpha+theta) to be the most sensitive in two prior studies, only that index was used in this experiment. Once again, RMSE scores for the tracking task were analyzed to examine the effects of both workload and feedback conditions on tracking performance.

As in the previous two studies, these investigators found that the system made more switches between automatic and manual modes in the negative as opposed to positive feedback conditions. In addition, Prinzel et al. (in press) observed the same type of interaction between feedback condition and task mode for the value of the engagement index (i.e., the index was higher in automatic than in manual mode under negative feedback with the opposite pattern observed under positive feedback). Once again, tracking performance was superior under negative as opposed to positive feedback.

Regarding workload, more switches between task modes were made in the high workload condition. Participants also rated the high workload condition higher on the TLX, but there were no differences between feedback conditions. Thus, the system appears to be sensitive to manipulations in task load.

ABSOLUTE CRITERION
AND WINDOW SIZE

The original system described by Pope et al. (1995) used the slope of the engagement index derived from successive measurements to implement changes between task modes. Hadley, Mikulka, Freeman, Scerbo, and Prinzel (1997) reported a potential flaw in this procedure. Specifically, under this method the system can potentially shift modes of operation any time there is a change in arousal. The system does not, however, take into account the absolute level of engagement. Thus, for example, in situations where the operator's level of engagement is substantially below his or her mean for the session, the system could change task modes after only a slight increase in the value of the index even though the overall level of engagement is still quite low. To circumvent this problem, Hadley et al. redesigned the system to implement changes based upon an absolute level of engagement. Specifically, these investigators had their subjects perform the task for 5 minutes prior to the experimental trial. They then determined the mean value of the engagement index and used this value as the criterion for task allocations in the subsequent trial. Thus, changes between task modes were made relative to this absolute criterion instead of the slope of the engagement index from successive measurements.

Hadley et al. (1997) also examined the effect of window size in computing the engagement index. In all of the studies described earlier, the index was based upon 40 seconds of EEG activity. Hadley et al. argued that an improvement in system sensitivity might be obtained with a smaller window width. Thus, they compared system operation with window sizes of 40 and 4 seconds.

Hadley et al. (1997), once again, observed an interaction between feedback condition and task mode using the absolute criterion for task allocations. The value of the engagement index was higher for the automatic than for the manual mode under negative feedback with the opposite pattern observed under positive feedback. This effect, however, was moderated by window size. Specifically, the differences between task mode and feedback condition were greater under the 4-second window. In addition, more switches between modes occurred under negative as opposed to positive feedback and the narrower window generated more switches than the 40-second window. These investigators also found that tracking performance was bolstered by the narrower window.

The findings of Hadley et al. (1997) are important for several reasons. First, they demonstrate the use of EEG band ratios as an index of engagement is quite robust. Modifying the system to change tasks based upon an absolute criterion of engagement did not alter the pattern of results obtained in previous studies. Second, these finding show that a narrower window may not only improve the sensitivity of the system to changes in engagement, but may facilitate performance as well.

CONCLUSION

The system developed by Pope et al. (1995) provides one example of how psychophysiological measures can be employed in an adaptive interface. In particular, they have shown that EEG power band ratios are a promising way to assess engagement. The subsequent work by Prinzel and his colleagues confirmed these results. Moreover, this body of work has demonstrated that the level of the engagement index, particularly beta/(alpha+theta), validates system operation. The index reaches a higher level in the negative feedback condition when the operator is passively monitoring the display than when he or she is engaged and actively performing the task. In addition, Prinzel et al. (in press) shows that the negative feedback condition does indeed lead to better performance, which is paramount for any prospect of practical applications.

Although the results to date show much promise for creating adaptive automation around psychophysiological input, clearly more work needs to be done. At present, a systematic examination of electrode placements has not been conducted. Thus, it is not known whether the current configuration is optimal, whether input from four electrodes is necessary, or whether the importance of a given electrode site is affected by the specific task or tasks performed by the operator. Also, the system in its present configuration only allows for two modes of operation, automatic and manual. It remains to be seen whether the range of values produced by the engagement index can be divided and used to allocate tasks among multiple modes of automation. Further, it would also be important to know whether the findings obtained with a task allocation strategy in the present system would also be observed with a single task partitioning strategy. Despite the need for additional work, the research thus far has been successful in demonstrating a system capable of coupling the level of automation to the level of operator engagement and facilitating performance with a psychophysiologically based adaptive interface.

Although the need for continued investigation is clear, it does not detract from another important aspect of the work of Pope et al. (1995). These investigators stated that the methodology they employed can serve as a useful tool for assessing candidate indices of engagement. In particular, comparing system behavior under negative and positive feedback conditions allows the evaluation of potential indices for their ability to generate stable or unstable operating conditions, respectively. Recently, Byrne and Parasuraman (1996) argued that this approach was essential to evaluating potential parameters in an adaptive automation framework.

From an applied perspective, this line of work lends additional support to the feasibility of adaptive technology (see Scerbo, 1996). Although the biocybernetic system described earlier shows promise, it is still confined largely to the laboratory. Many issues would have to be addressed before a system such as this could actually be applied to work in the real world. The most obvious hurdle concerns the technical problems associated with EEG recording in the field and enabling the system to interface with a variety of application platforms. However, the current pace of improvements in physiological recording may diminish the importance of this issue in the next 10 to 20 years. Aside from technical concerns, there are other more social issues that beg consideration. Adaptive interfaces usurp some of the authority that operators have over their systems. Consequently, issues such as reliability and trust become much more critical as well as who is ultimately responsible for system operations. Concerns such as these have led some researchers to begin addressing issues in adaptive automation from a team perspective (see Scerbo, in press). In the end, the feasibility of adaptive technology may rest more with our willingness to accept and work with "electronic teammates" than in advancements in physiological recording equipment.

ACKOWLEDGMENTS

The authors wish to acknowledge the assistance of our colleagues at the NASA Langley Research Center for work described in this chapter. In particular, we wish thank Dr. Alan T.Pope for his continued support as well as his comments.

REFERENCES

Beatty, J., & O'Hanlon, J.F. (1979). Operant control of posterior theta rhythms and vigilance performance: Repeated treatments and transfer of training. In N.Birbaumer & H.D. Kimmel (Eds.), *Biofeedback and self-regulation* (pp. 235–246). Hillsdale, NJ: Lawrence Erlbaum Associates.

Byrne, E.A., & Parasuraman, R. (1996). Psychophysiology and adaptive automation. *Biological Psychology, 42,* 249–268.

Comstock, J.R., Jr., & Arnegard, R.J. (1992). *The multi-attribute task battery for human operator workload and strategic behavior research* (Memo. No. 104174). National Aeronautics and Space Administration, Hampton, VA.

Cunningham, S., Scerbo, M.W., & Freeman, F.G. (in press). The electrocortical correlates of daydreaming during vigilance tasks. *Journal of Mental Imagery.*

Davidson, R.J. (1988). EEG measures of cerebral asymmetry: Conceptual and methodological issues. *International Journal of Neuroscience, 39,* 71–89.

Gale, A., & Edwards, J. (1983). The EEG and human behavior. In A.Gale & J.Edwards (Eds.), *Physiological correlates of human behavior* (pp. 99–127). New York: Academic.

Giambra, L.M. (1993). The influence of aging on spontaneous shifts of attention from external stimuli to the contents of consciousness. *Experimental Gerontology, 28,* 485–492.

Gomer, F.E. (1981). Physiological monitoring and the concept of adaptive systems. In J. Moraal & K.-F.Kraiss (Eds.), *Manned systems design: Methods, equipment, and applications* (pp. 271–287). New York: Plenum.

Hadley, G., Mikulka, P.J., Freeman, F.G., Scerbo, M.W., & Prinzel, L.P. (1997). An examination of system sensitivity in a biocybernetic adaptive system. *Proceedings of the 9th International Symposium on Aviation Psychology* (pp. 305–309). Columbus: Ohio State University.

Hancock, P.A., & Chignell, M.H. (1987). Adaptive control in human-machine systems. In P.A.Hancock (Ed.), *Human factors psychology* (pp. 305–345). Amsterdam: North-Holland.

Hart, S.G., & Staveland, L.E. (1988). Development of NASA-TLX (Task Load Index): Results of empirical and theoretical research. In P.A.Hancock & N.Meshkati (Eds.), *Human mental workload* (pp. 139–183). Amsterdam: North-Holland.

Kramer, A.E (1991). Physiological metrics of mental workload: A review of recent progress. In D.L.Damos (Ed.), *Multiple-task performance* (pp. 279–328). London: Taylor & Francis.

Lubar, J.E (1991). Discourse on the development of EEG diagnostics and biofeedback for attention-deficit/hyperactivity disorders. *Biofeedback and Self-Regulation, 16,* 201–225.

Lubar, J.E, Swartwood, M.O., Swartwood, J.N., & O'Donnell, P.H. (1995). Evaluation of the effectiveness of EEG neurofeedback training for ADHD in a clinical setting as measured by changes in T.O.V.A. scores, behavioral ratings, and WISC-R performance. *Biofeedback and Self-Regulation, 20,* 83–99.

Morrison, J.G., & Gluckman, J.P. (1994). Definitions and prospective guidelines for the application of adaptive automation. In M.Mouloua & R.Parasuraman (Eds.), *Human performance in automated systems: Current research and trends* (pp. 256–263). Hillsdale, NJ: Lawrence Erlbaum Associates.

Natani, K., & Gomer, F.E. (1981). *Electrocortical activity and operator workload: A comparison of changes in the electroencephalogram and in event-related potentials* (Tech. Rep. E2427), McDonnell Douglas Corporation.

Offenloch, K., & Zahner, G. (1990). Computer aided physiological assessment of the functional state of pilots during simulated flight. *Proceedings of the NATO Advisory Group for Aerospace Research and Development, 490,* 9–1–9–9.

O'Hanlon, J.F., & Beatty, J. (1977). Concurrence of electroencephalographic and performance changes during a simulated radar watch and some implications for the arousal theory of vigilance. In R.R.Mackie (Ed.), *Vigilance: Theory, operational performance and physiological correlates* (pp. 189–201). New York: Plenum.

Parasuraman, R. (1983). Vigilance, arousal, and the brain. In A.Gale & J.Edwards (Eds.), *Physiological correlates of human behavior* (pp. 139–143). New York: Academic.

Parasuraman, R., Bahri, T., Deaton, J.E., Morrison, J.G., & Barnes, M. (1992). *Theory and design of adaptive automation in aviation systems* (Progress Report No. NAWCADWAR92033–60). Warminster, PA: Naval Air Warfare Center, Aircraft Division.

Parasuraman, R., Mouloua, M., Molloy, R., & Hilburn, B. (1996). Monitoring of automated systems. In R.Parasuraman & M.Mouloua (Eds.), *Automation and human performance: Theory and applications* (pp. 91–115). Mahwah, NJ: Lawrence Erlbaum Associates.

Pope, A.T., Bogart, E.H., & Bartolome, D. (1995). Biocybernetic system evaluates indices of operator engagement. *Biological Psychology, 40,* 187–196.

Prinzel, L.J., Freeman, F.G., Scerbo, M.W., Mikulka, P.J., & Pope, A.T. (in press). A closed-loop system for examining psychophysiological measures for adaptive automation. *International Journal of Aviation Psychology.*

Prinzel, L.J., Scerbo, M.W., Freeman, F.G., & Mikulka, P.J. (1997). Behavioral and physio logical correlates of a bio-cybernetic, closed-loop system for apative automation. In M. Mouloua & J.M.Koonce (Eds.), *Human-automation interaction: Research and practice* (pp. 66–75). Mahwah, NJ: Lawrence Erlbaum Associates.

Rouse, W.B. (1976). Adaptive allocation of decision making responsibility between supervisor and computer. In T.B.Sheridan & G.Johannsen (Eds.), *Monitoring behavior and supervisory control* (pp. 295–306). New York: Plenum.

Rouse, W.B., Geddes, N.D., & Curry, R.E. (1987–1988). An architecture for intelligent interfaces: Outline of an approach to supporting operators of complex systems. *Hu-man—Computer Interaction, 3,* 87–122.

Scerbo, M.W. (1996). Theoretical perspectives on adaptive automation. In R.Parasuraman & M.Mouloua (Eds.), *Automation and human performance: Theory and applications* (pp. 37–63). Mahwah, NJ: Lawrence Erlbaum Associates.

Scerbo, M.W. (in press). Adaptive automation: Working with a computer teammate. In L.J. Hettinger & M.W.Haas (Eds.), *Psychological issues in the design and use of virtual, adaptive environments.* Mahwah, NJ: Lawrence Erlbaum Associates.

Sterman, M.B., Schummer, G.J., Dushenko, T.W., & Smith, J.C., (1987). *Electroence-phalographic correlates of pilot performance: Simulation and in-flight studies* (pp. 1–16). Neuilly sur Seine, France: AGARD.

Streitberg, B., Röhmel, J., Herrmann, W.M., & Kubicki, S. (1987). COMSTAT Rule for vigilance classification based on spontaneous EEG activity. *Neuropsychobiology, 17,* 105–117.

Wiener, E.L. (1989). *Human factors of advanced technology ('glass cockpit') transport aircraft* (Tech. Rep. 117528). Moffett Field, CA: NASA Ames Research Center.

Wiener, E.L., & Curry, R.E. (1980). Flight-deck automation: Promises and problems. *Ergonomics, 23,* 995–1011.

Wilson, G.F., & O'Donnell, R.D. (1988). Measurement of operator workload with the neuropsychological test battery. In P.A.Hancock & N.Meshkati (Eds.), *Human mental workload* (pp. 63–93). Amsterdam: Elsevier.

Woods, D.D. (1996). Decomposing automation: Apparent simplicity, real complexity. In R. Parasuraman & M.Mouloua (Eds.), *Automation and human performance: Theory and applications* (pp. 3–17). Mahwah, NJ: Lawrence Erlbaum Associates.

Yamamoto, Y., & Isshiki, H. (1992). Instrument for controlling drowsiness using galvanic skin reflex. *Medical & Biological Engineering & Computing, 30,* 562–564.

Chapter 12

Slow Brain Potentials as a Measure of Effort? Applications in Mental Workload Studies in Laboratory Settings

Gabriele Freude
Peter Ullsperger
Federal Institute for Occupational Safety and Health,
Berlin, Germany

The use of information and communication technologies makes work tasks physically easier and minimizes exposure to traditional hazards. On the other hand, work is increasingly accompanied by information overload, time pressure, high levels of sustained attention and long working hours that may affect employees' well-being and health. The assessment of the various aspects of mental workload resulting from human-computer interaction remains a fundamental prerequisite for preventing potential work-related risks.

Kramer, Sirevaag, and Braune (1987) have described mental workload as the cost of performing a task in terms of a reduction in the capacity to perform additional tasks that use the same processing resource. Sanders (1983) elaborated a cognitive-energetical model of task performance that is based on a multiple resource view that combines both structural and energetical components. The structural level describes the flow of information through various processing stages from stimulus to response. At the energetical level, the physiological mechanisms of arousal, effort, and activation are distinguished. Whereas the arousal and activation components are hypothesized to be determined by involuntary factors, effort is under voluntary control, mediates response selection, and coordinates the arousal and activation subsystems (Mulder, 1986; Pribram & McGuinness, 1975). Kok (1997) pointed out that these energetical mechanisms provide the gain for the data processing system and have direct ties not only with processing stages, but also with typical state variables like fatigue, effects of drugs, and so on. Hockey (1997) goes further than previous approaches and has developed a generalized control model that provides mechanisms for dynamic regulatory activity underlying adaptive responses to environmental demands. In this model, effort is located centrally as a coordinating process, adjusting the balance of input and output operations. *Effortful regulation* refers to the attempt to maintain a particular task state under overload, external distraction, or stress.

In the following, we are concerned with the measure of preparatory slow brain potentials (SP; for review, see e.g., Birbaumer et al, 1990; Rockstroh et al., 1989) as an index of voluntarily controlled effort. Following are several examples from laboratory studies of the relationship between SP measures and effort expenditure varied by such factors as task demand, time pressure, practice, breaks, and working hours. We begin with a particular component of preparatory slow brain potentials, the movement-related readiness potential or Bereitschaftspotential (BP; for review, see e.g., Deecke, 1984; Freude & Ullsperger, 1987a), which has particular importance for application in modern workplaces.

FUNCTIONAL ASPECTS
OF THE BEREITSCHAFTSPOTENTIAL

When working in the modern computer workplace, the operator usually communicates with the computer by means of a keyboard, whereas the output of the computer is presented on a video display. Key presses on the computer keyboard or computer mouse always accompany computer work to obtain information, to enter digital and alphanumeric information into the computer, and to issue processing commands. These goal-directed and task-relevant key presses can be used as triggers to analyze the BP in a more or less ecologically valid situation without the need for any additional or artificial stimuli. Mostly, the movements to perform the key presses are highly trained and do not systematically vary in their motor and ballistic properties. However, the actual task demands or the expected effects and consequences of key presses may differ considerably.

In most BP experiments, subjects were asked to perform simple voluntary movements of their index fingers (Kornhuber & Deecke, 1964, 1965) at self-paced intervals. Onsets of the electromyographic activity or movements provided the triggers for backward averaging of the bioelectrical brain activity. A strict motor interpretation of the BP was initially favored for two reasons. There was a direct relationship between the negative-going BP shift and movement initiation, and BP negativity was greater over the hemisphere contralateral to the movement consistent with functional somatotopic organization of the motor cortex in the brain.

However, other findings provided evidence against the exclusive motor interpretation of BP. As early as 1965, Kornhuber & Deecke stressed the significance of intentional task involvement and motivation. Deecke (1976) stated, that "BP may reflect a general facilitation process, preactivating those brain regions which will be needed under the special experimental situation under study" (p. 91). The generation of a voluntary movement is usually conduced with a certain goal and with expectancy about the effect of the movement. Therefore, the possibility that the BP would overlap other components of slow brain potentials such as the contingent negative variation (CNV; Walter et al., 1964) or stimulus preceding negativity (SPN; Brunia, 1993; Damen, Freude, & Brunia, 1996) is not surprising. These SPs are known to be particularly sensitive to psychological influences such as intention, attention, and expectation-con-structs that are meaningful during goal-directed mentally demanding tasks.

The BP elicited during computer work consists of at least two major overlapping potentials. One potential is associated with motor preparation, whereas the other is related to preparation for the anticipated task (e.g., facilitation of information processing in the sensory system). The preparatory processes preceding the active movement affect not only movement execution, but also post-movement processing stages that facilitate performance by increasing the speed and accuracy of task processing (Donchin et al., 1984; Requin, Lecas, & Bonnet, 1984). During the preparatory interval, operators can intentionally allocate necessary resources for task processing by effort mechanisms. This is especially the case when operators have previously experienced a mismatch between the task demands and the amount of resources necessary to attain the particular goal. Consistent with the concept of cerebral potentiality and cerebral performance (Rockstroh et al., 1989), preparatory negativity is seen as a reflection of voluntarily influenced energy supply necessary for effective information selection, information processing and motor output in the immediate future.

Numerous experimental studies provide evidence that the amplitude and to pography of preparatory SPs reflect the type of task (e.g., mental computation, scanning of short-term memory, mental rotation, concept formation, planning and execution of motor responses) and the task demands (McCallum, 1993; McCallum, Cooper, & Pocock, 1988; Rösler, Heil, & Glowalla, 1993). In particular, SPs are found over those parts of the cortex where, according to neurological evidence, structures essential for solving the task are located (Lang et al., 1989; Rösler, Heil, & Glowalla, 1993). However, the location of the SP does not necessarily imply that the potential is generated in brain tissue directly beneath the electrode location. In case of foot movements, for example, BP is not more pronounced contralateral to the movement side, but a paradoxical ipsilateral BP occurs (Boschert, Hink, & Deecke, 1983; Brunia, 1984; Brunia, Voorn, & Berger, 1985).

The finding that negativity increases with task demands and that greater negativity is associated with better performance (McCallum, 1993; Rösler, Heil, & Glowalla, 1993) confirms the assumption that negative slow waves reflect the activity of the involved brain regions. In other words, negative slow waves may be interpreted as a manifestation (cf. Hockey, Coles, & Gaillard, 1986) of intentional active effort to perform a given task. Increased intentional involvement effects an increased readiness to spend additional effort in the task execution. This notion was confirmed by studies of the influence of motivation (McAdam & Seales, 1969), attention (McAdam & Rubin, 1971), and the intentional involvement in task processing (Freude & Ullsperger, 1987b; Freude et al., 1988; Kristeva, 1977; McAdam & Seales, 1969).

Our approach (Freude et al., 1988, 1989; Freude, Ullsperger, & Molle, 1995) is based on the notion that the preparatory negativity that occurs prior to self-paced key presses on the computer keyboard reflects mainly the preparation to the anticipated mental task. Because BP is considered to occur in combination with other SP components, we use the broader term of SPs in the following section.

SLOW BRAIN POTENTIALS AS INDEX OF MENTAL EFFORT

In this section, we review examples of the application of SPs in laboratory studies evaluating effort expenditure in the context of mental workload assessment. Studies are also presented on the effects of different work-related factors, such as task difficulty and time pressure, on SPs. Two further effort-related studies will also be presented that examine early human error detection and that evaluate an operator's level of training and skill acquisition.

Effects of Task Difficulty

McCallum et al. (1988,1993) were the first to apply SPs in the field of workload assessment during video display unit (VDU) work. They started from the no tion that a subject's "task involvement has always presented itself as one of the major determinants of the amplitude of these shifts" (McCallum & Pleydell-Pearce, 1993, p. 165). As discussed in the preceding paragraph, the dimensions of task involvement, attention, and motivation are related to effort. McCallum and Pleydell-Pearce (1993) conducted a series of studies in which the level of effort expenditure and workload were manipulated during monitoring tasks over a protracted period of time. The mental tasks employed were a visual display presentation that the subjects interacted with via manual responses. While subjects had to track a moving letter across the screen, various manipulations were introduced to vary the level of workload related to the speed and distance traveled by the tracked letter, the smoothness or irregularity of the movement, and the number of letters to be memorized. The main result was that higher tracking difficulty was associated with a higher negative SP shift. In line with this result, Lang et al. (1989) reported for a visual learning paradigm that mirror tracking is accompanied by higher negative cerebral potentials than simple tracking. These results are in general agreement with a number of studies examining the effects of various task-difficulty manipulations on SP shifts (for review, see Rockstroh et al., 1989).

Applying a dual-task approach (for review, see, e.g., Kramer, 1991), McCallum and Pleydell-Pearce (1993) increased workload by introducing secondary tasks. SP amplitudes at frontal (Fz) and occipital (Oz) leads were more negative during tracking combined with secondary tasks than during tracking alone. Adding a secondary task to the primary tracking task increased SP negativity, which may be interpreted as a sign of increased effort expenditure due to a higher cognitive demand and an increased motor activity.

Effects of Time Pressure

Freude et al. (1988) found a significant effect of time pressure on SPs during cognitively demanding computer work. Adult subjects had to solve arithmetic problems under graduated time pressure. They could choose the duration of subsequent task presentation by pressing one of three keys on a computer keyboard corresponding to three steps of task duration. As soon as a key had been pressed, the task appeared on the computer display and disappeared after the time interval corresponding to the selected duration. As the SP waveforms in Fig. 12.1 show, SP was found to be significantly more negative when the tasks were solved

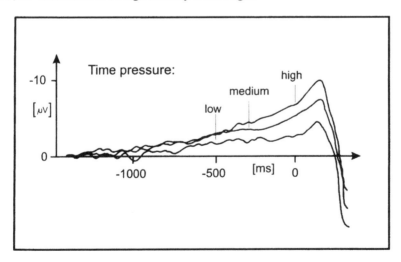

FIG. 12.1. Slow brain potentials prior to mental arithmetic performed under three different steps of time pressure (variation of task presentation time). SP negativity increases when a short duration of task presentation is expected (key press at 0 ms).

under higher time pressure pointing to higher effort expenditure in preparation to the anticipated higher task demand.

Two further studies (Freude & Ullsperger, 1994; Freude, Ullsperger, & Mölle, 1995) focused on evaluation of time-dependent changes of effort expenditure during VDU work lasting several hours. The task requirements differed in both experiments with respect to the task (monotonous text processing vs. highly demanding mental arithmetic) and the function of the key press itself (result input vs. calling up tasks).

Monotonous Spell Checking Task

Subjects had to perform a text processing task over a period of 4 hours. SP was analyzed separately for four subsequent 50-minute work periods that were interrupted by 10-minute breaks. The influence of duration of computer work on the SP was examined by comparing the four work periods as well as the time immediately after breaks and the time toward the end of each work period. Subjects checked a text for misspelled words using the spell-checking function of a text processor. Subjects had to decide whether highlighted words in the text were to be replaced by words presented at the status line and then press a corresponding right- or left-hand key. Subjects had to press the right-hand key with the right index finger if the word fit the context and if it was correctly spelled.

Figure 12.2 shows the significant time-dependent SP changes at parietal and frontal electrodes. For the parietal lead (Fig. 12.2, upper panel), it is evident that SP amplitudes are more negative in the first part of each work period, that is, in the work period immediately after the break. Toward the end of each work period SP decreased and became even positive.

On the other hand, the frontal SP tended to change to the opposite direction across time (Fig. 12.2, lower panel). SP negativity was more pronounced in the period immediately after the 10-minute break and increased in the second part of each work period.

Functional aspects of brain regions should be considered when interpreting the time-dependent SP effects at parietal and frontal leads (cf. Creutzfeldt, 1995). It might be speculated that the decrease in parietal SP goes along with a reduction in attention toward the end of the 50-minute period of spell checking. On the other hand, the SP increase over the frontal cortex toward the end of each working period might be seen in the context of effort regulation, a function that is associated with the frontal cortex. The results suggest that the invested effort increased toward the end of each working period to meet the task requirements during this rather monotonous activity of spell checking.

Demanding Mental Arithmetic

Key presses were used in the spell-checking task for entering results and terminated each trial. In contrast, in the Freude, Ullsperger, and Mölle (1995)

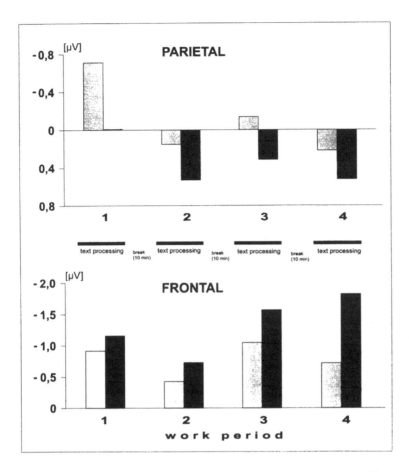

FIG. 12.2. Mean SP amplitudes (time interval: −500 ms to 0 ms; 0-moment of key press) for the first part (after breaks, lightly shaded bars) and the last part (toward the end of each work period, darkly shaded bars) of each work period. Upper part: parietal lead-SP values decrease from the beginning to the end of each work period. Lower part: frontal lead-SP increases from the first to the second part of each work period.

study, mental arithmetic was required and the key presses were used for calling up the arithmetical tasks. The experimental design including the sequence of activities required from the subjects is illustrated in Fig. 12.3.

Subjects continuously attained their performance limits in the course of the 3 hours long experiment (three 50-minute work periods) through a performance-dependent adjustment procedure that controlled the duration of task presentation. The subjects task involvement was achieved by feedback presentation: They were informed about the correctness of the task solution and were therefore enabled to adjust their effort accordingly.

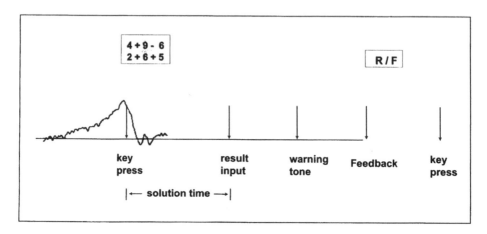

FIG. 12.3. Experimental design for analyzing time-dependent SP changes during performance-adjusted mental arithmetic. After solving each task, subjects got feedback information about correctness of calculation (R: correct; F: incorrect).

The performance data indicate improvement of arithmetic performance assumed to be an effect of practice. In comparison to the first work period, the number of calculated tasks increased by 28% in the third work period. Calculation accuracy, however, increased only by 7%.

When interpreting SPs in relation to practice effects, one should expect a SP decrease in the course of practicing the tasks. However, the negativity of SPs increased across time from the first to the last work period. Under the particular condition of the adaptive adjustment of task load, all effort-lowering effects of practice were continuously compensated for by decreasing the duration of task presentation after a correct calculation. Therefore, SP increase over the course of the experiment may be interpreted as reflecting increased effort expenditure to ensure the required performance outcome.

Considering the performance and the SP data, we determined the efficiency of mental work by calculating the ratio between performance outcome (number of calculated tasks) and the effort invested (as reflected by SP negativity). As shown in Fig. 12.4, the efficiency in the second and third work period is less than half that of the first work period. Thus, a disproportionate increase in effort due to the fatiguing influence associated with the long-lasting task demands may be assumed.

Task-Preceding SPs and Subsequent Performance Outcome

The prediction of errors is considered an important human factor concern (Wickens, 1990). Numerous experimental studies show that SP and performance parameters correlate significantly, where higher SP negativity indicates better performance.

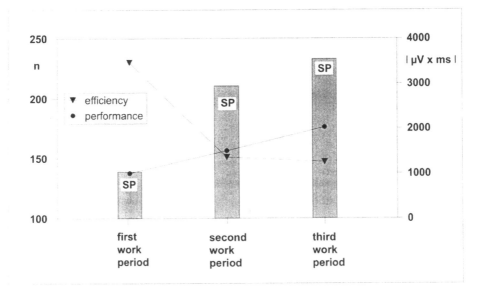

FIG. 12.4. Estimation of the efficiency of mental arithmetic for each work period (performance measure: n=number of calculated tasks).

Because SP negativity precedes the overt behavior and performance outcome, respectively, it should be possible to predict operators performance outcome on the basis of the preceding SP negativity.

Freude et al. (1989) investigated SP negativity in a pattern recognition task. Pattern recognition performance was significantly correlated with SPs measured prior to the tasks:

Higher SP amplitudes were found prior to correct and fast responses and lower SP amplitudes prior to incorrect and slow ones. This result is in accordance with the general conclusion of McAdam and Rubin (1971) that larger SPs are associated with certain and correct perceptions and smaller SPs are obtained when subjects reported "no idea" of the stimulus presented. Quite obviously, visual information processing was affected by cortical negativity. This assumption was confirmed by analyzing slow potential shifts generated prior to incorrect recognition where positive-going shifts occurred, particularly in the early stages of SP This result is congruent with the hypothesis that slow positive shifts are related to a lower excitability when performing cognitive tasks (Bauer & Nirnberger, 1981; Trimmel, 1987; Weber & Bauer, 1986). We assume that positive-going shifts in the early phases of task-preceding SP may be seen as an inappropriate allocation of attentional processing resources.

Hohnsbein, Falkenstein, and Hoormann (1996) investigated subjects with large spontaneous differences in performance accuracy, as defined by their error rates in a two-alternative bimanual choice reaction time task. Subjects with fewer errors had large late SP shifts, whereas subjects with many errors showed virtually no late SP. These authors explain the high error rate of "poor" subjects as insufficient preparation that may have impaired controlled processing.

Effects of Practice on SP

Training or practice on tasks helps to reduce mental workload. Gaillard (1993) stated that training is one important way to reduce the workload and increase the availability of spare capacity for other activities. As numerous studies have indicated, continuous repetition of motor and mental activities are paralleled by a reduction in SP negativity. Kristeva (1977) reported a SP decrease with simple repetitive finger movements during a 1-hour experiment Taylor (1978) analyzed SP during acquisition of a skilled motor task (pressing a series of buttons in a certain order in rapid succession) and found SPs to increase with proficiency of the motor response, but to decline after acquisition of the skilled motor task. Eulitz (1991) investigated SPs during a problem solving task and found a SP decrease during the learning process. Fattaposta et al. (1996) reported a SP amplitude reduction with repetition of a skilled performance task that required subjects to stop the sweep of an oscilloscope trace within a defined area of a screen. SP decrease was accordingly interpreted as an effect of practice that makes the task less demanding and, consequently, less effortful. Freude, Ullsperger, and Erdmann (1998) used a complex visual monitoring task and found a SP decrease when comparing two subsequent experimental blocks, which also may reflect a decrease in effort expenditure due to practice. Furthermore, it could be shown that the level of practice has an effect on the SP-performance relationship. For example, a significant correlation between SP and performance was only found for the second experimental block after having sufficiently practiced the complex visual monitoring task.

CONCLUSIONS

Taken together, we suppose that slow brain potentials provide information about mental workload that cannot easily be obtained from other physiological parameters. Although we agree that mental workload is a multidimensional construct and that no single measurement technique can be expected to tap all relevant aspects of human mental workload, we believe that the slow brain potentials are a manifestation of invested effort and represent an important complementary workload measure (for general advantages and disadvantages of physiological measures for workload assessment, see Kramer, 1991). Slow brain potentials indicate the preparatory mobilization of resources and increased cortical activity under demanding conditions.

As Hockey (1997) pointed out, performance may be protected at the expense of increased effort and energetical costs. Alternatively, the performance goals can be reduced without further costs. We demonstrated that by combining an effort measure with a corresponding performance measure an efficiency index of mental work may be obtained; for example, the decrease in expended effort due to practice or the increase due to fatigue can be evaluated. Thus, information about cost/performance tradeoffs in highly demanding conditions can be made on the basis of slow brain potentials. Provided that further research confirms the sensitivity and reliability of the slow brain potentials as a manifestation of effort, they will certainly augment the batteries of measures tapping important dimensions of mental workload.

REFERENCES

Bauer, H., & Nirnberger, G. (1981). Concept identification as a function of preceding negative or positive spontaneous shifts in slow brain potentials. *Psychophysiology, 18,* 466–469.

Birbaumer, N., Elbert, T., Canavan, A.G.M., & Rockstroh, B. (1990). Slow potentials of the cerebral cortex and behavior. *Physiological Reviews, 70*(1), 1–41.

Boschert, J., Hink, R.F., & Deecke, L. (1983). Finger movement versus toe movement-related potentials: Further evidence for supplementary motor area (SMA) participation prior to voluntary action. *Experimental Brain Research, 52,* 73–80.

Brunia, C.H.M. (1984). Contingent negative variation and readiness potential preceding foot movement. *Annals of the New York Academy of Sciences, 425,* 403–406.

Brunia, C.H.M. (1993). Stimulus preceding negativity: Arguments in favor of nonmotoric slow waves. In W.C.McCallum & S.H.Curry (Eds.), *Slow potential changes in the human brain* (pp. 147–161). New York: Plenum.

Brunia, C.H.M., Voorn, F.J., Berger, M.P.F. (1985). Movement-related slow potentials: II. A contrast between finger and foot movements in left-handed subjects. *Electroencephalography and Clinical Neurophysiology, 60,* 135–145.

Creutzfeldt, O.D. (1995). *Cortex cerebri.* New York: Oxford University Press.

Damen, E.J.P., Freude, G., & Brunia, C.H.M. (1996). The differential effects of extremity and movement side on the scalp distribution of the readiness potential (RP) and the stimulus-preceding negativity (SPN). *Electroencephalography and Clinical Neurophysiology, 99,* 508–516.

Deecke, L. (1976). Opening remarks on motor aspects. In W.C.McCallum & J.R.Knott (Eds.), *The responsive brain* (pp. 91–98). Bristol, England: Wright.

Deecke, L. (1984). Movement-associated potentials and motor control: Introduction. *Annals of the New York Academy of Sciences, 425,* 398–401.

Donchin, E.M., Heffley, E., Hillyard, S.A., Loveless, N.E., Maltzman, J., Öhman, A., Rösler, F., Ruchkin, D., & Siddle, D. (1984). Cognition and event-related potentials: II. The orienting reflex and P300. In J.Karrer, J.Cohen, P.Tueting, & S.Koslow (Eds.), *Event-related potentials in man* (pp. 39–57). New York: New York Academy of Sciences.

Eulitz, C. (1991). *Zum Indikatorwert langsamer Hirnpotentiale und P300 für die Untersuchung von Stimuluserwartung und -bewertung in einer Lernanforderung* [On the indicator value of slow brain potentials and P300 for the analysis of stimulus expectation and evaluation during learning]. Unpublished doctoral dissertation, Humboldt-Universität, Berlin.

Fattaposta, F., Amabile, G., Cordischi, M.V., Di Venanzio, D., Foti, A., Pierelli, F., D'Alessio,

C.D., Parisi, A., & Morrocutti, C. (1996). Long-term effects on a new skilled motor learning: An electrophysiological study. *Electroencephalography and Clinical Neurophysiology, 99,* 495–507.

Freude, G., & Ullsperger, P. (1987a). Das Bereitschaftspotential (BP)-Bewegungskorrelierte Änderungen in der bioelektrischen Hirnaktivität [The Bereitschaft's potential (BP-) movement related changes in bioelectrical brain activity]. *Psychiatrie, Neurologie und Medizinische Psychologie, 39*(8), 449–459.

Freude, G., & Ullsperger, P. (1987b). Changes of the Bereitschaftspotential in the course of muscular fatiguing and non-fatiguing hand movements. *European Journal of Applied Physiology, 56,* 105–108.

Freude, U., & Ullsperger, P. (1994). *Analyse psychischer Beanspruchung anhand von Parametem der bioelektrischen Hirnaktivität* [Analysis mental strain by means of parameters of the bioelectrical brain activity]. Schriftenreihe der Bundesanstalt für Arbeitsmedizin. Fb 01 HK 041. Bremerhaven, Germany: Wirtschaftsverlag NW.

Freude, G., Ullsperger, P., & Erdmann, U. (in press). Slow brain potentials in a visual monitoring task-timing accuracy and pre-movement negativity. *International Journal of Psychophysiology.*

Freude, G., Ullsperger, P., Krüger, H., & Pietschmann, M. (1988). The Bereitschaftspotential in preparation to mental activities. *International Journal of Psychophysiology, 6,* 291–297.

Freude, G., Ullsperger, P., Krüger, H., & Pietschmann, M. (1989). Bereitschaftspotential and the efficiency of mental task performance. *Journal of Psychophysiology, 3,* 377–385.

Freude, G., Ullsperger, P., & Mölle, M. (1995). Application of Bereitschaftspotential for evaluation of effort expenditure in the course of repetitive display work. *Journal of Psychophysiology, 9,* 65–75.

Gaillard, A.W.K. (1993). Comparing the concepts of mental load and stress. *Ergonomics, 36*(9), 991–1005.

Hockey, G.R. (1997). Compensatory control in the regulation of human performance under stress and high workload: A cognitive-energetical framework. *Biological Psychology, 45,* 73–93.

Hockey, G.R.J., Coles, M.G.H., & Gaillard, A.W.K. (1986). Energetical issues in research on human information processing. In G.R.J.Hockey, A.W.K.Gaillard, & M.G.H.Coles (Eds.), *Energetics and human information processing* (pp. 3–21). Dordrecht, The Netherlands: Martinus Nijhoff.

Hohnsbein, J., Falkenstein, M., & Hoormann, J. (1996). Performance differences in reaction tasks are reflected in event-related brain potentials. *Psychophysiology in Ergonomics, 1*(1), 38–39.

Kok, A. (1997). Event-related potential (ERP) reflections of mental resources: a review and synthesis. *Biological Psychology, 45,* 19–56.

Kornhuber, H.H., & Deecke, L. (1964). Hirnpotentialänderungen beim Menschen vor und nach Willkürbewegungen, dargestellt mit Magnetbandspeicherung und Rückwärtsanalyse [Brain potential changes in man before and after voluntary movements demonstrated by means of tape recording and backward analysis]. *Pflügers Archiv, 281,* 52.

Kornhuber, H.H., & Deecke, L. (1965). Hirnpotentialänderung bei Willkürbewegungen und passiven Bewegungen des Menschen: Bereitschaftspotential und reafferente Potentiale [Brain potential changes during voluntary and passive movement in man: Bereitschafts potential and reafferent potentials]. *Pflügers Archiv, 284,* 1–17.

Kramer, A.F. (1991). Physiological metrics of mental workload: A review of recent progress. In D.L.Damos (Ed.), *Multi-task performance* (pp. 279–323). London: Taylor & Francis.

Kramer, A.F., Sirevaag, E.J., & Braune, R. (1987). A psychophysiological assessment of operator workload during simulated flight missions. *Human Factors, 29,* 145–160.

Kristeva, R. (1977). Study of the motor potential during voluntary recurrent movement. *Electroencephalography and Clinical Neurophysiology, 42,* 588.

Lang, W., Zilch, O., Koska, Ch., Lindinger, G., & Deecke, L. (1989). Negative cortical DC shifts preceding and accompanying simple and complex sequential movements. *Experimental Brain Research 74,* 99–104.

McAdam, D.W., & Rubin, E.M. (1971). Readiness potential, vertex positive wave, contingent negative variation and accuracy of perception. *Electroencephalography and Clinical Neurophysiology, 30,* 511–517.

McAdam, D.W., & Seales, D.M. (1969). Bereitschaftspotential enhancement with increased level of motivation. *Electroencephalography and Clinical Neurophysiology, 27,* 73–75.

McCallum, W.C. (1993). Human slow potential research: A review. In W.C.McCallum, & S.H.Curry (Eds.), *Slow potential changes in the human brain* (pp. 1–12). New York: Plenum.

McCallum, W.C., Cooper, R., & Pocock, P.V. (1988). Brain slow potential and ERP changes associated with operator load in a visual tracking task. *Electroencephalography and Clinical Neurophysiology, 69,* 453–468.

McCallum, W.C., & Pleydell-Pearce, C. (1993). Brain slow potential changes associated with visual monitoring tasks. In W.C.McCallum & S.H.Curry (Eds.), *Slow potential changes in the human brain* (pp. 165–189). New York: Plenum.

Mulder, G. (1986). The concept of mental effort. In G.R.J.Hockey, A.W.K.Gaillard, & M.G.H.Coles (Eds.), *Energetics and human information processing* (pp. 175–198). Dordrecht, The Netherlands: Martinus Nijhoff.

Pribram, K., & McGuinness, D. (1975). Arousal, activation and effort in the control of attention. *Psychological Review, 82,* 116–149.

Requin, J., Lecas, A., & Bonnet, J. (1984). A model of motor preparation. In S.Kornblum & J. Requin (Eds.), *Preparatory states and processes* (pp. 260–284). Hillsdale, NJ: Lawrence Erlbaum Associates.

Rockstroh, B., Elbert, T., Canavan, A.G.M., Lutzenberger, W., & Birbaumer, N. (1989). *Slow cortical potentials and behavior.* Munich, Germany: Urban & Schwarzenberg.

Rösler, E, Heil, M., & Glowalla, U. (1993). Monitoring retrieval from long-term memory by slow event-related brain potentials. *Psychophysiology, 39,* 170–182.

Sanders, A.F. (1983). Toward a model of stress and human performance. *Acta Psychologica, 53,* 61–97.

Taylor, M.J. (1978). Bereitschaftspotential during the acquisition of skilled motor task. *Electroencephalography and Clinical Neurophysiology, 45,* 568–576.

Trimmel, M. (1987). Contingent negative variation (CNV) influenced by preceding slow potential shifts. *Electroencephalography and Clinical Neurophysiology, 66,* 71–74.

Walter, W.G., Cooper, R., Aldridge, V.J., McCallum, W.C., & Winter, A.L. (1964). Contingent negative variation: An electric sign of sensorimotor association and expectancy in the human brain. *Nature, 203,* 380–384.

Weber, G., & Bauer, H. (1986). Informationsverarbeitung und Leistungsniveau in Abhängigkeit kortikaler Gleichspannungsschwankungen [Information processing and performance level in dependence on cortical steady potential shifts]. *Zeitschrift für experimentelle und angewandte Psychologie, 33*(1), 164–176.

Wickens, C.W. (1990). Applications of event-related potential research to problems in human factors. In J.W.Rohrbaugh, R.Parasuraman, & R.Johnson (Eds.), *Event-related brain potentials. Basic issues and applications* (pp. 301–309). New York: Oxford University Press.

Chapter 13

Ocular Measures of Fatigue and Cognitive Factors

Erik J.Sirevaag
John A.Stern
Washington University

The gaze control system consists of the mechanisms concerned with the acquisition of visually presented information. Elements of this system include the eyes and associated musculature, the pupils, the eyelids, head and neck muscles, and the neural structures responsible for coordinating the complex task of producing a focused image of a potential information source upon the fovea. By virtue of the control exerted over the flow of visual input, this system is an excellent reflector of aspects of information processing. However, the sensitivity of the system to information processing is not restricted to the selective acquisition of visual information. Auditory information, as well as task difficulty and complexity, can also influence the control of gaze. We first review the literature dealing with issues of fatigue and information processing as they impact on and are reflected in the gaze control system, and we conclude with a discussion of issues that must be considered when attempting to implement measures of oculomotor activity in applied settings.

What can these systems tell us about cognitive processes in general, and "mental fatigue," an admittedly vague concept, in particular? Everyone knows when they feel fatigued, but an objective measure of this construct has yet to be developed. We will assume that work-induced fatigue develops as a function of time-on-task, and that the rate at which fatigue effects appear is a function of the complex interaction between a variety of both subject and task variables.

Subject variables include *state* factors (i.e., sleep history, drug intake, and biological rhythms) and *trait* attributes (i.e., the ability to focus and maintain attention). Task variables include the nature (perceptual/central/motor) as well as the magnitude (or difficulty level) of the demands placed on the operator. A third category of mediating variables emerges as the result of the interaction between a given operator and a particular task. This category includes factors such as operator expertise and the perceived consequences of a breakdown in performance. One task facing the engineering psychologist is to determine the degree to which a given measure is both sensitive to and diagnostic of different variables present in this complex equation.

SACCADIC EYE MOVEMENTS

Eye movement and fixation position (point of regard) data have been successfully employed for decades in order to determine the frequency, duration, and sequential probabilities (link analysis) of looking at particular displays (Fitts, Jones, & Milton, 1950; Senders, Elkind, Grignetti, & Smallwood, 1964). Such data have stimulated a number of attempts to develop quantitative models of the factors underlying instrument panel monitoring behavior (Carbonell, 1966; Kvalseth, 1977; Moray, Neil, & Brophy, 1983; Moray, Richards, & Low, 1980; Senders, 1955,1983). Analyses of visual scanning patterns have also been used to quantify mental effort (Krebs & Wingert, 1976), to contrast the behavior of expert and novice helicopter pilots (Stern & Bynum, 1970), and to characterize optimal scanning strategies in contexts ranging from military environments (Clare, 1979) to the field of radiology (Gale & Worthington, 1983). All of these analyses assume that because changes in the position of the eye control the flow of visual information into the nervous system, eye movement data can be used to generate inferences concerning an operator's strategic, high-level decision-making processes.

We now review data from laboratory studies of fatigue and attention, as well as the literature on eye movements during reading, which indicate that cognitive task demands and operator state variables (such as fatigue) also exert an influence on eye movements. Do saccadic eye movements change as a function of time-on-task? Some of the first systematic observations were made by Dodge (1917) who used a camera to record eye movements as subjects continuously shifted their gaze between two points separated by 60° of visual arc. Dodge observed that, after a series of trials, saccades were slower, terminated less accurately, and became less regular. He attributed these effects to fatigue. A number of investigators have extended these results to saccades subtending a variety of visual angles (40°: Becker & Fuchs, 1969; 30°: McFarland, Holway, & Hurvich, 1942; 20°: Schmidt, Abel, Dell'Osso, & Daroff, 1979; 17°: Specht, 1941).

Bahill and Stark (1975) and Robinson (1981) both report additional time-on-task effects. The frequency of *glissades* (saccades terminating in a slow drifting eye movement) increases with time-on-task. Inefficient initial saccades requiring that another saccade be produced in order to move the eyes to a target location (a phenomenon known as *double saccading*) are also more frequent during later task segments. Schroeder and Holland (1968) evaluated changes in saccades while subjects performed a vigilance task over a 40-minute period. Saccade frequency decreased significantly over time; however no other tempo ral/morphological properties of the saccades were studied.

Saccade frequency has also been shown to significantly decline during the course of a 2-hour air traffic control simulation (Stern, Boyer, Schroeder, Touchstone, & Stoliarov, 1994); saccade amplitude remained constant. Although saccade velocity also decreased over time, McGregor and Stern (1996) subsequently attributed this decrease to a concomitant increase in blink frequency. When the analysis of saccade velocity was restricted to saccades occurring only in the absence of a blink, there was no time-on-task effect on the velocity measure. Saccades during a blink are generally of longer duration than those made in the absence of a blink. We say "generally" because normal velocity saccades can also be observed during a blink. Thus, the phenomenon is not attributable to biological limitations such as friction of the eyeball against the lid. We suspect that this relationship is centrally programmed because saccade duration appears to adjust to blink duration. Slower velocity saccades are seen in conjunction with longer lid closure durations.

A more recent study (Wang & Stern, 1997), in which subjects performed a cognitively demanding vigilance task for a 1-hour period, demonstrated that the velocity of saccades that move the eyes to a location in response to the presentation of task relevant information at peripheral locations (termed *reactive saccades*) does not vary with time-on-task. However, the distributions of saccades that either (a) move the eyes back to a central location following the abstraction of this information *(return saccades),* or (b) move the eyes peripherally in anticipation of the presentation of a stimulus *(anticipatory saccades)* do incorporate an increased frequency of slower velocity saccades during later task segments. To summarize, the velocities of externally paced reactive saccades appear impervious to time-on-task effects. The average velocities of the self-paced return and anticipatory saccades also remain relatively stable; however, an increased frequency of especially slow self-paced saccades occurs during later task segments.

Wang and Stern (1996) found significant increases in *glissadic eye movements* (saccades characterized by either slow initiation times or long termination periods) as a function of time-on-task. As was the case for slow velocity saccades, glissades were limited to self-paced return and anticipatory saccades. These results suggest that the central processor can overcome any effects of fatigue on the mechanisms responsible for generating saccadic movements. Such effects are restricted to self-paced saccades, where the speed and accuracy of eye movements are less crucial than when the eye is moving in response to externally presented information.

In an early study (Hoffman, 1946), students read for a 4-hour period uninterrupted by the presentation of comprehension tests. Under these circumstances, Hoffman obtained significant differences in a number of oculomotor variables as a function of time-on-task. Blink rate increased significantly and the time required to read a line became longer and more variable. Subjects skipped entire sentences and paragraphs and tended to stop reading for several seconds at a time, allowing their eyes to roam over the pages without reading.

A subsequent study (using the same equipment) on reading and visual fatigue was conducted by Carmichael and Dearborn (1947). College and high school students were asked to read for a 6-hour period. Subjects read hard copy text and also used a microfiche reader. The reading matter included material that was "intrinsically interesting" as well as material best described as "dry" (an 18th century treatise on economics). Although subjects changed their attitude toward performing the task over the 6-hour period, no changes in oculomotor activity or reading comprehension were obtained. Carmichael and Dearborn suggest that the crucial difference between their experiment and the earlier one was that frequent comprehension checks required their readers to pay closer attention to the text. They also point out that subjects in the Hoffman study were paid less for their participation. Because of these two factors, Carmichael and Dearborn concluded that subjects were less motivated to actually read the material in the Hoffman study, leading to the markedly different results.

We favor an equally parsimonious explanation, which was considered and rejected by Carmichael and Dearborn. The authors reviewed the literature on performance decrements associated with "homogeneous" versus "varied" tasks and concluded that varied tasks were associated with reduced fatigue levels as compared to homogenous tasks. Their testing paradigm constituted a more varied task than the one used by Hoffman because the comprehension tests administered every 25 pages allowed subjects to rest, however briefly, and change tasks 10 or 12 times during the 6-hour period. A number of authors have reported that short rest breaks allow for the maintenance of high levels of efficiency in vigilance tasks (Bergum & Lehr, 1962; Colquhoun, 1959). For example, Sussman (1970) found that a brief rest period after a few hours of task performance led to an almost complete recovery in the performance of a tracking task. Additional research is required to disentangle whether task variety or motivational factors were responsible for the differences between these two otherwise very similar studies of reading behavior.

Do saccadic eye movements and the time between such movements *(fixation pauses)* change as a function of cognitive task demands or strategies? There is a wealth of literature clearly demonstrating such effects. For example, when required to abstract detailed (as opposed to general) information from text, there are significant changes in the pattern of eye movements dependent upon the skill of the reader (Stern, 1978). Competent readers (as indexed by a reading speed and comprehension test) demonstrate a small increase in fixation pause duration, but no change in the number of forward going fixations per line of text. Less competent readers, on the other hand, demonstrate no change in fixation pause duration, but restrict the size of the informational chunk taken in at one time by increasing the number of fixation pauses per line of text.

In situations where the location and timing of stimuli are predictable, subjects are able to move their eyes to foveate an anticipated source of information well in advance of the actual delivery of this information. Put another way, when expecting an imperative stimulus with highly predictable spatiotemporal characteristics, subjects are able to generate anticipatory saccades. Conversely, saccadic activity associated with nonpredictable stimuli is more reactive in nature. We have found that aspects of anticipatory saccades vary significantly when subjects expect that an impending stimulus is likely to require a response. The expectation of an imperative stimulus increases the probability that an anticipatory saccade will occur. Furthermore, anticipatory saccades preceding imperative stimuli are also more accurate (requiring smaller amplitude post-stimulus corrective saccades or no corrective saccades to foveate the stimulus) than those produced in anticipation of nonimperative stimuli.

These are but a few examples of how the analysis of eye movements can aid the engineering psychologist wishing to characterize the manner in which visual information is abstracted from the environment. Interestingly, the operator is seldom aware of (or at least is unable to verbalize) the nature of these processes. Thus, readers generally report that they read more slowly and make more regressive eye movements when reading for detailed as compared to general information. They do not, however, report that they either spend more time on each informational chunk or take in smaller informational chunks.

EYE BLINKS

What can blinks tell us about fatigue and aspects of information processing? Stern, Boyer, and Schroeder (1994) reviewed the literature dealing with blink rate and fatigue. There is little to add to that review. They concluded that an increased blink rate associated with increased time-on-task is a well established phenomenon. Predictably, given the conflicting patterns of saccadic results detailed earlier, Hoffman (1946) obtained a significant increase in blink rate while subjects read continuously for a 4-hour period; Carmichael and Dearborn (1947), with Hoffman collecting the data, found no change in blink rate over a 6-hour reading period. Once again, it is difficult to know whether this discrepancy should be attributed to motivational or task-pacing differences between the two studies.

A series of articles by Luckiesh (1943, 1946, 1947) reporting an increase in blink frequency as a sign of visual fatigue were severely criticized by Tinker (1943a, 1943b, 1945, 1946) and Bitterman and Soloway (1946). However, the paradigms used to refute Luckiesh's findings employed procedures in which data collection was periodically interrupted by having subjects perform a second task in which their comprehension of the reading material was tested. Once again, studies finding an effect of fatigue upon an oculomotor parameter entailed the continuous performance of a visually demanding task; studies in which the reading task was periodically interrupted failed to elicit fatigue effects.

Stern et al. (1994) point out that fatigue is only one of a host of variables affecting blink rate. The timing and frequency of stimulus presentation is also critically important. Thus, when stimuli are presented at predictable intervals, blink production tends to become externally paced and greater interstimulus intervals (within limits) are associated with decreased blink rates. Blink rate is also sensitive to factors such as visual information-processing load and perceived risk; tasks that are more demanding lead to a greater inhibition of blinking than less demanding tasks. Therefore, the blink rate of a driver in a complex urban environment will be considerably lower than when the same individual is driving on a rural interstate. In a military context utilizing a flight simulator, Stern and Skelly (1984) found that blink rates associated with flying over friendly skies were higher than when the flights took place over enemy territory. Similarly, Wilson (1993) obtained significant differences in blink rate for different segments of an air-to-ground training mission.

For this reason, Stern, Boyer, Schroeder, Touchstone, and Stoliarov (1996) have suggested that aspects of blinking other than rate may be more diagnostic of time-on-task effects. For example, blink closure duration (or the period of time during which the pupil is obscured by the lid during a blink) demonstrates significant time-on-task effects under many conditions. In the Stern et al. (1996) study involving 2 hours of performance in an air traffic control simulation, significant increases in blink closure duration were obtained. Lid closures less than 300 milliseconds were categorized as "normal blinks," those ranging between 300 and 500 milliseconds were termed "long closure duration blinks," and closures longer than 500 milliseconds were labeled "eye closures." All three of these measures indicate that the amount of time the eye is closed increases as a function of time-on-task. These authors also suggested an additional measure of relevant blink activity, namely the frequency of blink flurries. *Blink flurries,* defined as the occurrence of a minimum of three blinks with interblink intervals less than 1 second, were also found to occur more frequently with increased time-on-task.

Morris and Miller (1996) evaluated oculometric variables in pilots operating a flight simulator after a night of sleep deprivation and found several relationships between time-on-task, oculomotor measures, and flight performance. Blink rate, blink duration, long closure duration blink rate, blink amplitude, and saccade rate all correlated significantly with flight performance. A stepwise regression analysis was performed to evaluate how well the oculomotor variables predicted performance errors. The best univariate predictor was blink amplitude (R-square=0.36). The addition of the long closure duration blink rate to the regression equation increased the R-square value to 0.54. Including a third oculomotor variable, blink closure duration, further increased the R-square to 0.61. These authors suggest that the decreases in blink amplitude found to predict performance errors were most likely related to eyelid "droop" associated with sleepiness. These results strongly suggest that oculometric variables can be used to index aspects of performance inefficiency.

When the temporal characteristics of stimuli that operators are required to process are predictable, the timing of blinks with respect to these stimuli can be useful in determining how long it has taken the subject to perceive and interpret the information presented. For example, blink latency following the visual presentation of items to be memorized is significantly affected by the number of items. A larger set of items leads to a longer delay in blinking than a smaller set (Stern & Dunham, 1990). We have interpreted these and other results dealing with the timing of blinks following stimulus presentation as suggesting that individuals blink at times where the intake and processing of information can be momentarily inhibited.

An additional finding is that poststimulus blink latencies get shorter during later task segments. This is as true for the timing of blinks following nonvisual stimuli requiring a simple discrimination (Tone a is longer than Tone b) as it is for more involved decision-making tasks (such as being required to make a manual response following the presentation of three consecutively presented even integers). Our interpretation of this effect is that the ability to inhibit blinking is compromised by "fatigue" effects. However, it is difficult to ascribe this effect solely to fatigue. The possibility that the task becomes easier (i.e., makes fewer demands on the operator) as a function of increased exposure cannot easily be dismissed. Similarly, subjects may become more "comfortable" with the constraints imposed on them by the recording environment during later task segments, and the observed reductions in blink latencies may simply reflect this process of adaptation.

HEAD MOVEMENTS

What can head movements coupled to eye movements tell us about aspects of cognition and fatigue? Eye—head coordination in the acquisition of visual information is not a simple function of the angular distance gaze must shift in order to acquire information. The nature of the information to be acquired, the predictability of the location of the information, and task difficulty all affect both the proportion of gaze shift accomplished by the head movement and the timing of head movements with respect to the movement of the eyes. Bizzi, Kalil, Morazzo, and Tagliaso (1972) found in the monkey, and Zangemeister and Stark (1982) found in humans, that the timing of head movements with respect to eye movements is significantly affected by the predictability of target movements. With unpredictably moving targets, eye movements precede head movements (termed *classical gaze shifts*). When targets appear at predictable locations, the head frequently begins to move before the eye (labeled *predictive gaze shifts*). During classical gaze shifts, head movements are initiated approximately 20 to 40 milliseconds after saccade initiation. Predictive gaze shifts have much more variable head and eye movement initiation patterns.

Moschner and Zangemeister (1993) conclude that eye-head coordination is controlled by "high level preview control mechanisms" and suggest that the subject's intention is an important factor in accounting for these relationships. Successful prediction of target location and the associated eye-head movement coordination allows for decreasing the latency with which target information is acquired and enhances the accuracy with which gaze is shifted to the target lo cation. In most of these studies, the eccentricity of targets was relatively large, usually in excess of 40°.

A different literature has addressed the roles played by head and eye movements during reading. Fischer (1924) observed that during reading there are almost always head movements. Of these, left head movements (presumably associated with shifting gaze from the end of one line to the beginning of the next) are most apparent; movements to the right are slower and less easily observed. According to Fischer, a report by Erdmann and Dodge indicated that unpracticed readers are more likely to make upper body and head movements than practiced readers. Netchine and collaborators (1981, 1984, 1987) have studied head and eye movements in children and adults as they read silently and out loud. They report that for both age groups the likelihood of head movements is greater during oral than silent reading, and that the proportion of gaze control accomplished with head movements is smaller in adults than in children. When text difficulty was manipulated, these authors found that the likelihood of head movements increased in both groups as difficulty increased.

What effect does inhibiting head movements have on reading? Salel and Gabersek (1975) reported that restraining the head had an adverse effect on reading performance. These results are reminiscent of studies demonstrating that the inhibition of subvocalization (in persons who subvocalize during reading) may significantly increase reading speed at the cost of impaired comprehension (Hardyck & Petrinovich, 1970). Thus, head movements during reading may index an aspect of reading skill, but should not be treated as a cause of reading-skill difficulties. Readers will usually express surprise upon being made aware of head movements. According to Netchine, these head movements are predictive in nature (they are initiated before the eye movement starts and completed after the saccade has terminated) and must be centrally programmed since the associated eye movements do not include a compensatory saccade to adjust for the change in head position.

Mourant and Grimson (1977) evaluated gaze shifts in automobile drivers as they made lane changes and merged with highway traffic. These authors evaluated three young novice, three young experienced, and three mature drivers and came to the conclusion that the occurrence of predictive gaze shifts increases with driving skill development. They suggest that drivers should be taught to make predictive head movements. Given the adverse consequences of constraining head movements reported by Salel and Gabersek (1975), additional research is required to determine whether a policy of training predictive head movements is, in fact, advisable.

Combined analyses of head and eye movement data have also been enployed in the evaluation of user interfaces in complex human-machine systems. For example, a study by Sirevaag et al. (1993) examined the processing demands imposed by two different communication formats (digital and verbal) in a high fidelity stimulation of an advanced multifunction helicopter. Analyses of videotaped head and eye movement data indicated that the digital display format increased the time pilots spent in a heads down (instrument scanning) posture. Increased task shedding (failure to perform certain mission critical subtasks) during digital communication segments was also confirmed via the analysis of the videotaped data.

PUPIL DIAMETER

We now turn to a review of the literature in which changes in the diameter of the pupil are associated with cognition, fatigue, and sleepiness. It is somewhat surprising that relatively little attention has been paid to this measure over the years. One possible reason for this is the difficulty involved in quantifying the diameter of the pupil (a difficulty that has largely been eliminated by the advent of improved image processing technologies and algorithms). The knowledge that pupil diameter changes reflect variables other than the amount of light falling on the eyes has been suggested since antiquity. Women in the middle ages used belladonna, a drug that dilated the pupils, to make themselves look more attractive. The earliest systematic studies evaluated the relationship between pupil diameter and affective attributes. Hess and Goodwin (1974) present data demonstrating that larger pupils are associated with the judgment that a pictured face is happier and/or older when compared to the same face with smaller pupils.

Excellent reviews of the impact of cognitive manipulations on pupil diameter can be found in Beatty (1982) and Janisse (1977). The literature is quite clear; pupil diameter increases as cognitive load increases. For example, committing six numbers to memory produces greater dilation than memorizing just two numbers. Requiring subjects to repeat the numbers committed to memory in a backward order leads to an even greater dilation than the instruction to simply repeat them. We restrict our discussion to papers not covered by the aforementioned reviews.

Most studies since 1982 have confirmed the direct relationship between pupil diameter and task difficulty. Just and Carpenter (1995) utilized a lexical decision task employing sentences of varying complexity and concluded that sentences that required more thought before a response (as reflected in increased response latencies) were associated with greater increases in pupil diameter than sentences to which responses could be made more rapidly. Similarly, Schluroff et al. (1986) found that pupil dilation effects were responsive to syntactic complexity. Finally, Backs and Walrath (1992) demonstrated that pupil diameter was sensitive to the level of visual information processing demands.

A number of studies have evaluated what happens when task requirements exceed the capacity of the operator. For example, Granholm, Asarnow, Sarkin, and Dykes (1996) developed a short-term memory task in which subjects recalled lists of digits. Memory load was varied from low (5 digits) to moderate (9 digits) to excessive (13 digits). They concluded that pupil dilation increased systematically as a function of load until "the limits of available resources" were reached. As these limits were approached, pupil dilation became asymptotic and then declined as resource capacity was exceeded.

Other literature has explored the relationship between the pupil and processes related to anticipation (or expectancy). Richer and Beatty (1987) reported increases in pupil diameter associated with increases in the number of response alternatives. For stimuli requiring a response (Go trials), pupil dilation was linearly related to response uncertainty (reflected by the number of alternative responses); however, for stimuli not requiring a response (No Go trials), pupil dilation was inversely related only to the overall response probability.

A slightly different paradigm, in which only response probability was manipulated (from 25% to 50% to 75%), was employed by van der Molen, Boomsma, Jennings, and Nieuwboer (1989). These authors report that an increase in the probability of Go trials produced a small decrease in pupil dilation responses and a significant increase in peak dilation for No Go trials (although the amplitude of the pupillary response for Go trials was significantly greater than that for No Go trials). This pattern of results is further complicated by the findings of Richer and Beatty (1985) who determined that self-initiated finger movements produced increases in pupil diameter as a function of both movement complexity and force requirements.

Less ambiguous evidence supporting a relationship between anticipatory processes and pupil diameter is provided by Qiyuan, Richer, Wagoner, and Beatty (1985). These authors demonstrated that task-evoked pupillary dilations were sensitive to the surprise value of an event. Importantly, the omission of an expected stimulus produced as large a dilation as the occurrence of an unanticipated stimulus. Although the literature mentioned previously supports the hypothesis that pupil diameter is sensitive to anticipation, more studies in which subject expectancies are explicitly manipulated in a variety of ways are clearly needed.

Other laboratories have examined the relationship between pupillary changes and fatigue or sleepiness. Lowenstein and Loewenfeld (1964; see also Loewenfeld, 1993) draw several conclusions based primarily on incidental observations:

1. Pupil diameter decreases significantly as a subject goes from wakefulness to sleep.
2. The pupil in sleep-deprived, tired subjects is smaller than the pupil of subjects studied under nonsleep-deprived, alert conditions.
3. Spontaneous pupil diameter changes (fluctuations) are seen in tired subjects under conditions of darkness.

Yoss, Moyer, and Hollenhorst (1970) attempted to develop more objective procedures for the evaluation of these spontaneous fluctuations in pupil diameter and the resulting measures were related to narcolepsy as well as to feelings of fatigue. These spontaneous pupillary oscillations are referred to as *pupillary hippus* by Yoss et al. (although Loewenfeld, 1993, took exception to this term). The general procedure for quantifying pupillary hippus is to measure pupil diameter with the subject looking at a dim light spot for a 15-minute period while seated in a dark room. Under these conditions, the pupil reportedly remains large and stable in alert subjects.

As subjects become bored or fatigued, their eyelids droop, heads nod, and pupils constrict. Waves of dilation, reportedly associated with spontaneous arousal, occur and fade with the deepening of drowsiness. One study, which empirically related the occurrence of rhythmic fluctuations in pupil diameter with a measure of sleepiness (the Stanford Sleepiness Scale), examined three narcoleptic patients and three control subjects. Pressman et al. (1980) report that pupillary hippus (with a cycle length of 90–120 minutes) was seen only in the narcoleptic subjects. These authors also report that the narcoleptics generally exhibited smaller pupil diameters than controls, but that at the peak of hippus their pupils were as large as those of normal subjects. Interestingly, subjective reports of alertness were maximized when the pupils were large. The lowest subjective alertness ratings were obtained when the pupil was at the nadir of the hippus cycle. Fluctuations were not seen in the control subjects.

These "fatigue" waves are different from those described by Lowenstein and Lowenfeld (1964) who reported two types of waves: (a) slow waves, from 4 to 40 seconds in duration, with amplitudes less than 0.5 mm, and (b) faster waves, approximately 1 second in duration, with amplitudes no larger than 0.3 mm. Lowenstein hypothesizes that these waves may be related to efforts on the part of the subject to remain awake; with spontaneous recovery from sleep or drowsiness, the pupil dilates rapidly, followed by a slower constriction. He further suggests that these waves disappear as subjects fall asleep. In patients suffering from chronic fatigue, Lowenstein observed exaggerated fatigue waves. Similar waves also occurred more frequently in older subjects.

The study with the highest relevance to the area of engineering psychology was conducted by Yoss, Moyer, Carter, and Evans (1970), who examined 50 commercial airline pilots, 32 of whom were well rested whereas 18 were inadequately rested. Pilots were required to sit in the dark for 15 minutes and pupil diameter was continuously measured. The pupillary fluctuations were then rated by the investigators for alertness using a 4-point rating scale. Of the rested pilots, 12% received ratings of marginal or unsatisfactory; for the inadequately rested pilots, 50% received such ratings.

McLaren, Erie, and Brubaker (1992) utilized spectral analysis for the evaluation of hippus in pupillograms and correlated these analyses with physician ratings of pupillary indications of alertness using the 4-point scale developed by Yoss. Correlations between these two measures was highest (0.77) when they compared spectral energy in the 0.02 and 0.04 Hz frequency bands with the ratings. The correlation decreased when higher frequency bands were used. For example, the correlation between 0.25–0.27 Hz activity and the physician ratings dropped to 0.49. These authors also evaluated changes in pupil diameter over the 15-minute period. A measure reflecting the slope of pupil diameter decrease *(miosis)* over the 15-minute period, when used in combination with the spectral power in the low frequency bands, was highly correlated ($r=0.92$) with the physician ratings.

Newman and Broughton (1991) evaluated differences in pupillary activity between 10 narcoleptics and matched controls. No differences were obtained for pupil diameter under baseline conditions. Pupillary light reflexes and the pupillary orienting responses also failed to discriminate the groups. Narcoleptics, however, demonstrated significantly more spontaneous fluctuations in the dark than normals. Furthermore, Stanford Sleepiness Scale scores did not correlate with pupillary oscillations for either the control or the narcoleptic subjects.

An abstract by Pressman et al. (1980) starts with the following observations: "The measurement of pupillary size and activity is a well established indicator of alertness and sleepiness. Large and stable pupils are associated with alertness while constricted pupils and hippus are associated with decreasing alertness including pathological sleepiness" (p. 218). If true, an index of pupil diameter would constitute a promising candidate for a general alertness or fitness for duty measure. Our review of this literature leaves us somewhat less than satisfied that the relationship between pupillary fluctuations and levels of alertness is well established. Furthermore, we find that a variety of other factors also influence the pupil. Nociceptive stimulation induces rhythmic changes in pupil diameter (Tassorelli, Micieli, Osipova, Rossi, & Nappi, 1995), and respiratory fluctuations have also been related to pupil diameter oscillations (Yoshida, Yana, Okuyama, & Tokoro, 1994). Ohtsuka, Asakura, Kawasaki, and Sawa (1988) report that respiratory inspiration is associated with mydriasis and the expiratory phase with miosis. Withdrawal from heroin can produce spontaneous pupil diameter fluctuations (Grunberger et al., 1990), whereas physical illnesses, such as diabetes, significantly reduce pupillary fluctuations during darkness (Hreidarsson & Gundersen, 1988). Thus, although pupil diameter (and hippus) may be somewhat sensitive to the general level of alertness, the sensitivity to a variety of other variables limits the diagnostic utility of this measure.

All of the studies reviewed earlier evaluated pupil diameter changes as subjects sat in a dark room and focused on a dim light. A few studies have explored such changes under conditions of relatively constant levels of illumination as subjects performed a variety of cognitive tasks. Libby, Lacey, and Lacey (1973) report a significant decrease in pupil diameter over the course of an experiment lasting somewhat longer than 40 minutes. Peavler (1974) studied fatigue in telephone operators and was able to relate subjective feelings of fatigue to pupil diameter. Individuals reporting greater feelings of fatigue exhibited larger constrictions of the pupil (as indexed by pre- and postwork measures of pupil diameter) than operators experiencing less fatigue. A study by Peavler and Nellis (summarized in Janisse, 1977) evaluated pupil diameter in typists before and after 75 minutes of consecutive typing under ambient lighting conditions. Pupil diameter was significantly reduced late as compared to early in task performance. In experimental sessions lasting somewhat longer than 1 hour, Jaschinski et. al. (1996) reported significant miosis over time.

The literature on pupillary responses within the past 10 years is consistent in demonstrating that time-on-task effects, increased cognitive load, response uncertainty, and required level of motor activity are reflected in pupil diameter changes. We have examined the relationship between time-on-task and pupil size. Comparing pupil diameter during the initial 15 minutes of task performance to that for the last 15 minutes of a 1-hour long vigilance task, 11 of 14 subjects demonstrated a reliable reduction in pupil diameter over time. Thus, there is reasonably robust evidence indicating that pupil diameter decreases with increased time-on-task. However, as noted earlier, it is difficult to ascribe this affect solely to fatigue as alternative explanations related to learning and adaptation are also possible.

We have also studied changes in pupil diameter as a function of processing load and level of anticipation. We find that the pupil dilates as a function of the expectancy that a manual response will soon be required. In addition, the diameter of the pupil continues to increase during the enactment of the response. Finally, the diameter of the pupil tends to decrease at points during the task where subjects normally blink (even if no blink occurs). It is our hypothesis that blinks occur at points in time where one can take a time-out from taking in and processing information. Blinks are not obligatory at these points in time, but are most likely to occur here. Because we cannot measure pupil diameter during a blink, the number of blink-free trials where the pupil can be measured during this critical interval is smaller than during other points in time. However, we consistently find pupil constriction at these time points even though the number of trials available for pupil diameter measurement is often quite small. These pupillary responses appear to be impervious to time-on-task effects. They are as prominent during the last 15 minutes of task performance as during the first 15 minutes of a 60-minute task.

Pupil diameter measures appear to robustly reflect cognitive load. In our experience, currently available equipment can reliably measure pupillary dilations and constrictions on a single trial basis, obviating the need for signal averaging. We are presently examining whether changes in pupil diameter are predictive of local performance decrements. We would expect, for example, that pupil dilation in advance of subsequently missed targets (the occurrence of which the subject should have been able to anticipate) will be significantly smaller than on trials where the target is correctly identified.

CONCLUSION

Clearly, this review confirms that gaze control measures are sensitive to aspects of information processing in general, and fatigue, in particular. However, the plethora of factors mediating the control of gaze limits the degree to which these measures can be considered selectively diagnostic of specific psychological states and processes. This problem is often encountered when interpreting physiological measures and can be dealt with in at least two ways. First, researchers should not rely on a limited subset of measures to characterize complex processes. Rather, an assessment battery in which oculomotor measures augment information obtained from other sources, such as subjective reports, overt performance measures, and other physiological systems should be employed. Second, experiments should be well designed and controlled to isolate the factors of interest The difficulties in identifying the factor responsible for the divergent findings obtained by Hoffman (1946), Carmichael and Dearborn (1947), Luckiesh (1946), and Tinker (1946) illustrate how relatively minor differences in experimental design can critically impact obtained results.

A major strength of oculomotor monitoring, of special importance to the engineering psychologist, is that measurements can be obtained using unobtrusive sensors. Hopefully, in the not-to-distant future, relatively inexpensive technologies for remotely monitoring blinks, saccades, pupil diameter, and head position will be available. High bandwidth is another great advantage of gaze control measures. Because these metrics can be obtained frequently and in the absence of overt responses they can be used to address a variety of important issues.

One such issue concerns whether fatigue evolves as a linear (or monotonic) function of time-on-task or is better described as producing intermittent periods of impaired performance that may increase in frequency and duration as a function of time-on-task. There is some empirical support for the latter position. In subject-paced tasks, periods characterized by unusually slow responses (*performance blocks* defined as local response latencies deviating from the mean by more than two standard deviations) increase in both frequency and duration as a function of time-on-task (Bills, 1931). Bills reported that sleep loss and subjective reports of fatigue were associated with increases in the frequency of such performance blocks. Response latencies for events preceding the block were also generally longer than the average, whereas those following a block were generally shorter. These blocks typically lasted less than 1 second and occurred at rates as low as 3 per minute. Similarly, Carmichael and Dearborn (1947) had students read for 6 hours and found that the major signs of fatigue were tempo rary blocks (i.e., brief periods of interruption in the reading of text).

This state of affairs clearly poses a problem for studies in which fatigue effects are examined by averaging across performance intervals subtending many minutes (as is often the case). Although averaging increases measurement reliability, it may also miss or at least obfuscate the underlying structure of how fatigue-related performance decrements develop over time. Future exploration of the microarchitecture of moment-to-moment changes in the control of gaze may provide answers to this question and many others of relevance to the fields of engineering psychology and human performance assessment.

REFERENCES

Backs, R.W., & Walrath, L.C. (1992). Eye movement and pupillary response indices of mental work load during visual search of symbolic displays. *Applied Ergonomics, 23,* 243–254.

Bahill, A.T., & Stark, L. (1975). Overlapping saccades and glissades are produced by fatigue in the saccadic eye movement system. *Experimental Neurology, 48,* 95–106.

Beatty, J. (1982). Task evoked pupillary responses, processing load, and the structure of processing resources. *Psychological Bulletin, 91,* 276–292.

Becker, W., & Fuchs, A. (1969). Further properties of the human saccadic system: Eye movements and correction saccades with and without visual fixation points. *Vision Research, 9,* 1247–1258.

Bergum, B.O., & Lehr, D.J. (1962). Vigilance performance as a function of interpolated rest. *Journal of Applied Psychology, 46,* 425–430.

Bills, A.G. (1931). Blocking: A new principle of mental fatigue. *American Journal of Psychology, 43,* 230–245.

Bitterman, M.E., & Soloway, E. (1946). The relation between frequency of blinking and effort expended in mental work. *Journal of Experimental Psychology, 36,* 134–136.

Bizzi, E., Kalil, R.E., Morasso, P., & Tagliasco, V. (1972). Central programming and peripheral feedback during eyehead coordination in monkeys. In J.Dichgans & E.Bizzi (Eds.), *Cerebral control of eye movements and motion perception* (pp. 220–232). Basel, Switzerland: Karger Verlag.

Carbonell, J.R. (1966). A queueing model for many-instrument visual sampling. *IEEE Transactions on Human Factors in Electronics, HFE-7,* 157–164.

Carmichael, L., & Dearborn, W.F. (1947). *Reading and visual fatigue.* Boston: Houghton Mifflin.

Clare, J.N. (1979). Recognition as an extension to the detection process. In J.N.Clare & M.

A.Sinclair (Eds.), *Search and the human observer* (pp. 279–283). London: Taylor & Francis.

Colquhoun, W.P. (1959). The effect of a short rest pause in inspection efficiency. *Ergonomics, 2,* 367–376.

Dodge, R. (1917). The laws of relative fatigue. *Psychological Review, 24,* 89–113.

Fischer, F.P. (1924). Concerning head movements associated with visual information acquisition. *Archive fur Opthamologie, 113,* 394–416.

Fitts, P.M., Jones, R.E., & Milton, J.L. (1950). Eye movements of aircraft pilots during instrument landing approaches. *Aeronautical Engineering Review, 9,* 1–5.

Gale, A.G., & Worthington, B.S. (1983). The utility of scanning strategies in radiology. In R. Groner, C.Menz, D.F.Fisher, & R.A.Monty (Eds.), *Eye movements and psychological functions: International views* (pp. 169–191). Hillsdale, NJ: Lawrence Erlbaum Associates.

Granholm, E., Asarnow, F.R., Sarkin, A.J., & Dykes, K.L. (1996). Pupillary responses index cognitive resource limitations. *Psychophysiology, 33,* 457–461.

Grunberger, J., Linzmayer, L., Fodor, G., Presslich, O., Praitner, M., & Loimer, N. (1990). Static and dynamic pupillometry for determination of the course of gradual detoxification of opiate addicted patients. *European Archive of Psychiatry and Clinical Neuroscience, 240,* 109–112.

Hardyck, C.D., & Petrinovich, L.F. (1970). Subvocal speech and comprehension level as a function of the difficulty level of reading material. *Journal of Verbal Learning and Verbal Behavior, 9,* 647–652.

Hess, E.H., & Goodwin, E. (1974). The present state of pupillometrics. In M.P.Janisse (Ed.), *Pupillary dynamics and behavior* (pp. 209–248). New York: Plenum.

Hoffman, A.C. (1946). Eye movements during prolonged reading. *Journal of Experimental Psychology, 36,* 95–118.

Hreidarsson, A.B., & Gundersen, H.J. (1988). Reduced pupillary unrest. Autonomic nervous system abnormality in diabetes mellitus. *Diabetes, 37,* 446–451.

Janisse, M.P. (1977). *Pupillometry: The psychology of the pupillary response.* New York: Wiley.

Jaschinski, W., Bonaccker, M., & Alshuth, E. (1996). Accommodation, convergence, pupil diameter and eye blinks at a CRT display flickering near the fusion limit. *Ergonomics, 39,* 152–164.

Just, M.A., & Carpenter, P.A. (1995). The intensity dimension of thought: Pupillometric indices of sentence processing. In J.M.Henderson, M.Singer, & F.Ferreira (Eds.), *Reading and language processing.* Mahwah, NJ: Lawrence Erlbaum Associates.

Krebs, M.J., & Wingert, J.W. (1976). *Use of the oculometer in pilot workload measurement* (Report No. NASA CR-144951). Washington, DC: National Aeronautics and Space Administration.

Kvalseth, T. (1977). Human information processing in visual sampling. *Ergonamics, 21,* 439–454.

Libby, W.L., Jr., Lacey, B.C., & Lacey, J.I. (1973). Pupillary and cardiac activity during visual attention. *Psychophysiology, 10,* 270–294.

Loewenfeld, I.E. (1993). *The pupil: Anatomy, physiology and clinical applications.* Ames IA/ Detroit, MI: Iowa State University Press/Wayne State University Press.

Lowenstein, O., & Loewenfeld, I.E. (1964). The sleep-waking cycle and pupillary activity. *Annah New York Academy of Science, 117,* 142–156.

Luckiesh, M. (1943). Some comments on Dr. Tinker's review of "Reading as a visual task." *Journal of Applied Psychology, 27,* 360–362.

Luckiesh, M. (1946). Comments on criteria of ease of reading. *Journal of Experimental Psychology, 36,* 180–181.

Luckiesh, M. (1947). Reading and the rate of blinking. *Journal of Experimental Psychology, 37,* 266–268.

McFarland, R.A., Holway, A.H., & Hurvich, L.M. (1942). *Studies of Visual Fatigue.* Boston: Harvard Graduate School of Business Administration.

McGregor, D.K., & Stern, J.A. (1996). Time-on-task and blinks: Effects on saccade duration. *Ergonomics, 39,* 649–660.

McLaren, J.W., Erie, J.C., & Brubaker, R.F. (1992). Computerized analysis of pupillograms in studies of alertness. *Investigative Ophthalmology and Visual Science, 33,* 671–676.

Moray, N., Neil, G., & Brophy, C. (1983). *The behaviour and selection of fighter controllers* (Tech. Rep.). London: Ministry of Defense.

Moray, N., Richards, M., & Low, J. (1980). *The behaviour of fighter controllers* (Tech. Rep.). London: Ministry of Defense.

Morris, T.L., & Miller, J.C. (1996). Electrooculographic and performance indices of fatigue during simulated flight. *Biological Psychology, 42,* 343–360.

Moschner, C., & Zangemeister, W.H. (1993). Preview control of gaze saccades: Efficacy of prediction modulates eye-head interaction during human gaze saccades. *Neurological Research, 15,* 417–432.

Mourant, R.R., & Grimson, C.G. (1977). Predictive head movements during automobile mirror sampling. *Perceptual and Motor Skills, 44,* 283–286.

Netchine, S., Greenbaum, C., & Guihou, M.-C. (1984). Cephalic and ocular components of gaze displacement during oral and silent reading in children and adults. In A.G.Gale & F. Johnson (Eds.), *Theoretical and applied aspects of eye movement research* (pp. 223–230). Amsterdam: Elsevier.

Netchine, S., Pugh, A.K., & Guihou, M.-C. (1987). The organization of binocular vision in conjunction with head movements in French and English readers of 9 and 10 years. In J. K.O'Reagan & F.Levy-Schoen (Eds.), *Eye movements: From physiology to cognition* (pp. 333–342). Amsterdam: Elsevier.

Netchine, S., Solomon, M., & Guihou, M.-C. (1981). Composants oculairs et cephaliques de l'organisation des deplacements du regards chez les jeunes lecteurs [Cephalic and ocular components of gaze displacement in children while reading]. *Psychologie Francaise, 26,* 110–124.

Newman, J., & Broughton, R. (1991). Pupillometric assessment of excessive daytime sleepiness in narcolepsycataplexy. *Sleep, 14,* 121–129.

Ohtsuka, K., Asakura, K., Kawasaki, H., & Sawa, M. (1988). Respiratory fluctuations of the human pupil. *Experimental Brain Research, 71,* 215–217.

Peavler, W.S. (1974). Individual differences in pupil size and performance. In M.Janisse (Ed.), *Pupillary dynamics and behavior* (pp. 159–176). New York: Plenum.

Pressman, M.R., Spielman, A.J., Korczyn, A., Rubenstein, A., Weitzman, E.D., & Pollak, C. (1980). Pupillometry in normals and narcoleptics throughout the course of a day. *Sleep Research. Los Angeles Brain Information Service/Brain Research Institute, 9,* 218a.

Qiyuan, J., Richer, F., Wagoner, B.L., & Beatty, J. (1985). The pupil and simulus probability. *Psychophysiology, 22,* 530–534.

Richer, F., & Beatty, J. (1985). Pupillary dilation in movement preparation and execution. *Psychophysiology, 22,* 204–207.

Richer, F., & Beatty, J. (1987). Contrasting effects of response uncertainty on the task-evoked pupillary response and reaction time. *Psychophysiology, 24,* 258–262.

Robinson, D.A. (1981). Control of eye movements. In V.B.Brooks (Ed.), *Handbook of physiology, Section I: The nervous system: Vol. II. Motor control, Part 2* (pp. 1275–1320). Bethesda, MD: American Physiological Society.

Salel, D., & Gabersek, S.J. (1975). Cinq aspects due comportement de l'enfant en cours d'apprentissage de la lecture [Five aspects of behavior in children while studying reading]. *Revue d'EEG et de Neurophysiologie Clinique, 5,* 345–350.

Schluroff, M., Zimmerman, T.E., Freeman, R.B., Hofmeister, K., Lorscheid, T., & Weber, A. (1986). Pupillary responses to syntactic ambiguity of sentences. *Brain and Language, 27,* 322–344.

Schmidt, D., Abel, L., Dell'Osso, L.F., & Daroff, R.B. (1979). Saccadic velocity characteristics: Intrinsic variability and fatigue. *Aviation, Space and Environmental Medicine, 50,* 393–395.

Schroeder, S.R., & Holland, J.G. (1968). Operant control of eye movements during human vigilance. *Science, 161,* 292–293.

Senders, J.W. (1955). Man's capacity to use information from complex displays. In H. Quastler (Ed.), *Information theory in psychology.* Glencoe, IL: Free Press.

Senders, J.W. (1983). *Visual scanning processes.* Tilburg, Netherlands: University of Tilburg Press.

Senders, J.W., Elkind, J.E., Grignetti, M.C., & Smallwood, R.P. (1964). *An investigation of visual sampling behavior of human observers* (NASA-CR-434). Cambridge, MA: Bolt, Beranek, & Newman.

Sirevaag, E.J., Kramer, A.F., Wickens, C.D., Reisweber, M., Strayer, D.L., & Grenell, J.F. (1993). Assessment of pilot performance and mental workload in rotary wing aircraft. *Ergonomics, 36*(9), 1121–1140.

Specht, H. (1941). Fatigue and hours of service of interstate truck drivers. Eye movements and related phenomena. *U.S. Public Health Bulletin, 265,* 209–225.

Stern, J.A. (1978). Eye measurements, reading and cognition. In J.W.Senders, D.F.Fisher, & R.A.Monty (eds.), *Eye movements and the higher psychological functions* (pp. 145–155). Hillsdale, NJ: Lawrence Erlbaum Associates.

Stern, J.A., Boyer, D.J., & Schroeder, D. (1994). Blink rate: A possible measure of fatigue. *Human Factors, 36*(2), 285–297.

Stern, J.A., Boyer, D.J., Schroeder, D., Touchstone, M., & Stoliarov, N. (1994). Blinks, saccades, and fixation pauses during vigilance task performance: I. Time-on-task. DOT/FAA/AM9426. FAA. *Office of Aviation Medicine.*

Stern, J. A., Boyer, D.J., Schroeder, D., Touchstone, M.R., & Stoliarov, N. (1996). Blinks, saccades, and fixation pauses during vigilance task performance: II. Gender and time-of-day. DOT/FAA/AM/96/9. FAA. *Office of Aviation Medicine.*

Stern, J.A., & Bynum, J.A. (1970). Analysis of visual search activity in skilled and novice helicopter pilots. *Aerospace Medicine, 41,* 330–335.

Stern, J.A., & Dunham, D.N. (1990). The ocular response system. In J.T.Caccioppo & L.G. Tassinary (Eds.), *Principles of psychophysiology: Physical, social, and inferential elements* (pp. 513–553). Boston: Cambridge University Press.

Stern, J.A., & Skelly, J.J. (1984). The eyeblink and workload considerations. *Proceedings*

Human Factors Society, 28th Annual Meeting (pp. 942–944). Santa Monica, CA: Human Factors Society.

Sussman, D. (1970). *An investigation of factors affecting driver alertness* (Tech. Rep. No. FH-11–7313). Buffalo, NY: Cornell Aeronautical Laboratories.

Tassorelli, C., Micieli, G., Osipova, V., Rossi, F., & Nappi, G. (1995). Pupillary and cardiovascular responses to the cold pressor test. *Journal of Autonomic Nervous System, 55*, 45–49.

Tinker, M.A. (1943a). A reply to Dr. Luckiesh. *Journal of Applied Psychology, 27*, 469–472.

Tinker, M.A. (1943b). Review of "Reading as a visual task" by M.Luckiesh & F.K.Moss. *Journal of Applied Psychology, 27*, 116–118.

Tinker, M.A. (1945). Reliability of blinking frequency employed as a measure of readability. *Journal of Experimental Psychology, 35*, 418–424.

Tinker, M.A. (1946). Validity of frequency of blinking as a criterion of readability. *Journal of Experimental Psychology, 36*, 453–460.

van der Molen, M.W., Boomsma, D.I., Jennings, J.R., & Nieuwboer, R.T. (1989). Does the heart know what the eye sees? A cardiac/pupillometric analysis of motor preparation and response execution. *Psychophysiology, 26*, 70–80.

Wang, L., & Stern, J.A. (1996). Effects of stimulus duration and task experience on gaze shift. *Ergonomics, 39*, 140–150.

Wang, L., & Stern, J.A. (1997). *Oculometric evaluation of subjects performing a vigilance task: The Bakan continuous performance task.* Unpublished manuscript.

Wilson, G.F. (1993). Air-to-ground training missions: A psychophysiological workload analysis. *Ergonomics, 36*, 1071–1087.

Yoshida, H., Yana, K., Okuyama, F., & Tokoro, T. (1994). Time varying properties of respiratory fluctuations in pupil diameter of human eye. *Methods Internal Medicine (MVI), 33*, 46–48.

Yoss, R.E., Moyer, N.J., Carter, E.T., & Evans, W.E. (1970). Commercial airline pilot and his ability to remain alert. *Aerospace Medicine, 41*, 1339–1343.

Yoss, R.E., Moyer, N.J., & Hollenhorst, R.N. (1970). Pupil size and spontaneous pupillary waves associated with alertness, drowsiness, and sleep. *Neurology, 20*, 545–547.

Zangemeister, W.H., & Stark, L. (1982). Gaze latency: Variable interaction of head and eye movement latency. *Experimental Neurology, 75*, 389–406.

Chapter 14

The Use of Psychophysiology for Evaluating Stress-Strain Processes in Human-Computer Interaction

Wolfram Boucsein
University of Wuppertal, Germany

During the 1960s, when visual display terminals (VDTs) came widely into use, computers were predominantly tools for data input and coding tasks. During this time, the use of psychophysiological recordings was restricted to examining factors directly affecting physiology, such as body position during work. Early reviews of adverse effects of computerization by Grandjean (1980) and by Sauter, Gottlieb, Jones, Dodson, and Rohrer (1983) mentioned a decrease in critical fusion frequency (CFF) together with complaints about eye strain as well as about neck, shoulder, and arm pain in VDT workers. Although recommendations and checklists have been available since the 1980s to ensure ergonomic setup of VDT workplaces (e.g., Grandjean, 1987; Grandjean & Vigliani, 1980), typical complaints are still an issue in VDT work. Besides physiological measures such as critical flicker frequency that require an interruption of the ongoing task, unobtrusive measures, such as electrooculographic (EOG) and electromyographic (EMG) activity or even pupil size, have been widely used to determine ergonomic features of the VDT system (e.g., Åborg, Fernström, & Ericson, 1995; Taptagaporn & Saito, 1990; Yamamoto & Kuto, 1992).

The most dramatic changes, both in the nature of tasks and in the organization of work, emerged from the shift from the physical workplace that has desk tops, filing cabinets, drawing boards, and so on, to the virtual workplace that has everything in a single computer. This enables the operator to perform a wide variety of tasks, and even to communicate, without leaving his or her work-place. Instead of using computers as a mere tool, human-computer interaction (HCI) emerged as a new mode of performing a wide variety of jobs. Office automation is perhaps the most obvious example of pervasive spread of HCI. However, the transition process from industrial to information-age office designs has been slow and incomplete in most large institutions, superimposing elements of computer technology such as VDTs on relatively unchanged work structures (Howell, 1991). Work has become more demanding, leading to an increase in mental workload on a lot of jobs. On the other hand, partitioning tasks has become a source of boredom and dissatisfaction, diminishing the human role in data entry tasks or operating a system with reduced degrees of freedom.

Despite the obvious advantages of office automation, such as relief from repetitive work, several disadvantages soon became evident. Subjective complaints reported comprised monotony and feelings of boredom resulting from partitioning formerly complex tasks into repetitive subtasks, a decrease in decision latitude, low levels of experienced job autonomy, and excessive external control (see Lundberg, Melin, Evans, & Holmberg, 1993). In addition to those complaints already mentioned, ergonomic research in HCI focused on psychophysiological changes that could be related to the structure of the human-computer dialogue.

This chapter deals with three topics in HCI where the use of psychophysiology has become a formidable research tool: quantifying workload imposed by computerization of workplaces, determining adverse affects of temporal delays in HCI caused by features of the computer system, and measuring the psychophysiological benefit of rest breaks during computer work.

Only a small fraction of the studies reported have been conducted outside the laboratory. The main reason for the predominance of laboratory research in psychophysiology may be the problem of bringing various physiological measures to the field. However, the now available multichannel ambulatory monitoring systems (see chap. 5, this volume) will increase the number of physiological systems that can be monitored continuously. Hypotheses that have been developed and for which evidence has been found within experimental studies that include the differential sensitivity of cardiorespiratory and electrodermal variables for determining mental versus emotional strain (Boucsein, 1992, 1993) need to be tested in the field as well.

DETERMINING THE AMOUNT OF MENTAL WORKLOAD IN HUMAN-COMPUTER INTERACTION

One of the most important issues raised during the introduction of computers at various workplaces that can be solved by the application of psychophysiological methods is to what extent mental workload may be increased by VDT work (for a general discussion of the psychophysiology of mental workload, see chap. 2, this volume). Psychophysiological research in this area of application has been frequently performed by means of endocrine and cardiovascular measures, but other variables, such as spectral analysis of the electroencephalogram (EEG), are used as well.

In a study by Johansson and Aronsson (1984), two groups of administrative workers with different amounts of VDT work (low=10% or high=more than 50%) were compared with respect to their workload. During the morning, the group with high VDT work had elevated levels of epinephrine compared to the low VDT group performing mainly secretarial work. This effect reversed during the afternoon. The differences in epinephrine excretion paralleled the difference in distribution of workload during the day, because high VDT workers tried to get their work done early as a precaution against possible computer system breakdowns, whereas low VDT employees were normally rushed with the end of the work day approaching. Norepinephrine levels yielded no difference between the groups. Interestingly, the high VDT group showed not only significantly higher epinephrine values throughout, but also a delayed deceleration of the epinephrine level during their leisure time after work as compared to the control group with less VDT work. The results of this study suggest that the group with high VDT work is more prone to the development of stress-related disorders because they displayed higher epinephrine levels at work and a delayed unwinding at home, a result that could not be obtained from subjective data. Accelerated and long-lasting epinephrine excretion is known as an indicator of stress reactions that can lead to health damaging effects (Selye, 1956). Enhancement of urine epinephrine levels after work was also reported by Gao et al. (1990) who found a tendency for epinephrine to increase after simple data entry work (only integers) compared to a more complicated data entry task (numbers with decimals) performed for 150 minutes by 29 students randomly assigned to one of the working conditions. The norepinephrine level yielded a slight, but nonsignificant, tendency to decrease after the more complicated data entry work, whereas diastolic blood pressure (during standing) increased significantly.

A simulated realistic data entry task that lasted 120 minutes was used by Floru, Cail, and Elias (1985) to investigate relationships between performance and physiological measures in 12 subjects trained in the task on a previous day. Whereas the average heart rate (HR) decreased from the beginning to the end of the task, which is typical for such kind of laboratory settings, the power in occipital EEG bands yielded significant intra- and intersubject correlations with performance changes during the task of the following kinds: Beta activity increased with increasing performance whereas both alpha and theta activity increased with performance decrement. However, a performance decrement observed during the last 15 minutes of the first hour of work was accompanied by a slight increase in beta activity and a diminution of alpha power. A continuing increase in theta activity during this period was interpreted as an indicator of a critical imbalance.

The usability of EEG theta activity as an indicator of fatigue due to mental load in HCI has been demonstrated by Yamamoto, Matsuoka, and Ishikawa (1989). In two pilot studies, the so-called frontal-midline theta waves (i.e., theta waves occurring from the center of the frontal region to the parietal region) showed a positive correlation with performance during a highly concentration-demanding VDT task in some of their subjects. They suggested that the occurrence of increased frontal-midline theta activity could be used to determine the point when an operator needs relief from highly demanding VDT work.

In a laboratory study with 19 computer novices, Pinkpank and Wandke (1995) compared a menu-language with a command-language technique in a computer dialogue task. When it was signaled to the subjects that the latter technique should be used in the task, the P300 amplitude was increased, indicating a greater amount of mental effort. In addition, P300 amplitudes tended to be larger for task-inadequate techniques and for the technique not preferred by the individual. The 0.1 Hz spectral component in the electrocardiogram was significantly reduced during the execution of the more complicated command technique, indicating the greater amount of mental effort required. During the execution of the most difficult tasks, a considerable reduction of the 0.1 Hz component emerged when an inadequate menu technique was imposed on the user.

Wastell (1990) also reported clear evidence for reduced heart rate variability (HRV) reflecting the amount of mental effort involved in different laboratory HCI tasks, such as using a text or a line editor with full effort or typing under time pressure compared to relaxed use of the text editor or just copy typing.

On the other hand, psychophysiological measures may not be sensitive enough for differentiating between VDT task measures, such as task complexity under short-term laboratory conditions. Lundberg et al. (1993) could not find significant differences in HR, blood pressure (BP), urinary epinephrine, norepinephrine, or cortisol levels in 30 male students between a monotonous data entry task and a stimulating learning task performed for 90 minutes on consecutive days in counterbalanced order. However, recovery to baseline level after 1 hour of task completion was slower in the monotonous condition for five of the six variables, reaching significance for epinephrine. Another laboratory study by Tanaka, Fukumoto, Yamamoto, and Noro (1988) was not supportive of epinephrine being a valid indicator of workload induced by HCI. Twenty-three subjects with a wide range of computer experience and ages performed 2 hours of searching for target words on a VDT with a 2-minute break every 15 minutes. A paper search task was used as the control condition. Only norepinephrine showed significant changes, despite the fact that work speed was higher in the control task compared to the VDT task. Because no changes in norepinephrine excretion emerged when large instead of small letters were used for VDT work, the authors concluded that it was not the use of a computer as such, but the increased difficulty reading small letters that was responsible for the norepinephrine increase.

The interpretation of psychophysiological differences between different types of work, such as working with and without a computer, should be performed with caution. Differences in posture or the amount of physical work may considerably contribute to the results, as was shown in a study by Springer, Müller, Langner, Luczak, and Beitz (1990). Thirty-three students of mechanical engineering performed a set of eight design tasks at a computer aided design (CAD) system and a drawing board for 120 minutes. Various physiological measures such as HR, electrodermal activity (EDA), and neck muscle EMG were increased at the drawing board. This was, however, not due to a relief from mental strain using CAD where performance actually was worse, but a result of standing at the drawing board versus sitting at the CAD system.

STRESS-INDUCING PROPERTIES
OF COMPUTER SYSTEM RESPONSE TIMES

Inadequate temporal structures of work, such as involuntary delays in the work flow have been known as major stressors from early investigations on workers at assembly lines performed in the 1920s. When time-sharing computer systems came into use during the late 1960s, delays in the work flow between computer and operator caused by system response times (SRTs) were consistently found to be associated with reports of annoyance and low user satisfaction (Carbonell, Elkind, & Nickerson, 1968). A computer SRT is defined as the time it takes from the moment a user terminates his or her input until the computer presents the result of the operation or displays readiness for another input (Shneiderman, 1992). The computer does not accept any further user commands during the SRT, the termination of which is signaled by a prompt, a blinking cursor, the appearance of a pull-down menu, or a new window. With the introduction of computers in all kinds of workplaces, the possibility of stress being induced by SRTs generated considerable interest because it became apparent that long SRTs were not only a source of added costs but also of stress-strain processes (Boucsein, Greif, & Wittekamp, 1984; Martin, 1973; Shneiderman, 1984; Youmans, 1983).

Although computers have become significantly faster since then, the adverse effects of SRTs still persist in the form of inadequate temporal structures during HCI. Hardware speed is often jeopardized by bulky software packages, powerful input and output devices, extensive network functions, and huge data banks with simultaneous access for many users. Furthermore, the possibility of multitasking in Windows-based systems raises the additional problem of adequate scheduling of work flow with regard to SRTs that may result from secondary tasks. In general, SRTs and even more system breakdowns, which may be labeled as infinite SRTs have been shown to be potent stressors in VDT work (Johansson & Aronsson, 1984).

Stress-inducing properties of SRTs have been investigated by our group since 1982. In most of the studies performed in our laboratory, we used an easy to learn detection task. A space surrounded by identical letters had to be targeted in a line of random letters and spaces presented in the center of a visual display. One study (Thum, Boucsein, Kuhmann, & Ray, 1995) used a more complicated task that individually standardized mental strain. SRTs were varied systematically between 0.5 and 8 seconds. Additional time pressure was induced by using incentives for working as fast and as correctly as possible or by announcing overtime work in the case of incorrect solutions. Skin conductance level (SCL), the frequency and mean amplitude of nonspecific electrodermal responses (NS.EDR), systolic and diastolic blood pressure (SBP and DBP) as well as mean arterial BP, HR, HRV, respiration rate, and frontalis EMG were used as physiological recordings. Task completion time, numbers and relevance of keystrokes, cursor movements, and errors were taken as performance measures. Ratings of mood and bodily symptoms served as subjective measures.

Table 14.1 summarizes the results from six laboratory experiments with 242 subjects altogether. Unexpectedly, our general findings were that negative physiological, behavioral, and subjective consequences resulted not only from rather long but also from rather short SRTs. If incentives for working fast such as time pressure were present and SRTs were rather short (top of the third column of Table 14.1), our subjects developed considerable strain, mainly reflected by increases of cardiorespiratory activity (i.e., SBP, DBR, and respiration rate), EMG frontalis activity and NS. EDR frequency, and a decrease in HRV They showed signs of impaired performance, such as an increase in error rate, cursor movements, and irrelevant keystrokes. They also had more subjective complaints of the sort made in computerized workplaces. When no such incentives as time pressure were imposed on the subjects under the same SRT conditions, physiological and subjective effects were less computer-task specific. Only rather slight increases in SBP, DBR, and HR were observed. General subjective arousal and the number of task-related bodily symptoms in general were increased. Although not instructed to do so, our subjects spontaneously worked at high speed and displayed high error rates (top of the fourth column of Table 14.1). Because our subjects were not experienced computer users, they may have been rushed by the tasks being presented too fast for their personal tempo.

A different picture emerged when SRTs were rather long. At first glance, it may seem that rather long SRTs simply resulted in subjects rationally adopting a more relaxed and careful work style than when incentives for working fast such as time pressure were present (bottom of the third column of Table 14.1). This means that the subjects made fewer errors, presumably because error correction was much more time consuming. However, any increased relaxation was more perceived than real, because our subjects experienced their working situation as being uncomfortable, as shown by an increase in negative emotions and headache symptoms, and their increased SCL, NS.EDR frequency or mean

TABLE 14.1
Results from Six Studies on System Response Times

SRT	Category	Incentive (Time Pressure) Present	Incentive (Time Pressure) Absent
	physiological	SBP,[b,e] mean arterial BP[f] and/or DBP[e] increased HRV decreased[e] Respiration rate increased[e] EMG frontalis power increased[e] NS.EDR frequency increased[f]	SBP and DBP increased[a] HR increased[d]
short (0.5–2 seconds)	behavioral	Error rate increased[b,e,f] Number of cursor movements,[b] working speed,[e] and irrelevant keystrokes[f] increased	Error rate increased[a,d] Working speed increased[d]
	subjective	More headache and eye symptoms reported[b]	General arousal increased[d] Number of task-related symptoms increased[a]
	physiological	SCL increased[b] NS. EDR frequency increased[b] NS. EDR amplitude increased[e] SBP and DBP decreased[e] Respiration rate decreased[e]	Increasing amount of EDA during SRT[c] NS. EDR frequency initially increased but decreased later[d]
long (8 seconds or longer)	behavioral	Error rate low[b,e]	Error rate low[a,d] Number of cursor movements increased[c] Task completion time prolonged[c]
	subjective	Negative motivation and emotions increased[e] More symptoms of headache reported[f]	General well-being enhanced at the beginning of work[c]

[a]Schaefer, Kuhmann, Boucsein, and Alexander (1986), $N = 20$;
[b]Kuhmann, Boucsein, Schaefer, and Alexander (1987), $N = 68$;
[c]Kuhmann (1989), $N = 48$;
[d]Kuhmann, Schaefer, and Boucsein (1990), $N = 24$;
[e]Thum, Boucsein, Kuhmann, and Ray (1995), $N = 40$;
[f]Kohlisch and Kuhmann (1997), $N = 42$.

NS.EDR amplitude suggested an augmentation of emotional strain. Results without incentives such as time pressure (bottom of the fourth column of Table 14.1) were similar with respect to a low error rate. However, subjects used their resources less effectively, increasing the number of cursor movements and the time needed for task completion. Furthermore, enhanced subjective well-being, which had been reported at the beginning, faded during the course of work when emotional tension developed as indicated by an increasing amount of EDA during SRTs as long as 8 seconds (Kuhmann, 1989). The decrease in EDA after an initial increase reported by Kuhmann, Schaefer, and Boucsein (1990) corresponds to an application of SRTs within a task instead of between tasks. In general, effects of long SRTs on EDA are more pronounced under time pressure compared to conditions where no incentives for working fast are present.

The adverse effects of rather short SRTs consistently found in several studies of our group have been challenged by Schleifer and Okogbaa (1990). In their own study, 45 female professional typists performed an alphanumerical data entry task in response to a series of prompts displayed on a VDT screen on 4 consecutive days. Their subjects worked either under short and invariable (350 milliseconds) or long and variable (3–10 seconds) SRTs. Half of the subjects received a monetary incentive for working faster and a penalty for performing more errors compared to their individual baseline. Their results indicated that time pressure imposed by monetary incentives led to increased cardiovascular strain on the last 2 days (increased SBP and DBR, decreased HRV). No significant effects of SRTs or interactions between incentive and SRT conditions were found in the physiological measures recorded. This is not surprising since the typists used as subjects may have been very much acquainted with the combination of short SRTs and monetary incentives. On the other hand, given results obtained by our group, adverse effects of long SRTs may not show up in cardiorespiratory variables, but in electrodermal variables instead, which were not recorded by these authors.

An obvious explanation for the stress-inducing effects of short SRTs as found by our group is the increase in workload due to higher work density, which is in accordance with elevated cardiorespiratory activity (top of the third and fourth columns of Table 14.1). However, since increased cardiorespiratory activity can be regarded as a main marker variable for physical strain (Boucsein, 1993), it should be verified that our results were not mediated by an increased amount of physical activity required under short as compared to long SRTs.

To verify this, motor demands and mental load were varied as independent factors in two experiments performed by Kohlisch and Schaefer (1996). The task used in the first study was keeping a left moving cursor within a target area in the middle of the VDT screen by means of compensatory keystrokes. Motor activity required was manipulated in four steps by cursor speed (interkeystroke intervals being 150,300,600, or 1,200 milliseconds) while mental load was varied by the required accuracy for keystroke synchronization (narrow or wide target area). In a completely balanced within-subjects design with 24 subjects, 20-second training and 90-second tests were conducted for each combination of conditions. Heart period (instead of HR), NS.EDR frequency, and mean interkeystroke intervals were recorded. Both cardiovascular and electrodermal measures remained unaffected by motor activity when mean interkeystroke intervals were 300 milliseconds or longer, which is in accordance with Carriero's (1975) finding that tapping below 333 milliseconds did not exert considerable influence on these physiological variables. This result was confirmed in a second study with 42 subjects using an additional arithmetic task for the induction of mental load, where no significant effects of motor activity below 360 milliseconds were found. The reported range for mean interkeystroke intervals in HCI is between 891 milliseconds and 5.01 seconds (Rauterberg, 1992) with a mean of 453 milliseconds for data entry tasks (Henning, Sauter, Salvendy, & Krieg, 1989). Because those intervals had been as high as 880 milliseconds in the Kuhmann, Boucsein, Schaefer, and Alexander (1987) study, an increased motor demand cannot explain the psychophysiological changes observed under short SRTs, leaving increased mental workload as the most likely explanation.

In order to keep metabolic demands due to mental workload constant while varying SRTs, Thum et al. (1995) performed a study with 40 subjects using an adaptive computer task developed by Kuhmann (1979). This task consisted of a randomly generated 6×6 matrices of 36 two-digit numbers where the subjects had to decide whether one, both, or none of the two target numbers were present. An algorithm continuously varied the presentation time of the matrices, becoming shorter after a correct response and longer after two subsequent mistakes, thus ensuring that all subjects achieved the same percentage of correct responses. Feedback was given after each task. The first of three 7-minute trials was performed with 1.5 second SRTs, the two others with 0.5 second and 4.5 second in balanced order. Half of the subjects were given monetary incentive for good performance. HR, SBP, DBR, respiration rate, EDA, and frontalis EMG were recorded, and task features (e.g., difficulty in comparison with a reference trial) as well as emotional states during the task were rated after each trial. As compared to the medium SRT of 1.5 seconds, both short (0.5-second) and long (4.5-second) SRTs significantly increased HRV while HR was reduced. In addition, short SRTs increased DBR, while long SRTs increased NS.EDR amplitude and reduced SBP and respiration rate. Compared to 0.5 seconds, the 4.5 second SRTs increased NS.EDR amplitude and HRV, but reduced SBP, DBR, and respiration rate. This complex outcome shows that both SRTs shorter and longer than a certain medium range may cause considerable psychophysiological strain. However, effects of short SRTs on HR and HRV as displayed in the upper part of Table 14.1 may be due to mental workload since they no longer appear when this factor is controlled, as in the Thum et al. (1995) study.

However, given that various psychophysiological studies have shown detrimental effects of very short SRTs, the general recommendation "faster is better" (cf. the title of Smith, 1983) does not seem to be appropriate for HCI, except for programmers (Lambert, 1984). Therefore, the recommendation is to determine an optimal SRT for the task in question by means of a psychophysiological study.

The criterion for optimal would be no marked increases in cardiovascular activity, low NS.EDR frequency, no increased general muscle tension, best performance, and the lowest reports of pain symptoms. With the particular task and sorts of subjects used by our group, the optimal SRT turned out to be 5 to 6 seconds (Kuhmann, 1989). That optimum would probably be shorter with experienced computer users who are familiar with the task (such as the typists used by Schleifer & Okogbaa, 1990), as well as with other kinds of tasks (e.g., about 1.5 seconds for the task used by Thum et al., 1995) or even much longer as in the case of Barber and Lucas (1983) where the optimal SRT was 12 seconds for establishing telephone circuits in interstate networks. However, the main point is not what the optimum is, but that what is optimal be determined by using adequate psychophysiological variables in addition to performance criteria. This concept of an optimal SRT is in accordance with Shneiderman's (1992, p. 277) notion that there is a preferred SRT for a given user and task and both shorter and longer SRTs may generate debilitating effects on the effectiveness of HCI.

A deliberate investigation of the concept of an optimal SRT was conducted by Kohlisch and Kuhmann (1997) with 42 subjects performing the detection task used by our group under three conditions of 10-minute duration each in a balance repeated measurement design (1 seconds=short, 5 seconds=medium, and 9 seconds=long SRTs). An incentive for working quickly and accurately was established by announcing unpaid overtime work if the subjects performance did not meet a combined speed-accuracy criterion checked by the computer. Measures of speed and accuracy, HR, BP, and EDA were recorded, and subjective ratings of mood, arousal, and bodily symptoms were obtained. Performance decrement and cardiovascular strain clearly emerged under the short SRTs; the percentage of errors, irrelevant keystrokes, and mean arterial BP were significantly elevated compared to the other two conditions. Unexpectedly, the NS.EDR frequency was significantly higher under the short as compared to the long SRT condition. However, the amount of headache reported was significantly elevated under 9 seconds as compared to 5-second SRTs. Therefore, Kohlisch and Kuhmann (1997) considered 5 seconds as an optimal SRT for the task and subjects investigated.

Theoretical approaches for explaining the adverse effects of prolonged SRTs date back to Miller's (1968) review on response times in HCI. Based on communication theory, Miller recommended an upper limit of 2 seconds for intratask and 15 for intertask SRTs in order to avoid a breakdown in the human-computer dialogue. Furthermore, computer users may become increasingly concerned with prolonged SRTs because the penalty for errors increases, which further slows down their performance and increases the error rate (Shneiderman, 1992). Decreases in the user's motivation and work satisfaction have also been found in prolonged SRTs (e.g., Schleifer & Amick, 1989; Thum et al., 1995; Treuerniet, Hearty, & Planas, 1985) and thus may contribute to performance decrements. Because temporal uncertainty may also be an important stress-inducing factor in HCI (Boucsein et al., 1984), it has been suggested that not only the length but also the variability of SRTs may be critical for the amount of resulting strain by means of inducing uncertainty concerning the temporal course of HCI (Carbonell et al., 1968).

However, no dramatic effects were found when SRT variability was introduced (for a summary, see Shneiderman, 1992, chap. 7.6). For example, in two studies that introduced 0 or 80% variability of SRTs as a second experimental factor besides 2- and 8-second SRT duration, neither significant main effects of variability nor interactions with duration were found (Kuhmann, Boucsein, Schaefer, & Alexander, 1987; Schaefer, Kuhmann, Boucsein, & Alexander, 1986). Therefore, another study was performed by Schaefer (1990) with 48 subjects to resolve this issue. After a 20-minute training with 1 second SRTs for all subjects, four independent groups performed the detection task used in our earlier studies with either 2-, 4-, 6-, or 8-second fixed SRTs for another 20 minutes. The following six blocks of 20 minutes each used a so-called stop-reaction time procedure. In this method, a sequence of events presented with constant intervals is stopped from time to time. The subject has to press a key as soon as he or she notices such an unpredictable event. They were told that system breakdowns would occur in irregular sequence instead of the ordinary SRTs and a "restart key" would enable them to continue their work. However, to prevent subjects from speeding up their work by pressing this key, the restart needed 8 seconds. No significant effects of SRTs on HR, SBP, or DBP were found. The plot of mean stop-reaction times against SRTs fitted a positively accelerated power function (exponent 3.39). An alternative linear model had to be rejected, apparently due to the marked increase of stop-reaction times under the 8-second SRT condition. A similar result was obtained for the standard deviations of the stop-reaction times. This points to a deterioration of temporal sensitivity for SRTs between 6 and 8 seconds duration, because means and standard deviations of stop-reaction times were in agreement with Weber's Law up to 6 seconds, but not with 8-second SRTs. Impaired predictability resulted in performance decrement under the 8-second SRT condition, where subjects needed more time to complete their tasks because more cursor movements were performed. As a consequence, the failure of our earlier studies to determine a separate effect of SRT variability was reinterpreted insofar as temporal uncertainty could have been the critical factor for the adverse effects not only of the variable, but also the constant 8-second SRT condition, since subjects may have not been able to subjectively distinguish those conditions.

Another possible explanation for the adverse effects of inadequate SRTs on the performance of mental tasks may be derived from the fact that SRTs are normally used for preparing the subsequent work step. Detrimental effects of SRTs can be expected if their anticipation does not correspond to their actual duration. In general, it will be more likely for such preparatory problems to appear in intertask

SRTs than in intratask SRTs. Therefore, an intratask SRT paradigm was used in the Schaefer and Kohlisch (1995) study determining mental processes during SRTs by means of EEG recordings from Fz and Pz. Their 54 subjects performed a series of 360 target recognition tasks serving as a model for database retrieval. After presentation of an artificial target word on the screen, SRTs of either 1, 2, or 4 seconds were given to one third of the subjects prior to the appearance of a recognition list. The matching of SRT with the subject's expectation was varied as an additional within-subjects factor as follows. The standard SRT was reduced by 25% in 30 trials and increased by 25% in another 30 trials, which were randomly interspersed between standard trials. Latencies and amplitudes of N100 and P300 following SRT termination were obtained for each condition. N100 and P300 latencies were larger and recognition time was longer when SRTs were unexpectedly short, whereas P300 amplitudes were increased and working speed was reduced for the subsequent task when SRTs were unexpectedly long. Besides the distortions in the information processing, this result stresses the importance of a stable and predictable SRT in HCI.

An additional study on intratask SRTs has been performed by Kuhmann, Schaefer, and Boucsein (1990) with 24 subjects working on the modified detection task used by our group. Two rows of random letters with spaces as used in the other studies were combined into one task, after which the subjects had to report whether a target (i.e., a space surrounded by two identical letters) was detected or not in either line. After a 15 minute training trial with constant 1-second intratask SRTs, subjects performed four 10-minute trials in counterbalanced order with intratask SRTs of 2 seconds or 8 seconds, crossed with 0 or 80% variability as a second factor. Again, SRT variability did not have any effect on physiological and performance measures, only significantly less subjective well-being and more symptoms of general excitement were reported under variable as compared to constant SRTs. In accordance with the results from intertask SRTs, cardiovascular activity (i.e., HR) was increased while working under short intratask SRTs. However, an increase in NS.EDR frequency observed under long SRTs at the beginning reversed during the course of the trials. As could be expected from the hypothesis outlined earlier, that detrimental effects of SRTs would show up if they were introduced within a task, SRTs mainly affected performance in the second part of the task. Under short SRTs, subjects worked significantly faster on the second line, but their error rate increased as well. Again, performance decrements and physiological and subjective strain are likely to occur not only when SRTs exceed a certain optimal range, but also when SRTs are too short for the task in question. This has been found for both intratask and intertask SRTs.

A theoretical approach for determining the course of strain development during SRTs based on the cognitive stress theory of Lazarus (1966) has been formalized by Holling and Gediga (1987) using a probability model. Following the proposal of Carbonell et al. (1968), a linear nondecreasing function for psychological costs is assumed (e.g., anger associated with SRT duration). The subjective expectation of SRTs is regarded as a probability distribution. The integrated conditional expectation of the future costs constitutes a precise prediction of the time course for developing strain during the actual SRT. In order to analyze the predictions of the model, Holling and Gediga performed a laboratory study with 72 subjects working on the simple detection task used by our group for 6 trials of 12 minutes each. Four experimental conditions made of 2or 8-second SRT duration crossed by constant or variable SRTs were applied in a counterbalanced order. Ratings of subjective stress, weariness, and anger were obtained after each trial, and HR and BP were recorded. Their results were supportive of the strain-increasing property of SRT variability and highly supportive of 8-second SRTs being more stressful than 2-second SRTs. However, the assumption of a linear cost function had little predictive power for the course of strain developing during SRTs.

Stress induced by long SRTs may be reduced by feedback provided on the status of the task in progress. In a laboratory study with 48 subjects whose task was to correct errors found by the computer in seven files consisting of 100 forms each, Holling (1993) applied two versions of the task described for Holling and Gediga (1987) in a counterbalanced order. One version gave no information about the status of the task, whereas the other continuously displayed the number of the form actually being checked by the computer. Standardized differences between the Holling and Gediga (1987) measure of strain under both conditions were computed for EDA (SCL) and HR. Stress detected by EDA was significantly lower when feedback was given, paralleling an appropriate decrease in the subjective strain measures of anger and arousal, whereas feedback did not yield such a significant effect on HR.

Another possibility of reducing strain imposed on the operator by unavoidable long SRTs had been suggested as early as 1968 by Miller. He recommended artificially prolonging SRTs, allowing the user to behave like a time-sharing system himself. In modern computer operating systems providing multitasking features, long computer processing times can be utilized to perform another work step. Although strain imposed by forced waiting is avoided by filling SRTs with other tasks, additional effort is needed for an optimal scheduling of parallel tasks. This may produce new psychophysiological strain. In a recent study performed by Schaefer, Schäfer, and Boucsein (1997), 40 students worked on a mocked power plant control center setup procedure. A panel showed 36 displays to be set to particular initial values, some of which were to be requested from a virtual remote data bank by transmission lines. During the processing times for those requests, which were 10 seconds for half of the subjects and 30 seconds for the other half, the setup procedure could be either continued or additional requests could be started. As an additional within-subjects factor, feedback on the progress of opened requests was either not provided or given as a static or dynamic display. EDA, HR, peripheral blood volume, and respiration were continuously recorded, and keystrokes and request handling were written in log files. Subjective ratings were taken after each feedback condition. Results showed that the multitasking features were more often used when the processing times were long, where subjects also reported more arousal and pain symptoms. For short processing times, both frequency and amplitudes of NS.EDRs were significantly increased, indicating emotional strain. Cardiovascular measures did not show significant effects. Compared to the former results of the our group with single-task systems, the psychophysiological patterns seem to be reversed. Under multitasking conditions, long SRTs seem to be more convenient because they may be used for the performance of other work steps which is not the case with short SRTs. However, EDA may have indicated the inappropriateness of using multitasking procedures during very short processing times.

RECOVERY FROM STRAIN DURING BREAKS
UNDER DIFFERENT WORK/REST SCHEDULES

The most important features of rest breaks during work are to counteract the development of fatigue and, thus, improve performance and well-being. The duration of those breaks may range from seconds (microbreaks) to fractions of an hour. Furthermore, they may be taken spontaneously (concealed breaks), involuntarily (interruptions of work flow by machinery or by other persons), or scheduled by agreement between workers and employer. Not much research has been performed to quantify the outcome of breaks by means of psychophysiological recording.

In a study on the effects of microbreaks on the HR of 20 female data entry operators, Henning et al. (1989) allowed for a self-regulated break in the middle of each 40-minute work period, which were separated by 10-minute scheduled breaks. Six work periods were presented each day for 2 consecutive days. Mean microbreak length was 27.4 seconds. Regression analyses revealed that long microbreaks were predictive of low error correction rates and lower HR in the second half of the work period, implying that the degree of recovery was linked to the length of the microbreaks. On the other hand, low correction rates during the first half of the work period and work-related fatigue and boredom were associated with an increase in microbreak length, suggesting that the break period was self-adjusted relative to the mood state. However, workers tended to terminate microbreaks before performance recovery was complete. In another study with 30 two-person teams working together on a VDT work-substitution task for 46 minutes, Henning, Bopp, Tucker, Knoph, and Ahlgren (1997) manipulated the kind of feedback provided on the sufficiency of following the instruction to take team-managed 1 minute breaks for every 10 minutes of work. Feedback reduced the error rate but did not affect HR and HRV.

A comparison of break schedules during mental work by Rohmert and Luczak (1973) demonstrated that both short work/rest schedules (e.g., 25 minutes work/5 minutes break) and long work/rest schedules (e.g., 100 minutes work/20 minutes break) led to increased performance quality and decreased subjective and physiological stress symptoms after the breaks. However, during the course of the day, fatigue, recorded by physiological and performance measures, increased under the short break schedule, thus indicating a loss of resources. Unfortunately, it has not been shown that these different break effects were associated with different physiological recovery processes. If unfavorable break schedules for mental work will lead to increased blood pressure, muscle tension, pain symptoms, and negative emotional states, the probability of the development of stress-related disorders, such as hypertension, headache, and gastric ulcers may be increased in addition to the decreased performance quality.

In order to test psychophysiological effects of recovery during the breaks under different work/rest schedules in the field, Boucsein and Thum (1997) applied two different schedules during a complex and highly demanding computer task. Based on previous research on less complex, short cyclic mental tasks such as data typing, 10 minutes of break after each hour of work were introduced for most German computer work places. However, it is not clear that this sort of work/rest schedule remains appropriate for more complex tasks. One reason this schedule may be inappropriate is that the process of thinking is interrupted. Another reason could be that adequate recovery of stress-related physiological changes during complex work with display units may require more than a short break.

The Boucsein and Thum (1997) study was conducted at the European Patent Office in The Hague, The Netherlands. Eleven patent examiners who were members of a prototyping group using a computer workplace with two VDTs served as subjects. Instead of working with paper documents, these examiners had access to patents on a laser disk that could be displayed on the 24-inch computer screen. On the basis of a search through granted patents, the examiners had to write a report about the novelty of the current application. Each of the examiners performed his or her work under two different break schedules on different days, the order of which was balanced: a long break schedule with 15 minutes rest after 100 minutes of work and a short break schedule with 7.5 minutes rest after 50 minutes of work. The total break time on each day was 82.5 minutes for all subjects, including a lunch break.

Physiological recordings of HR, SCL, EMG (neck lead), respiration, pulse wave transit time (PWTT, which becomes shorter with elevated BP), and gross body movements were taken continuously using an ambulatory monitoring device (VITAPORT). Subjects indicated the beginning and end of each break with a marker. They were instructed to sit relaxed in their chair during the first 2 minutes of each break and during the first 2 minutes of the subsequent work period after each break, in order to obtain artifact-free physiological recordings of the recovery during the rest break. The three breaks selected for evaluation were at 11 a.m., 3 p.m., and 5 p.m. because the accumulated break time was identical at these points of time. Subjective variables were acquired eight times each day by an adjective checklist and ratings of bodily symptoms. No performance variables were acquired because the keyboard action did not correlate with either task difficulty or performance quality. The only indicator of performance quality is the final report about the patent application, the quality of which could not be measured objectively during the study because objections against the report can be filed even years after the report was written. For the physiological variables, three change scores were calculated as relaxation period after the break minus relaxation period at the beginning of the break. The amount of total break time under both conditions was exactly the same at all three points of evaluation. The amount of body movement measured by an accelerator was very low (on average 0.003 g force per 2 minutes) and did not differ significantly between the different break schedules and the points of measurement.

There was a general trend for HRV (calculated as variation coefficient of interbeat intervals) to decrease during the working time and a significant increase in HRV under short-break condition. Moreover, the decrease in HRV became more prominent during the long break in the late afternoon, indicating increased mental effort while struggling against fatigue. In the morning, there was an increase in HRV during short breaks, pointing to a more effective reduction of mental strain compared to the long-break schedule. The analysis of the SCL resulted in a significant break duration by working time interaction. Although an increase in the SCL during the 11 a.m. and 3 p.m. breaks and a decrease during the 5 p.m. break could be observed under the long-break schedule, the reverse pattern was found under the short-break schedule, namely a decrease in the morning and an increase in the late afternoon. This electrodermal pattern of results suggests that the long break schedule actually increases emotional strain during the 11 a.m. and 3 p.m. break, which is analogous to the laboratory results presented in Table 14.1, where de trimental effects of rather long SRTs have been found. In contrast, the longer break schedule during the 5 p.m. break was associated with a reduction in emotional strain. This could be due to the fact that the 5 p.m. break was very close to the regular end of the working day at 5:30 p.m. so that an interruption of thinking processes was less likely because the subjects knew that they would leave the workplace shortly after the end of this break. A comparison of the differential HRV and SCL results supports the working hypothesis that these measures reflect different aspects of strain during HCI that may be labeled as mental and emotional strain (Boucsein, 1993).

Preliminary recommendations derived from these results were as follows. In order to minimize loss of efficiency during the course of a workday, a prolongation of the break time in the late afternoon may be of great benefit. Because a 50 minutes work/7.5 minutes break schedule was associated with less mental strain, as indicated by an increased HRV, and also with less emotional strain during the 11 a.m. and 3 p.m. reading, as indicated by the SCL, such a short break schedule should be applied until the late afternoon. However, the more effective reduction of fatigue and stress symptoms under the long break schedule during the last point of measurement, as indicated by the decrease in the SCL, suggests that fewer but longer breaks should be taken in the late afternoon to counteract the loss of resources. Furthermore, subjects reported more subjective well-being under the short-break schedule compared to the long-break schedule.

Another aim of the Boucsein and Thum (1997) study was to compare psychophysiological changes during system breakdowns with those during regular breaks and consultations by other examiners, either by phone or in person. A total of 12 system breakdowns happened, all except one before noon, three under the short-break schedule and 9 under the long-break schedule. Because no significant differences between those particular short and long breaks were found, the system breakdowns were compared with the nearest regular break and the nearest consultation. Physiological variables before, during, and after the interruptions were compared for the three types of interruptions.

Significant interactions between the types of interruption and points of measurement were obtained for EMG and SCL. Neck muscular tension was lowest during regular breaks. Consultations showed a similar effect with higher muscular tension in general, whereas neck EMG was considerably increased during system breakdowns. Interestingly, the SCL showed a steady decay after system breakdowns, but a steady increase during the regular breaks. Cardiovascular measures, on the other hand, yielded significant differences between the types of interruption. HR was lowest and PWTT was longest during of system breakdowns, whereas the highest HR and the shortest PWTT appeared during consultations. Thus, cardiovascular reactions indicated some kind of physical/mental strain being most pronounced during consultations. Scheduled rest breaks, however, were the only type of interruption associated with a steady decline in HR, indicating a decrease in mental strain that persists after the break. The increase in SCL during consultations may have been a speech artifact. Unexpectedly, SCL yielded a steady decline during system breakdowns, which may be explained by the patent examiners as members of a prototyping group being rather familiar with such events, not worrying much about a system failure. On the other hand, the steady increase in SCL during the regular breaks points to the possibility of increasing emotional strain, as opposed to a decrease in physical and mental strain. This could be an effect of interrupting the complex reasoning task the examiners might have been experiencing.

Given these complex results obtained in a field setting with a highly demanding computer task, it may be suggested to not apply a rigid work/rest schedule at all for today's complex HCI. Instead, computer users should take flexible rest pauses depending on work demands and the situational context. However, care should be taken to have a rest/break schedule that allows for ample psychophysiological recovery in all phases of computer work.

CONCLUSIONS

Psychophysiological recording has been shown to be a valuable tool in evaluating stress-strain processes that may typically emerge during HCI. Because the amount of physical strain is relatively low during computer work, different psychophysiological marker variables may be chosen for detecting the amount of mental and emotional strain in this kind of research. Mental workload during VDT operations can be determined by means of changes in cardiac activity such as a decrease in HRV, by changes in EEG measures, such as an increased P300 amplitude or theta activity, and by an increase in catecholamine excretion. On the other hand, strain that emerges from the more emotion-related aspects of workload will appear as an increase in spontaneous electrodermal activity.

Another important goal that may be achieved by means of psychophysiological recordings is determining the optimal duration and schedule of necessary breaks during HCI. In particular, psychophysiological recovery during breaks can be used as a measure of predicting the feasibility of a given rest/break schedule for the prevention of VDT work-related diseases that may result from accumulated psychophysiological strain.

My opinion is that the use of psychophysiological tools should be backed up by appropriate modeling. The neurophysiologically based model provided in chapter 1 may serve as such an attempt, predicting that cardiovascular markers, such as HRV, can be used as easy-to-measure indicators for mental strain, whereas electrodermal variables form an equivalent for the emotional area of cognitive impairment. Such an approach, however, cannot be conducted with recordings from one single psychophysiological system but requires multivariate methodology (see Table 1.1 in chap. 1, this volume). Recommendations for an optimal design of both hard- and software may be developed using psychophysiological results from laboratory and field settings as shown in this chapter.

REFERENCES

Åborg, C., Fernström, E., & Ericson, M.O. (1995). Psychosocial and physiological effects of reorganizing data entry work—a longitudinal study. In A.Grieco, G.Molteni, B.Piccoli, & E.Occhipinti (Eds.), *Work with display units 94* (pp. 63–66). Amsterdam: Elsevier.

Barber, R.E., & Lucas, H.C. (1983). System response time, operator productivity, and job satisfaction. *Communications of the ACM, 26,* 972–986.

Boucsein, W. (1992). *Electrodermal activity.* New York: Plenum.

Boucsein, W. (1993). Psychophysiology in the computer workplace—goals and methods. In H.Luczak, A.Çakir, & G.Çakir (Eds.), *Work with display units 92* (pp. 135–139). Amsterdam: North-Holland.

Boucsein, W., Greif, S., & Wittekamp, J. (1984). Systemresponsezeiten als Belastungsfaktor bei Bildschirm-Dialogtätigkeiten [System response times as a cause for stress at dialogue-oriented visual display units]. *Zeitschrift für Arbeitswissenschaft, 38*(10 NF), 113–122.

Boucsein, W., & Thum, M. (1997). Design of work/rest schedules for computer work based on psychophysiological recovery measures. *International Journal of Industrial Ergonomics, 20,* 51–57.

Carbonell, J.R., Elkind, J.I., & Nickerson, R.S. (1968). On the psychological importance of time in a time sharing system. *Human Factors, 10,* 135–142.

Carriero, N.J. (1975). The effects of paced tapping on heart rate, skin conductance, and muscle potential. *Psychophysiology, 12,* 130–135.

Floru, R., Cail, F., & Elias, R. (1985). Psychophysiological changes during a VDU repetitive task. *Ergonomics, 28,* 1455–1468.

Gao, C., Lu, D., She, Q., Cai, R., Yang, L., & Zhang, G. (1990). The effects of VDT data entry work on operators. *Ergonomics, 33,* 917–924.

Grandjean, E. (1980). Ergonomics of VDUs: Review of present knowledge. In E.Grandjean & E.Vigliani (Eds.), *Ergonomic aspects of visual display terminals* (pp. 1–12). London: Taylor & Francis.

Grandjean, E. (1987). Design of VDT workstations. In G.Salvendy (Ed.), *Handbook of human performance* (pp. 1359–1397). New York: Wiley.

Grandjean, E., & Vigliani, E. (Eds.) (1980). *Ergonomic aspects of visual display terminals.* London: Taylor & Francis.

Henning, R.A., Bopp, M.I., Tucker, K.M., Knoph, R.D., & Ahlgren, J. (1997). Team-managed rest breaks during computer-supported cooperative work. *International Journal of Industrial Ergonomics, 20,* 19–29.

Henning, R.A., Sauter, S.L., Salvendy, G., & Krieg, E.F. (1989). Microbreak length, performance, and stress in a data entry task. *Ergonomics, 32,* 855–864.

Holling, H. (1993). Reduction of stress by the display of system operations in human computer interaction. In H.Luczak, A.Çakir, & G.Çakir (Eds.), *Work with display units 92* (pp. 129–134). Amsterdam: North-Holland.

Holling, H., & Gediga, G. (1987). Stress in waiting situations—predictions from a mathematical model. In E.E.Roskam & R.Suck (Eds.), *Progress in mathematical psychology, 1* (pp. 233–249). Amsterdam: Elsevier.

Howell, W.C. (1991). Human factors in the workplace. In M.D.Dunnette & L.M.Hough (Eds.), *Handbook of industrial and organizational psychology* (Vol. 2, pp. 209–269). Palo Alto, CA: Consulting Psychologists Press.

Johansson, G., & Aronsson, G. (1984). Stress reactions in computerized administrative work. *Journal of Occupational Behaviour, 5,* 159–181.

Kohlisch, O., & Kuhmann, W. (1997). System response time and readiness for task execution—the optimum duration of intertask delays. *Ergonomics, 40,* 265–280.

Kohlisch, O., & Schaefer, F. (1996). Physiological changes during computer tasks: Responses to mental load or to motor demands? *Ergonomics, 39,* 213–224.

Kuhmann, W. (1979, December). Realisation eines Verfahrens zur experimentellen Anforderungssteuerung bei einer einfachen Konzentrationsaufgabe [Operationalization of mental load by means of an adaptive algorithm]. Psychophysiologie und Verhaltensmodifikation, *Bericht über das IV. Kolloquium Psychophysiologische Methodik, Spitzingsee, Obb.* (pp. 18–21). Justus-Liebig-Universitat, Giessen, Germany.

Kuhmann, W. (1989). Experimental investigation of stress-inducing properties of system response times. *Ergonomics, 32,* 271–280.

Kuhmann, W., Boucsein, W., Schaefer, F., & Alexander, J. (1987). Experimental investigation of psychophysiological stress reactions induced by different system response times in human-computer interaction. *Ergonomics, 30,* 933–943.

Kuhmann, W., Schaefer, F., & Boucsein, W. (1990). Effekte von Wartezeiten innerhalb einfacher Aufgaben: Eine Analogie zu Wartezeiten in der Mensch-Computer-Interaktion [Effects of waiting times with simple problems: An analogy to waiting tims in human computer interaction]. *Zeitschrift für Experimentelle und Angewandte Psychologie, 37,* 242–265.

Lambert, G.N. (1984). A comparative study of system response time on program developer productivity. *IBM Systems Journal, 23,* 36–43.

Lazarus, R.S. (1966). *Psychological stress and the coping process.* New York: McGraw-Hill.

Lundberg, U., Melin, B., Evans, G.W., & Holmberg, L. (1993). Physiological deactivation after two contrasting tasks at a video display terminal: Learning vs. repetitive data entry. *Ergonomics, 36,* 601–611.

Martin, J. (1973). *Design of man-computer dialogue.* Englewood Cliffs, NJ: Prentice Hall.

Miller, R.B. (1968). Response time in man-computer conversational transactions. *Fall Joint Computer Conference, 33,* 267–277.

Pinkpank, T., & Wandke, H. (1995). Mental effort with the use of different dialogue techniques in human-computer interaction. *Zeitschrift für Psychologie, 203,* 119–137.

Rauterberg, M. (1992). An empirical comparison of menu-selection (CUI) and desktop (GUI) computer programs carried out by beginners and experts. *Behaviour and Information Technology, 11,* 227–236.

Rohmert, W., & Luczak, H. (1973). Ergonomische Untersuchung von Teilzeit-Schichtsystemen und Pausen bei informatorischer Arbeit [Ergonomical investigation of part time shift systems and recovery periods under conditions of information processing work]. *Internationales Archiv für Arbeitsmedizin, 31,* 171–191.

Sauter, S.L., Gottlieb, M.S., Jones, K.C., Dodson, V.N., & Rohrer, K.M. (1983). Job and health implications of VDT use: Initial results of the Wisconsin-NIOSH study. *Communications of the Association for Computing Machinery, 26,* 284–294.

Schaefer, F., (1990). The effect of system response times on temporal predictability of work flow in human-computer interaction. *Human Performance, 3,* 173–186.

Schaefer, F., & Kohlisch, O. (1995). The effect of anticipatory mismatch in work flow on task performance and event related potentials. In A.Grieco, G.Molteni, E.Occhipinti, & B. Piccoli (Eds.), *Work with display units 94* (pp. 241–245). Amsterdam: North-Holland.

Schaefer, F., Kuhmann, W., Boucsein, W., & Alexander, J. (1986). Beanspruchung durch Bildschirmtätigkeit bei experimentell variierten Systemresponsezeiten [Stress effects of experimentaly varied system response times in man computer interactions at visual display units]. *Zeitschrift für Arbeitswissenschaft, 40*(12 NF), 31–38.

Schaefer, F., Schäfer, R., & Boucsein, W. (1997). Psychophysiological work strain during multitasking human-computer interaction. *Psychophysiology, 34,* S79.

Schleifer, L.M., & Amick, B.C. (1989). System response time and method of pay: Stress effects in computer-based tasks. *International Journal of Human Computer Interaction, 1,* 23–39.

Schleifer, L.M., & Okogbaa, O.G. (1990). System response time and method of pay: cardiovascular stress effects in computer-based tasks. *Ergonomics, 33,* 1495–1509.

Selye, H. (1956). *The stress of life.* New York: McGraw-Hill.

Shneiderman, B. (1984). Response time and display rate in human performance with computers. *Computing Surveys, 16,* 265–285.

Shneiderman, B. (1992). *Designing the user interface: Strategies for effective human-computer interaction.* Reading, MA: Addison-Wesley.

Smith, D. (1983). Faster is better: A business case for subsecond response time. *In Depth, Computerworld (April 18),* 2–22.

Springer, J., Müller, T., Langner, T., Luczak, H., & Beitz, W. (1990). Stress and strain caused by CAD-work—results of a laboratory study. In L.Berlinguet & D.Berthelette (Eds.), *Work with display units 89* (pp. 231–238). Amsterdam: Elsevier.

Tanaka, T., Fukumoto, T., Yamamoto, S., & Noro, K. (1988). The effects of VDT work on urinary excretion of catecholamines. *Ergonomics, 31,* 1753–1763.

Taptagaporn, S., & Saito, S. (1990). How display polarity and lighting conditions affect the pupil size of VDT operators. *Ergonomics, 33,* 201–208.

Thum, M., Boucsein, W., Kuhmann, W., & Ray, W.J. (1995). Standardized task strain and system response times in human-computer interaction. *Ergonomics, 38,* 1342–1351.

Treurniet, W.C., Hearty, P.J., & Planas, M.A. (1985). Viewers' responses to delays in simulated teletext reception. *Behaviour and Information Technology, 4,* 177–188.

Wastell, D.G. (1990). Mental effort and task performance: Towards a psychophysiology of HCI. In Diaper, D.Gilmore, G.Cockton, & B.Shackel (Eds.), *INTERACT 90* (pp. 107–112). Amsterdam: Elsevier.

Yamamoto, S., & Kuto, Y. (1992). A method of evaluating VDT screen layout by eye movements analysis. *Ergonomics, 35,* 591–606.

Yamamoto, S., Matsuoka, S., & Ishikawa, T. (1989). Variations in EEG activities during VDT operation. In M.J.Smith & G.Salvendy (Eds.), *Work with computers: Organizational, management, stress and health aspects* (pp. 225–232). Amsterdam: Elsevier.

Youmans, D.M. (1983). *The effects of system response time on users of interactive computer systems.* Winchester, England: Hursley Human Factors Laboratory.

Chapter 15

Psychophysiological Analysis of Design Work: Ergonomic Improvements Derived From Stress-Strain Research

Holger Luczak
Johannes Springer
Institute of Industrial Engineering and Ergonomics, *Aachen University of Technology, Germany*

TASK AND PARADIGM

In spite of the positive interest of ergonomists in the application of psychophysiological methods, such methods have contributed little of a positive nature to the solution of problems in ergonomics and engineering design (Luczak, 1987; Plath & Richter, 1978). It is, however, interesting to note that the application of these methods in practical field situations delivers more useful results than laboratory research with artificial tasks. That is why the validity and reliability of a psychophysiological measurement technique may be taken as a given and its utility examined in an engineering psychophysiological context.

To demonstrate this approach, it is necessary to choose an exemplary task and a paradigm, that brings together the engineering and the psychophysiological perspectives. The task should be an information-processing or cognitive task with task components that can be supported by different technological means, one that has a high degree of representativeness for engineering. It should also be a significant task in terms of relevance to the working person. Relevance means both for the industrial application as well as for academic research about engineering work. Therefore, the task has to be able to generate a sufficient amount of interesting experimental and exploitable questions (Luczak, 1997).

With these background considerations, design work was chosen with computer aided design (CAD) systems as the respective task-support technology (Luczak & Springer, 1997).

Technically speaking, in the generation of a product, design is the functional unit in a company where customer orders specified as customer demands or technical requirements are transformed into a mostly graphic model of the product. This means part lists, part drawings, and composition drawings (see Fig. 15.1) as work results on the basis of the application of natural science and engineering knowledge about the product domain. The purpose of the CAD system in this context is to have the work result coded in a computer-compatible form, which means that the engineering data produced in the design department can be used with access to the respective data repository by all functional units in the company.

Organizationally speaking, the design process consists of phases, in which an order is subdivided into tasks and subtasks with specified intermediate or final work results (see Fig. 15.2). Beneath this phase-oriented sequential labor partition, according to the approaches of design methodology, a hierarchical labor partition can be found: From the academically qualified design engineer down to the draftsman a wide range of design responsibilities exists. Furthermore a competency-oriented labor partition into mechanical, electrical, hydraulic, and so on, problem solving and integration procedures is typical for design processes.

The CAD system has the function in this organizational context of rationalizing the operations management of information handling and of integrating different information chunks into a detailed work description. The output is a completely documented technical product.

Ergonomically speaking design work is information-processing work in its purest sense, consisting solely of human information-processing procedures, with different levels of cognitive control and the possibility of influencing the working process by identifying and supporting algorithmic tasks executed through software components. With respect to these continua of cognitive tasks four principal components can be identified;

- Creative components, like defining objectives of a product; setting goals for development procedures; identifying problem domains and initiating problem solving; and, where methods are unknown, developing problem solving methods,
- Strategic components, like problem and task solving along prescribed methods/procedures with algorithmic and perhaps iterative character, such as the application of iterative optimization tools or simulation procedures to test functional components of the product, mathematical model development, and calculation,

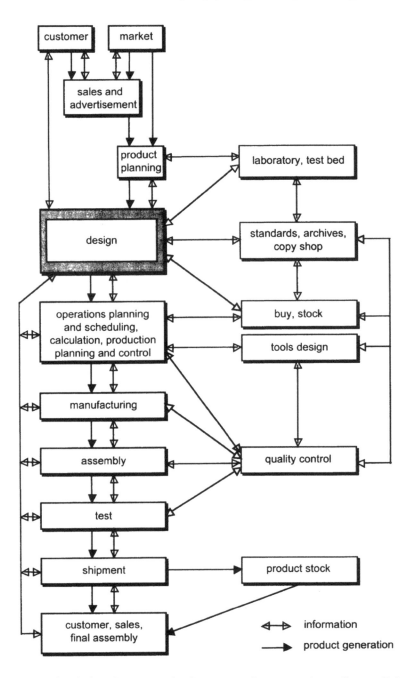

FIG. 15.1. The design department in the company's context (according to Beitz & Ehrlenspiel, 1984).

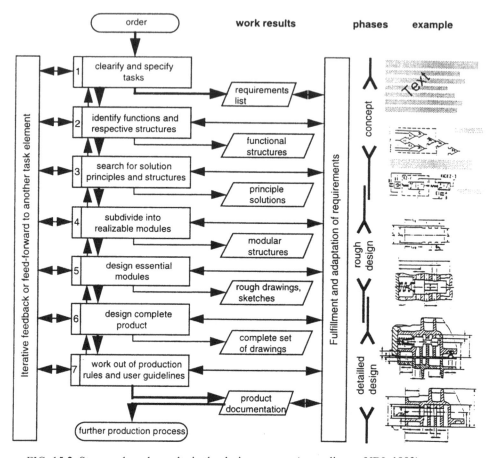

FIG. 15.2. Steps and work results in the design process (according to VDI, 1993).

- Routine components, like the application of simple algorithms, assigning and coding tasks, such as creation of parts lists, creation of dimensions, transforming sketches into drawings, and so on, and
- Automated components, like data entry, data transfer, data recall without loading effortful processes in the brain.

The paradigm used was an advanced variant of the stress-and-strain concept (Luczak, 1975; see Fig. 15.3) derived from the basic concept as introduced by Rohmert (1971). In this context, stress is an exterior influence on a designer, such as a difficult design task or high room temperature (Luczak, 1975). These stressors generate strain experienced by the designer. When analysing the

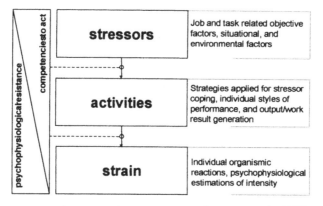

FIG. 15.3. The advanced stress-strain concept (according to Luczak, 1975).

stress-strain effects, the aim is to identify cause-effect relationships that are strong enough to be measured and tested statistically.

Typically, especially in engineering design, stressors can be varied by the designer through his activities. For example, if the designer has to solve a difficult task where difficulty is subjectively interpreted, he performs activities according to his individual problem solving strategies. The aim is to reduce the complexity and difficulty of the task *(task decomposition)*. Individuals utilize different strategies and activities to solve problems. This implies different strain reactions. Therefore, the stress-strain relationship is dependent on the individual competencies to act, as well as on the psychophysiological resistance of the individual (Luczak, 1975). To compare the strain reactions between individuals, the variation of the activities to be performed must be minimized by the experimental design. In other words, if the task is to evaluate stress-strain relationships, the nature of the task must be clearly defined.

However, if the aim of a research effort is the investigation of design activities and the improvement of tools to aid these activities, then design activities and tasks closely related to the design practice must be taken as a reference. The design practice is characterized by a high degree of freedom in the related activities. Therefore the activities in an experimental study cannot be predetermined either by the type of the activity or their existence and distribution over time. This leads to three major problems for the use of classical psychophysiological measures in the analysis of engineering design tasks (see Fig. 15.4):

- The imposition of different stressors, that is, a task characteristic or difficulty and a tool functionality (e.g., handling problem) lead to unknown strain effects. Moreover, different types of superpositions can occur (Luczak, 1982). When stressors are added, strain increases. Stressors that compensate each other reduce strain. Stressors can be indifferent,

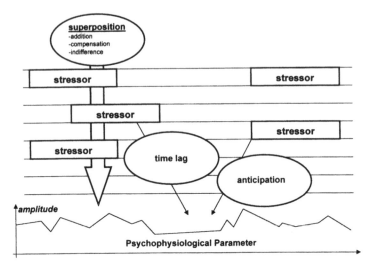

FIG. 15.4. Problems in the interpretation of stressors based on psychophysiological strain parameters.

suggesting that there is no interaction between the stressors in relation to the strain parameter mentioned. Moreover, the different stressors have different influences on different strain parameters, so the overall strain situation is indifferent.

- The strain parameter reaction can follow the occurrence of the stressor; therefore a positive time lag between the stressor and the strain parameter must be considered.
- The occurrence of a stressor can be anticipated, for example, a design problem that is anticipated before it can be described by rational arguments. Therefore, a possible strain reaction can be measured before the stressor can be identified as an action in response to the stressor. This leads to a negative time lag between the perceived stressor action and the strain reaction.

To give an example, Fig. 15.5 shows the results of cross-correlations for different time lags between a mean activity level and different strain and activity variables (motor activity of the right and left hand). The measurements were taken in a field study reported later. It is shown that most of the psychophysiological parameters show their maximal correlation at +1, which is a time shift by +20 seconds. Therefore, the psychophysiological response trails the stress situation by about 20 seconds. Nevertheless, the stress situation in Fig. 15.5 is described here at a

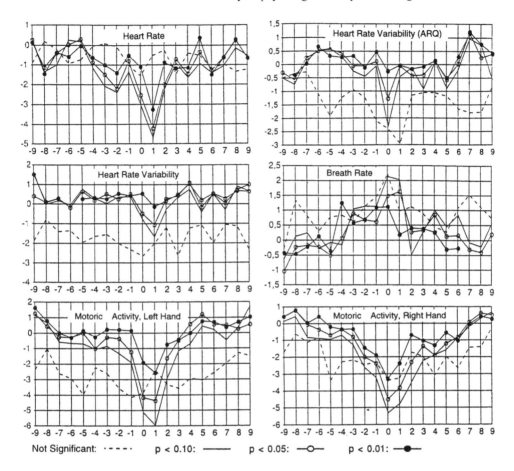

FIG. 15.5. Cross-correlations between the general activity, different stress indicators and motor activity (n=43 subjects; 80 hr standard-tasks; 141 hr project-tasks; time lag: −9=−3 min, +9=+3 min). The values are generated by a multiplication of the mean cross-correlation value (mean of 20 subjects) by the number of correlations found on the respective significance level.

mean activity level and does not differ between various activities that might be different in their strain effects and time lag.

In consequence, the concept is oriented to explain and interpret not only the effects of stable experimental conditions preset by a researcher, but gives room to a time-variant and parallel consideration of changes in stressor-variables, behavioural variables, activities for example, and psychophysiological indicators of strain.

RESEARCH OBJECTIVES

Psychophysiological stress-and-strain research in ergonomics is an approach that has problems in an absolute evaluation of working conditions: Zero points, maxima and minima, linearity, and so forth, of scales for stressors, activities, or strain are neither known nor easily definable in a reproducible manner.

Thus a relative evaluation approach using comparative studies is the efficacious way to determine the strength and rigor of the concept. Considering this, engineering psychophysiology research was centered around three problem fields:

- The comparative analysis of different degrees of technological development in work-system-design. Taking design work and CAD systems for the specification of this general problem, the question may arise whether or not design work takes advantage from the use of a CAD system in comparison to a conventional drawing-board. In consequence the experimental hypotheses to be tested may be formulated in simple words: More can be performed with less physiological resource consumption (Langner, 1991; Müller, 1992; Springer, Müller, Langner, Luczak, & Beitz, 1990).
- The comparative analysis of the task-support characteristics of different technological variants for work means. The marketing of CAD systems provides a variety of software and hardware solutions with different functionalities, different user interfaces, different technical equipment, and so on. Thus the experimental hypothesis for this type of problem may be formulated: Human performance and physiological costs depend on design and task-support characteristics of specific CAD technology (Langner, 1991; Luczak, Springer, Beitz, & Langner, 1992; Springer, 1992).
- A field oriented to the variety of different tasks in the different design phases and, respectively, the support of design aids. Especially for the early phases of engineering design, the use of design methods as well as complex information technology systems should be taken into account. For those design phases, CAD functionalities are based on complex artificial intelligence (AI) systems, data repositories, knowledge-based systems, and so on. Therefore, the type of study is a comparative analysis of the relative maxima and minima in mental or cognitive workload via psychophysiological indicators with reference to the cognitive structure of task execution. The hypothesis in this context may be: Cognitive engineering in terms of prescribed design-work-procedures with the use of AI-facilities can be improved or "optimized" by a psychophysiological resource consumption approach (Mühlbradt, Rückert, Springer, Beitz, & Luczak, 1994; Rückert, 1997).

COMPARISON OF LOW-LEVEL COMPUTER AIDS
WITH A DRAWING BOARD

Variation of Design Task Difficulty

A two dimensional CAD system should aid the designer in producing technical drawings. To compare different functionalities of CAD systems one needs to investigate whether or not the CAD system produces a real advantage over the conventional drawing board (e.g., in terms of design quality or execution time). This analysis should take into account the fact that different design tasks (based on type and difficulty) require different types of computer aids. Therefore, the set of design tasks in this study consisted of different levels of difficulty to vary the workload and were concerned with problems of mechanical engineering. The main aspect was geometrically oriented design, therefore the main activity was the making of technical drawings. Technical calculations and part lists were not to be included in this task because at the drawing board no calculation aids exist. The tasks were solvable by the subjects, students in mechanical engineering, within 120 minutes.

A set of design tasks selected according to these criteria was classified in an expert rating into three dimensions: *function-oriented* design; *production-oriented design*; and graphical representation (Müller, Springer, & Langner, 1989). The selection of eight design tasks of extreme factor-score combinations, two levels in three dimensions, yielded a factorial experimental design of the 2*2*2 type.

Design of the Laboratory Experiments

Thirty-three subjects participated in the experiments. They were students in mechanical engineering. Twenty subjects were needed for the homogeneous plan of the experiment (order of performing the tasks and of the tools used). The subjects' mean age was 23.35 years (standard deviation: 2.87) and the mean of the study duration was 4.1 semesters (standard deviation: 1.12). All participated in the engineering design exercises at the Technical University of Berlin, Germany and had an estimated 100 hours of experience in conventional technical drawing. All subjects participated in a 40-hour instruction course before the experiments to practice the use of the CAD system (Beitz, Langner, Luczak, Müller, & Springer, 1990).

In the experiments each subject had to perform the set of eight design tasks of varied difficulty alternatively at the CAD system (AutoCAD version 2.6) and at the drawing board. The order in which the separate tasks were assigned to the subjects was systematically manipulated.

Statistical procedures, such as analysis of variance and tests for independent groups, were used to determine the influence of the independent variables (applied tool and task difficulty in three dimensions) on the work results, working time, and strain experienced by the subjects.

Methods of Stress and Strain Measurements

The registration of the activities during the work session utilized video technology. One camera recorded the progress of work on the screen of the CAD system or on the drawing board, respectively; the second camera recorded the overall workplace. Both pictures were mixed, recorded on one tape, and coded with a list of items afterward. On the one hand, an analysis along these lines enabled us to create a profile of the actual stress components, on the other hand, it made possible the identification of strain outcomes, for example in identifying error occurrences.

In addition, various psychophysiological strain measures were continuously recorded during the whole work session to identify mental workload (Aasmann, Mulder, & Mulder, 1987; Luczak, 1987):

- From the electrocardiogram (ECG), heart rate, heart rate variability (Laurig et al, 1971; Luczak & Laurig, 1973), and the 0.10 Hz heart rate power spectra component were derived.
- Respiration rate was derived from thorax extension.
- The electromyogram (EMG) of the *musculus trapezius descendens* was measured as an indicator of a general strain level.
- Skin resistance was measured at the ball of the little finger to record electrodermal responses and their amplitude.
- Physical activity was measured with a three-axis accelerometer that allows interpretation of its influence on psychophysiological measures.

To analyze these measures, an online analysis was developed. After a first preprocessing step, the psychophysiological measures were recorded on an analog eight-track tape recorder. This record was used only as backup in case of a digitization failure. The same analog measures were preprocessed in a second step (filter, amplifier, etc.) and digitized online using an IBM-Compatible Personal Computer. The time-based synchronization between the video recording of activities and the psychophysiological measures was realized with the help of a video timer. The signal of the timer was mixed with the video recording, while its 1 Hz digital pulse was recorded on the analog eight-track tape recorder simultaneously with the psychophysiological data. An analog marker and a timer start pulse indicated the start of a work session both on the video- and analog-tape. Thus it was possible:

- To identify different strain situations and determine the corresponding stress and activity component, and

- To determine whether or not significant strain reactions on identified stress components occur.

Results

In a preliminary evaluation of the continuously measured psychophysiological variables, we obtained mean values and standard deviations of every parameter per work session. The tool used had an influence upon mean heart-rate, 0.10 Hz-power spectra component and the number and amplitude of skin-resistance reactions (see Fig. 15.6).

Apart from the 0.10 Hz component these parameters suggest a higher strain level at the drawing board. These results are in contradiction to the results of discontinuously measured psychophysiological parameters (flicker fusion frequency), as well as subjectively perceived strain, which were also measured in this study. This paradox was resolved by assuming a higher level of physical activity at the drawing board. Specifically, standing at the drawing board versus sitting at the CAD system causes an effect upon the psychophysiological measures. Increases in heart rate and fluctuations in skin resistance could be identified as artifacts caused by body movements. Moreover, the higher physical activity at the drawing board caused higher deviations of all psychophysiological

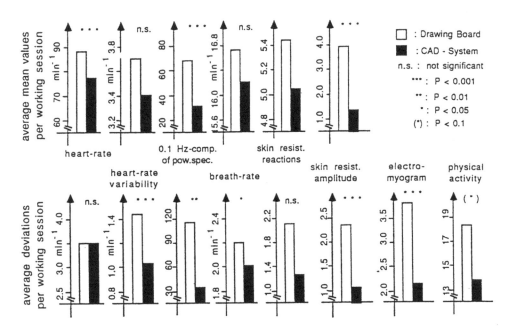

FIG. 15.6. Continuously measured psychophysiological strain parameters: Mean values and standard deviations for the drawing board and the CAD system.

parameters. However, a highly significant decrease of the 0.10 Hz component of the power spectra strongly indicated that mental workload at the CAD system was higher than at the drawing board. This result corresponds with the results of pre- and postcomparison of the subjective strain measures (Springer et al., 1990). However, it could not be decided whether the higher strain scores were the result of longer working time or a higher stress level caused by the CAD system (Langner, Müller, & Springer, 1990).

The difficulty of the working task had no influence on means and variance of the variables (see Table 15.1). Therefore, detailed evaluations of the psychophysiological parameters, especially the time series of activities and the combination of time series of strain measures and identified activities must demonstrate whether or not there is an influence of task difficulty upon the continuous measures of strain.

Cross-correlations between the use of different functions at the CAD system and the strain parameters indicate only some minor relationships at a 10% significance level (see Table 15.2). Assuming that the 32 correlations calculated in Table 15.2 were significant on the 10% level, 3 to 4 of the significant correlations may be due to chance.

In summary, psychophysiology gave no fruitful contribution to an improvement resulting from the use of a CAD system. Moreover, the comparison between different technological means, a conventional drawing board and a CAD system, by psychophysiological parameters was not possible. Different energetic situations, sitting and standing, hide every mental effect of the tool capabilities.

COMPARISON OF DIFFERENT CAD SYSTEMS IN FIELD EXPERIMENTS

Experimental Conditions

To compare different technological characteristics of CAD systems, the application of CAD systems in an industrial context was analyzed by field experiments (Springer, Langner, Luczak, & Beitz, 1991). Eleven companies from electromechanical engineering, mechanical engineering, and micromechanical engineering fields participated in this second study. The average size of the companies was 1,000 employees. Forty-three subjects participated in the study. The qualification levels ranged from engineering trainees, draftsmen, and technicians to design engineers at a university level (BSc, MSc). All subjects were well acquainted with the functionality and use of their CAD systems.

In contrast to the laboratory experiments, the whole work environment must be taken into account as stress factors. Therefore, events like disturbances by colleagues and telephone calls must be documented to interpret possible strain effects.

TABLE 15.1

Means and Explained Variance of Task Characteristics by the Psychophysiological Variables

	Functional Difficulty			Manufacturing Difficulty			Representational Difficulty			Tool CAD vs. Drawing Board		
	MW−	MW+	ZSSrel	MW−	MW+	ZSSrel	MW−	MW+	ZSSrel	MWB	MWC	ZSSrel
Hearts rate	14.0	13.2	.000	13.0	14.1	.007	12.9	14.2	.007	18.9	8.2	.378***
0.10 Hz heart rate variability	37.2	33.4	.013(*)	34.4	36.1	.002	37.3	33.2	.031**	43.1	27.3	.397***
Skin resist. responses (1/min)	5.14	5.33	.008	5.23	5.25	.002	5.26	5.22	.000	5.39	5.09	.069**
Breath frequency (1/min)	16.2	16.6	.011	16.5	16.3	.006	16.5	16.3	.000	16.7	16.1	.080***

MW−: Mean value for low difficulty; MW+: Mean value for high difficulty; MWB: Mean value at the drawing baord; MWC: Mean value at CAD system; ZSSrel: Amount of variance explained by the factor of task difficulty

***$p < .001$; **$p < .01$; *$p < .05$; (*)$p < .1$

TABLE 15.2
**Correlations Between the Time Line of the Use of Different CAD-Functionalities
and Strain Parameters**

CAD	Heart Rate	0.10 Hz Heart Rate Variability	Breath Frequency (1/min)	Skin Resistance Responses (1/min)
Task clarification	.12	.14	−.10	−.06
Rough design	.07	−.25*	−.03	.10
Detail design	.14	−.12	.06	.20(*)
Drawing	−.24*	.19(*)	.12	−.04
Definition	−.07	.09	−.08	−.23*
Generation	−.16	−.03	−.04	−.09
Manipulation	−.03	−.23(*)	−.02	−.11
Deletion	.02	−.11	.02	−.11

Note. Mean over whole duration.
***$p < .001$; **$p < .01$; *$p < .05$; (*)$p < .1$

The studies lasted 1 week per department. Before the study, the respective members of the staff and participating subdepartments (management, works council, design-department, CAD) were briefed in detail regarding the purpose, extent, and duration of the study.

The sessions took place in computer-aided workplaces in the design departments. These are generally the workplaces of the subjects. The whole measurement apparatus was installed on a mobile cart parked in the immediate work area. The psychophysiological parameters were transmitted by a long cable to the cart so that the subject was left with sufficient room to move.

For each subject, the completion of the study involved the process of being given a standard task (approximately 2 hours) as well as the solution to an actual problem (from the designers practical work) chosen by the subject (approximately 4 hours, these results are not presented in this report). The standard tasks (A, B, or C) were distributed in a manner that assured that each task was executed in its entirety at least once on each of the eight CAD systems. Every subject worked on all the assigned tasks without interruption (except for breakfast and lunch, meetings, etc.) in the course of 1 day.

Video and psychophysiological data of approximately 2 hours were usually recorded for every subject during the work on the standard task. In addition, data describing the environment conditions (temperature, humidity) and subject-related data resulting from questionnaires in the field of design and CAD experience were collected.

Results

For the elaboration of design environments, it is desirable to know what type of CAD functionality induces the least amount of strain on the designer. Obviously, the characteristics of those CAD systems that reduce strain are more appropriate for the designer than the other.

Figure 15.7 shows the mean correlations between the use of different functions and the heart rate variability for each CAD system. It also shows the results for those functions that should normally reduce stress by the reduction of needs for routine tasks. For example, dimensioning and hatching are functions that normally imply a reduced cognitive effort and are more routine operations. It can be seen that the generation and geometrical manipulation of the elements especially induces mental workload.

Comparing different CAD systems and, therefore, the different design of the functionalities, only a low relation between generation activities and mental strain was obtained for the SIGRAPH system. To explain this result one should know that the SIGRAPH system is based on parametric modeling: With the generation of elements, specific relations between the elements were generated simultaneously. Examples are measurement constraints (e.g., the distance has to be x mm) or generation constraints (e.g., the line is parallel to another line). With this functionality of the SIGRAPH system, manipulations that would be done later in the design process can be performed by the designer without large effort because all constraints connected with an element will induce manipulations of those elements that are connected by the constraint. Therefore, geometrical manipulation is a routine activity in this CAD system that induces a least amount of strain: Only a weak relationship between heart rate variability and geometrical manipulation was obtained (see Fig. 15.7).

By differentiating between the types of objects with which the designer is working, it can be shown that, independent of the type of the CAD system, the work with the basic elements results in an increase in the strain level. Many ergonomic deficiencies can be investigated by looking into the details of the generation process. The design of the menus, the type of dialogue, the description and explanation of parameters that must be defined, the need of completeness in parameter definition, and numerous others are deficiencies of the CAD systems. Therefore, many hints for the development of design improvements can be gamered from such an evaluation.

The relatively low correlations between working with complex elements and the strain level outlines the need for customization of the CAD systems. Complex elements that are customer specific are especially defined by either the company or the designer to reduce the effort in defining the complex geometrical topology with basic elements. The prescription of standard parts and their storage in data bases is an appropriate way for a reduction in the stress factors. In other words, if there are no prescriptions for standard parts a higher strain level results.

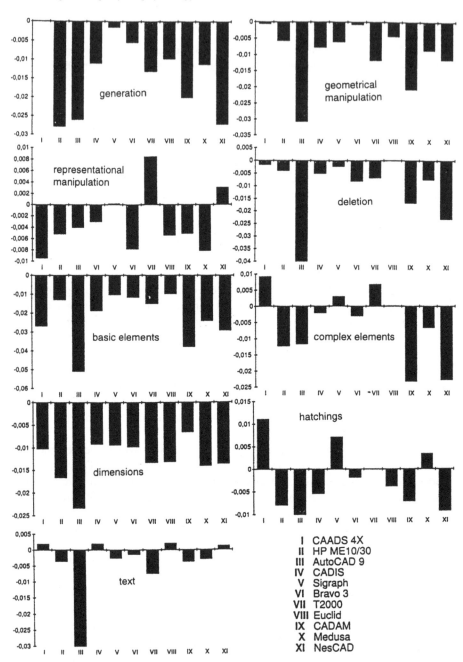

FIG. 15.7. Mean cross correlations compared for different CAD systems between heart rate variability (ARQ) and the use of different functions (generation, geometrical/representational manipulation, deletion) as well as the work on different objects (basic elements, complex elements, dimensions, hatchings, text), $N=43$.

In summary, the application of psychophysiological methods gave hints of where to focus attention for further design improvements of the CAD systems. Nevertheless, the situations and their related activities can be described only very roughly in terms of generation or working on different elements. A detailed insight into the problems the users had with the different functionalities was not possible because an investigation of the mental activities, mental structure, motivations, and problems experienced by the operators was impossible. This was taken into account in later studies.

COMPARISON OF ADVANCED COMPUTER AIDS WITH A DRAWING BOARD

Working Environment

A third approach was the comparison of different computer aids for design, which comprises conventional CAD and computer aided engineering (CAE) components, as well as artificial intelligence (AI) modules (e.g., for the selection of functional solutions in a specific problem domain).

During the experiments, which were again set in a laboratory, seven engineers worked with conventional tools including a drawing board, pencil, calculator, and catalogues. A second group consisting of six engineers worked using the design system KALEIT.

KALEIT was developed at the Institute for Machine Design to provide the engineer with a computer aid in every design phase (Beitz et al., 1992). It is divided into several modules, each module for one design step (Table 15.3), beginning with editors for the design task, the specifications, and the function structure (Kuttig, 1992). The preliminary and definitive layout can be drawn using a feature modeler based on the CAD system CATIA (from IBM-Dassault). For the selection and dimensioning of construction components, the designer can use a special knowledge-based system called WIKON during every design stage (Klein, 1992). With the exception of WIKON, all modules are integrated to X-windows with a common menu and can be handled in parallel.

Subjects and Design Task

The engineers participating in this third study graduated from the Technical University of Berlin, Germany, with Master's degrees in mechanical engineering. All had training in design methods, and those working with KALEIT knew at least one KALEIT module in advance. They were trained in the other modules during sessions prior to the experiments.

Every engineer was asked to solve the same design task, which consisted of designing a device to rotate a table in a washing machine. They had 6 hours to

TABLE 15.3
The Design System KALEIT (Design Analysis and Control System)

Human–Computer Interface	KALEIT-Menu		GEKO + CATIA		WIKON
Design step	Performance specifications	Function structure	Preliminary layout	Definitive layout	Information selection calculation
Design task					
Platform	IBM RISC 6000		IBM RISC 6000		SUN 386i

complete the task and were asked to use design methods according to Pahl and Beitz (1996; see Fig. 15.2).

Psychophysiological Approach

Keeping the results of the studies reported earlier and the discussion about the problems of psychophysiology with the analysis of nondetermined tasks in mind, further developments were needed to approach the psychophysiological measurement fruitfully. In relation to engineering design tasks, it is nearly impossible to investigate stressors by psychophysiological measurements without the help of the subjects who must interpret the strain reaction in terms of the casual stressors. On the other hand, working with subjective measurement methods alone (e.g., thinking-aloud methods) is stressful for the subjects because of a continuous bias of mental workload using thinking-aloud methods. Ex post continuous video interpretation are very time consuming. Accordingly, the subjects will not give a valid report of the stress situation or simply forget if they felt stressed or not.

Therefore, a combination of objective psychophysiological indicators for strain with an ex post analysis of the existing stressor combinations might be a fruitful approach to analyzing design activities. Once a sequence of activities was identified as stressful by psychophysiological strain parameters, the subjects

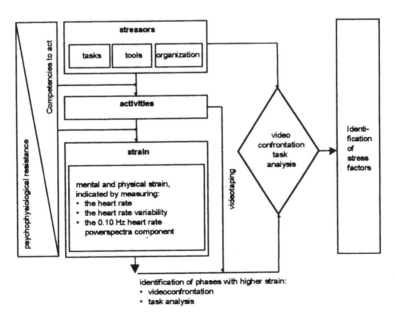

FIG. 15.8. The stress-and-strain concept, related measures and ex-post analysis was the stressor combination based on subjective statements.

were asked to give an interpretation of the situation (see Fig. 15.8). The subjects interpreted the situation after finishing the design task (ex post) to avoid the measurement bias inherent in thinking aloud methods.

In such subjective statements, detailed information was obtained about the nature of the design problems, the handling of tools (CAD, AI systems, conventional aids, etc.), and the environmental conditions (time pressure, disturbances, etc.). Because the design task was already finished, no time pressure exists to investigate all aspects of the stress situation. Because the situation was documented on video in different views (overall situation, CAD screen, desk views), the memory of the subject is maximally aided.

To measure the strain psychophysiologically, the following parameters were used in this study (Springer et al., 1990):

- Heart rate, which increases under stress, and
- Heart rate variability, which was measured specifically as ARQ (Laurig et al., 1971) as well as with the 0.10 Hz power spectra, and which decreases under informatory or physical stress (Luczak & Laurig, 1973).

Design and Data Analysis

To evaluate the strain experienced by the engineers while working on the design task, their heart rate was measured continuously. Every experiment was video taped so that the measured strain could be used to identify the stress factors. To avoid outside influence on the designer, the devices used for interpreting the measurements were located in a room adjoining the lab. In the lab room, only the designer and his or her tools, the video cameras, and the experiment manager were allowed. The experiment manager adjusted the video cameras when necessary and answered questions, for example, to clarify the design task or to help in unexpected situations with the computer systems.

During the experiment, the engineer's activities were categorized into a schema. Based on the videotape situation, the tools used by the designer, the representation mode and the functions of concern were categorized according to nearly the same schema used in the experiments discussed earlier.

Immediately after every experiment, a quick survey of the strain measurements was done and exceptional strain was identified. A critical phase was determined by the following procedure:

- An exponential regression by time was calculated for each psychophysiological parameter to correct the parameter by this regression. The aim was to exclude general time factors like adaptation or general fatigue.
- A z-transformation was calculated for each parameter.
- A time sequence was identified to be critical if at least one

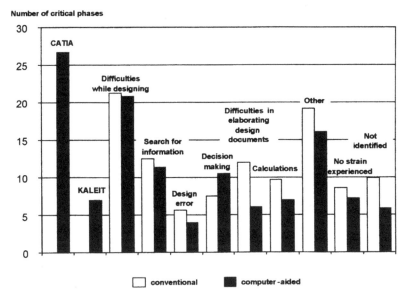

FIG. 15.9. Stress factors identified by the designers in the video confrontation (n=13 designers, N=208 critical phases).

psychophysiological parameter was out of the 2-level indicating higher strain.

Subjects were then confronted on videotape with these critical phases (strain-induced video-confrontation as described earlier). They were asked to explain the causes for the strain they had experienced. The analysis of these interviews provides information to help improve the computer systems as well as describing how the design task is performed. For that purpose, every critical period was identified by at least one stress factor according to what the engineers indicated as reasons for the strain.

Results

The overall incidence of the stress factors is represented in Fig. 15.9 (subfactors are not detailed). Overall, the 13 engineers were shown 208 critical phases (exceptional strain) to which 68 different stress factors were attributed 258 times.

In contrast to a previous study (Springer et al., 1990), the mean levels of the strain measurements showed no significant difference between conventional and computer-aided work designers. A possible reason for this is the prevalence of the design task as the main stress factor, masking the differences that may exist between the two work modes.

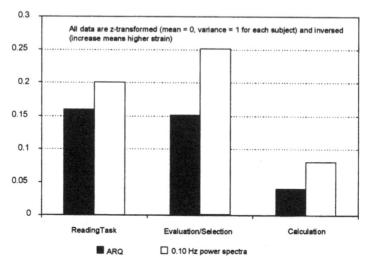

FIG. 15.10. Actions causing the highest strain.

Comparing the designers' activities with the strain measured, the following activities caused the highest strain (Fig. 15.10):

- Reading the design task,
- Evaluating and selecting different solutions, and
- Calculating.

These results are directly compatible with models of human information processing (Rasmussen, 1983). Calculation is, in most cases, a skill-and-rule-based activity and requires relatively little mental effort. In contrast to this, selection and evaluation, as well as the activities around task clarification, are more rule- and knowledge-based tasks that require higher mental effort. According to the stress-strain concept, these phases result in a higher strain level.

Although confrontation with the design task is a stress factor that is difficult to minimize by using computer aids, the designer can be relieved of calculations by using calculation programs adjusted to his or her needs. It is also possible to lessen the strain experienced when evaluating and selecting, as outlined in the following discussion.

Every designer generated several design variants in an individual order and degree of detail. From these variants, one had to be selected for detailing. Only one designer based his selection on a rational evaluation, which required the designer to establish the evaluation criteria himself and demanded rather detailed variants. Half of the engineers used no reproduceable evaluation and selection procedure. Through an analysis of their notepads and the interviews, the following evaluation criteria, specific to the given design task, could be found:

- Complexity (6 designers),
- Simplicity (6),
- Ease of assembly (5),
- Cost (4),
- Ease of manufacturing (3),
- Aesthetic (2),
- Ease of disassembly (2),
- Adapted to mechanical strain (2), and
- Number of parts, reliability, and ease of maintenance (1 each).

The other designers preferred to use formal evaluation methods (e.g., Pahl & Beitz, 1996) for selecting not only concept variants early in the design process, but also solution principles for subfunctions and detailed variants later in the design process. These methods provide a list of widely applicable evaluation criteria that the designer can use as support.

Thus, to minimize the stress factors *evaluation* and *selection,* the designer has to be relieved of the search for evaluation criteria. Because the computer-based design system KALEIT contains the design task and specification requirements, it is possible to envision an evaluation system that considers the design task, specifications, and drawings or notes made by the designer. Using its knowledge base and rules, it would generate evaluation criteria tailored to the specific construction. This would reduce mental efforts in the rule- and knowledge-based activities.

In a further step, the system could also propose an evaluation for the design variants, giving reasons for its evaluation in an explanation component. The designer could propose his or her own evaluation, which would be checked by the system for errors of judgment, thereby indicating to the designer where possible design errors may have occurred. Such a system was developed at the Institute for Machine Design as a version for selecting shaft-hub connections (Klein, Rückert, & Groeger, 1993).

Further analysis focused on the question of why some designers needed more time to perform the design task than others. The designers were divided into two groups according to the amount of time used. Interestingly, no significant difference was found in the quality of the definitive layout between the two groups. But the analysis of the video confrontation provides the following indications: The designers who needed more time to finish the task found it hard to determine the space for the device they had in mind, especially not to trespass the limits of the design area. They also discovered more design errors such as "assembly impossible," "device out of design area," or "function not fulfilled." Moreover, they mentioned "search for information," "decision making," and "difficulties in elaborating design documents" as a stress factor more often than the other designers. Strain experienced when searching for information can increase when the information cannot be found. Decision making may be

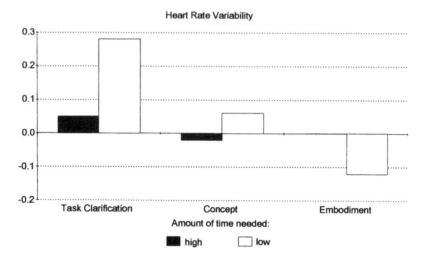

FIG. 15.11. Heart variability during the design phases, all data z-transformed (mean=0, variance=1 for each person), and inversed (increase means higher strain).

accompanied by higher strain when the criteria for making a decision are not clear or when the decision maker is aware of major flaws in the alternatives. The higher incidence of "difficulties in elaborating design documents" is either due to a lack of a sufficiently detailed mental model of the layout they wanted to represent in the design documents, or to an incapacity to transform the mental model to the required model in the design document. Thus, higher mental effort induces higher strain as well as lower performance.

All of these problems normally occur in the concept phase because the embodiment phase consists mainly of drawing the selected variant. This also explains the higher amount of time spent in the concept phase as well as the higher strain on the designers who needed more time to finish the design task than the other designers (see Fig. 15.11).

Limitations of the Study

To correctly interpret the results of this study, the following restrictions have to be taken into account:

- The study was done in a laboratory. Participants were all mechanical engineers, graduates from the Technical University of Berlin with training in design methods.
- The aim of the study was to identify and reduce stress factors in design work. The design processes in this study may not be representative of design processes found in the industry.

- The engineers were not required to verbalize their thoughts while designing. For the analysis, only the drawings, sketches, written notes, video tape, and interview could be used. Mental design processes in periods with lesser strain cannot be evaluated this way.

Thus, it is not possible to transfer the results directly to design processes in the industry.

CONCLUSION

Working with the CAD system was analyzed with psychophysiological measures, as well as activity-related and performance measurements. During the experiments, the most common and valid measurements, heart rate, respiration rate, and heart rate variability were measured (online) to evaluate stress and strain. Every experiment was videotaped to identify stress factors directly after the experiments. One major result of the first two studies was the lack of interpretation of strain reactions. Because the possible stressor can result from many different sources (task, design activities, CAD system, disturbances, etc.), an interpretation of the stressor related to a psychophysiological strain reaction is nearly impossible without the help of the subject. Therefore, a new method was developed called strain induced video-self confrontation.

The methods used in this final study to analyze computer-aids have been found to be very valuable. Many insights into how designers work were obtained. The interviews were evaluated using content analyses. Whereas psychophysiological indicators are common to the analyses of stress and strain, content analysis is not normally used in this context. This method, which is used mainly in the social sciences for the analysis of written material, allows the analysis of text by coding and structuring. This two-step procedure (measurement of psychophysiological indicators and the subsequent content analyses of the interviews) is thought to lead to a significant reduction in the amount of data, and at the same time, allow a deeper insight into the strategies used and problems encountered by the subjects.

The succession of phases gives an illustration of the overall strategy of problem solving chosen by the different subjects. With the help of the different phases of problem solving and the attached items, it was possible to create a qualitative matrix for each subject, presenting various kinds of problems and causes of strain in the various phases of the design process.

In general, the following design-oriented results were obtained from the various psychophysiological studies (Rückert & Springer, 1994a, 1994b):

- Most of the psychophysiological strain resulting from handling operations at the CAD system are due to ergonomic deficiencies of the systems.

- Stress can be minimized when the CAD system is designed to be as task appropriate as possible.
- The evaluation of the design activities and the video confrontation shows that the chosen strategy of problem solving is, in most cases, far from being a linear progression through the various phases. Individual problem solving strategies must be supported.
- Organizational factors like user qualification and training, support structures, data exchange with other applications, and so on are the most important factors for an efficient (and economic) CAD application in terms of performance and psychophysiological responses.

The results leave no doubt that a psychophysiological approach to the identification of problem areas for the design of working conditions provides a fruitful contribution to their improvement.

ACKNOWLEDGMENTS

This publication is dedicated to our honored colleague Wolfgang Beitz, Professor for Engineering Design at Technical University of Berlin. He died in November 1998 after an outstanding life dedicated to science. We are grateful for the many energizing discussions between Engineering Design and Ergonomics.

This research was funded over 6 years by a grant from the German Research Association. Thank you for giving us the opportunity to do this interdisciplinary research.

REFERENCES

Aasmann, J., Mulder, G., & Mulder, L.J. (1987). Operator effort and the measurement of heart rate variability. *Human Factors, 29*(2), 161–170.
Beitz, W., & Ehrlenspiel, K. (1984). Modellvorstellung für Entwicklung und Konstruktion [Models for product development and design]. *VDI-Z, 126*(7), 201–207.
Beitz, W., Langner, T., Luczak, H., Müller, T., & Springer, J. (1990). Evaluation of a compact CAD-course in laboratory experiments. *International Journal of Human-Computer Interaction, 2*(2), S.111–135.
Beitz, W., et al. (1992). Strukturen Rechnerunterstützter Konstruktionsprozesse. Abschlußbericht des Teilprojektes B2 des Sonderforschungsbereichs 203: Rechnerunterstützte Konstruktionsmodelle im Maschinenwesen [Structures of computer aided design processes. Final report of project B2 of reaserch group 203: Computer aided design models in mechanical engineering]. Berlin, Germany: TU Berlin.
Bortz, J., Lienert, G.A., & Boehnke, K. (1990). *Verteilungsfreie Methoden in der Biostatistik* [Distribution free methods in biostatistics]. New York: Springer-Verlag.
Klein, S. (1992). An example of knowledge-based decision making when selecting standard

components: Shaft-hub-connections. *Proceedings of the 4th International ASME Conference on Design Theory and Methodology DTM, Phoenix, AZ.*

Klein, S., Rückert, C., & Groeger, B. (1993). Erfahrungen aus der Anwendung eines wissensbasierten Systems zu Konstruktionsunterstützung—Ergebnisse eines Laborversuchs [Experiences with the application of a knowledge based system aiding engineering design—results of laboratory studies]. *VDI-Berichte Nr. 1079.* Düsseldorf, Germany: VDI-Verlag.

Kuttig, D. (1992, June). A model for describing functional modeling in the computer-aided design process. EDC-Workshop, *Understanding Function and Function to Form Evolution,* Cambridge, UK.

Langner, T. (1991). Analyse von Einflußfaktoren beim rechnerunterstützten Konstruieren [Analysis of influences in computer aided design]. Doctoral dissertation, University of Technology, Berlin. In W.Beitz (Ed.), *Schriftenreihe Konstruktionstechnik, Band 20.* Berlin, Germany: Verlag TU Berlin.

Langner, T., Müller, T., & Springer, J. (1990). Belastungs- und Beanspruchungsanalyse der Konstruktionsarbeit [Stress and strain analyses of engineering design work]. *Konstruktion, 42,* 157–164.

Laurig, W., Luczak, H., & Philip, U. (1971). Ermittlung der Pulsfrequenzarrhythmie bei körperlicher Arbeit [Calculation of heart rate variability during physical work]. *Internationale Zeitschrift für angewandte Physiologie einschließlich Arbeitsphysiologie, 30*(4), 40–51.

Luczak, H. (1975). Untersuchungen informatorischer Belastung und Beanspruchung des Menschen [Studies of mental stress and strain of the human being]. *Fortschritt-Berichte VDI-Z, Reihe 10, Nr.2.* Düsseldorf, Germany: VDI-Verlag.

Luczak, H. (1982). Grundlagen ergonomischer Belastungssuperposition [Basics of ergonomical superposition of workloads]. In W.Rohmert (Ed.), *Ergonomie der kombinierten Belastungen* (pp. 19–46). Cologne, Germany: O.Schmidt.

Luczak, H. (1987). Psychophysiologische Methoden zur Erfassung psychophysischer Beanspruchungszustände [Psychophisiological methods to measure psychophysical strain]. In C.F.Graumann et al. (Eds.), *Enzyklopädie der Psychologie, D-III-1* (pp. 185–259). Göttingen, Germany: Hogrefe.

Luczak, H. (1997). Task analysis. In G.Salvendy (Ed.), *Handbook of human factors and ergonomics* (2nd ed., pp. 340–416). New York: Wiley.

Luczak, H., & Laurig, W. (1973). An analysis of heart rate variability. *Ergonomics, 16*(1), 85–98.

Luczak, H., & Springer, J. (1997). Ergonomics of CAD Systems. In M.G.Helander, T.K. Landauer, & P.Prabhu (Eds.), *Handbook of human-computer interaction* (2nd ed.). Amsterdam: Elsevier.

Luczak, H., Springer, J., Beitz, W., & Langner, T. (1992). Betriebliche Feldexperimente zur Unterstützung von CAD-Systemen bei Konstruktionsprozessen [Field studies to the support of CAD systems in engineering design]. In A.-W.Scheer (Ed.), *Simultane Produktentwicklung* (pp. 233–275). Munich, Germany: gfmt Gesellschaft für Management und Technologie Verlags KG.

Mühlbradt, T., Rückert, C., Springer, J., Beitz, W., & Luczak, H. (1994). Analysis of stress and strain of design engineers solving mechanical problems—integrating a comparison of conventional and computer-aided design-work into the development of a design management system. *Proceedings of the 12th Triennal Congress of the International Ergonomics*

Association: Vol 5. Ergonomics and the Workplace (pp. 358–360). Mississauga, Ontario: Human Factors Association of Canada.

Müller, T. (1992). *Handlungsvollzug und psycho-physische Beanspruchung bei Konstruktionstätigkeiten* [Activities and psychophysical strain in engineering design]. Unpublished doctoral dissertation, Technical University of Berlin.

Müller, T., Springer, J., & Langner, T. (1989). Job analysis on design work. In K.Landau & W. Rohmert (Eds.), *Recent developments in job analysis.* London: Taylor & Francis.

Pahl, G., & Beitz, W. (1996). Engineering design. In K.Wallace (Ed.), *The design council* (2nd ed.). Berlin, Heidelberg, New York: Springer-Verlag.

Plath, H.-E., & Richter, P. (1978). Der BMS(I)-Erfassungsbogen—Ein Verfahren zur skalierten Erfassung erlebter Beanspruchungsfolgen [The BMS(I)-procedure—A strain measurement method based on subjective rating scales]. *Probleme und Ergebnisse der Psychologie, 65,* 45–67.

Rasmussen, J. (1983). Skills, rules, and knowledge: Signals, signs and symbols, and other distinctions in human performance models. *IEEE Transactions on Systems, Man and Cybernetics, SMC-13,* 257–266.

Rohmert, W. (1971). International Symposium on objectiv assessment of work load in air traffic control tasks. *Erganomics, 14,* 545–557.

Rückert, C. (1997). Untersuchungen zur Konstruktionsmethodik—Ausbildung und Anwendung [Studies of design methods—Education and application]. Doctoral dissertation, Technical University of Berlin. *VDI-Fortschritt Berichte Reihe 1, No. 293.* Düsseldorf, Germany: VDI-Verlag.

Rückert, C., & Springer, J. (1994a). Konstruktionsarbeit unter arbeitswissenschaftlichen und konstruktionsmethodischen Gesichtspunkten [Ergonomical and methodological aspects of design work]. *Konstruktion, 46*(1), 33–40.

Rückert, C., & Springer, J. (1994b). Testing design methodology through task performance. In T.K.Hight & F.Mistree (Eds.), *International Conference on Design Theory & Methodology, DTM '94* (pp. 161–168). New York: American Society of Mechanical Engineers.

Springer, J. (1992). Systematik zur ergonomischen Gestaltung von CAD-Software [Systematical and ergonomical design of CAD systems]. Doctoral dissertation, Technical University of Berlin. *VDI-Fortschritt Berichte Reihe 20, Nr. 60.* Düsseldorf, Germany: VDI-Verlag.

Springer, J., Langner, T., Luczak, H., & Beitz, W. (1991). Experimental comparison of CAD systems by stressor variables. *International Journal of Human-Computer Interaction, 3*(4), 375–405.

Springer, J., Müller, T., Langner, T., Luczak, H., & Beitz, W. (1990). Stress and Strain caused by CAD Work—Results of a laboratory study. In L.Berlinguet & D.Berthelette (Eds.), *Proceedings of the Work with Display Units Conference, Montreal* (pp. 231–238). Amsterdam: Elsevier.

Chapter 16

Stress and Health Risks in Repetitive Work and Supervisory Monitoring Work

Ulf Lundberg
Gunn Johansson[1]
Stockholm University

Assembly work and supervisory monitoring constitute two occupational activities essential to industrial production that represent different work conditions. For example, work on the assembly line of a car factory and process monitoring in a nuclear power plant differ in terms of demands for training and education, physical environments, consequences of human error, and so on. However, they also share a fundamental psychological characteristic in that both are associated with monotony (Johansson, 1991).

Both types of work are undergoing considerable change. In postindustrial so cieties, assembly work in its classical sense (Taylor, 1923) is being modified in order to allow more flexibility in production systems and more individual control and responsibility (e.g., Wall & Martin, 1987). Supervisory monitoring is being transformed by computerized control systems that tend to diminish requirements for activity by the operator and confine him or her to centralized control rooms at a distance from the actual production process.

This chapter reviews research on assembly work and other types of repetitive work, as well as supervisory monitoring using a psychobiological stress perspective. In analyzing these kinds of human activities, it is useful to distinguish between the stress perspective, which considers the total work environment including the physical, psychosocial, and organizational aspects, and a more

---

[1] Both authors are equally responsible for all parts of this chapter.

specific, task and performance-focused mental load perspective (Gaillard, 1993; see chap. 2, this volume). The focus here is on stress rather than mental load and performance. Comparisons with other occupational groups are offered, and long-term health consequences are discussed. Due to the high prevalence of musculoskeletal disorders in monotonous and repetitive work, special interest is focused on possible mechanisms linking work stress to muscle pain syndromes.

STRESS MODELS

Theoretical models of the stress process (Appley & Trumbull, 1986; Frankenhaeuser, Lundberg, Augustson, et al., 1989; Kahn & Byosiere, 1992) emphasize the imbalance between the individual's perception of the demands from the environment and his or her perceived resources to meet these demands. In terms of health impacts, a multifactorial etiology is assumed. In the development of the major health problems in postindustrial societies, work-related stress seems to play an important role, interacting with other psychosocial and physical environmental conditions, genetic factors, personality, and lifestyle in its influence on bodily functions.

The types of tasks covered in this chapter are usually organized in such a way that they expose the individual to an imbalance between demands and resources as implied earlier. For instance, Levi, Frankenhaeuser, and Gardell (1982) describe how stress is usually linked to overstimulation and overload, such as a very high work pace, conflicting demands, too much responsibility, or problems too complex to solve. They also describe how stress may occur due to underload or lack of stimulation. This may happen in very simple, repetitive jobs as well as in uneventful, monotonous monitoring tasks.

Another well-known and similar model of work-related stress is the job strain model suggested by Karasek and coworkers (Karasek, 1979; Karasek & Theorell, 1990). The model proposes that the combined load from high work demands and little influence over the pace and content of work is associated with elevated morbidity and poor well-being. Epidemiological studies (e.g., Johnson, Hall, & Theorell, 1989) show that work strain is significantly associated with elevated risk of coronary heart disease (CHD). More recently, low so cial support has been added to this model as a risk factor (Johnson & Hall, 1988). An additional model for work stress, the effort-reward model, has recently been proposed by Siegrist (1996). According to this model, the imbalance between effort (e.g., job involvement) and reward (e.g., status, appreciation, money) is the major determinant of job stress, which in a recent prospective study (Bosma, Peter, Siegrist, & Marmot, 1998) was found to predict CHD.

Biological functions, such as the autonomic nervous system, and the endocrine, cardiovascular, metabolic, and immune systems, play important roles in the organism's adaptation to the environment by protecting and restoring the body.

These psychophysiological responses are critical to survival (Seeman, Singer, et al., 1997; Selye, 1956), but may, under certain conditions, also have damaging health consequences. The allostatic load model (McEwen & Stellar, 1993; Seeman, McEwen, et al., 1997) proposes different ways in which long-term impact of physiological stress responses may increase health risks. Allostasis refers to the ability to achieve stability through change. A normal and economic response to a stressor requires activation of these physiological systems in order to cope with the stressor and then shuts off the allostatic response as soon as the stressor is terminated. However, over- or underactivity of the allostatic systems may add to the wear-and-tear of the organism. Different types of responses increasing the allostatic load can be distinguished: (a) activation of physiological systems that are too frequent and intense for rest and restitution to occur; (b) inability to shut off the stress response after the stress exposure or lack of adaptation to an environmental situation, causing overactivity and exhaustion of the systems; and (c) lack of adequate response in one system causing compensatory overactivation in other systems (McEwen, 1998).

Physiological responses and, in particular, the catecholamines and their concomitant impacts on other physiological functions, such as blood pressure, heart rate, and lipolysis, may serve as objective indicators of the stress process. However, these bodily responses are also assumed to link psychosocial stress to increased health risks.

Most research has focused on the influence of work stress on CHD. For example, longlasting elevated catecholamine levels are believed to contribute to the development of atherosclerosis and predispose for myocardial ischemia (Karasek, Russell & Theorell, 1982; Krantz & Manuck, 1984; Rozanski et al., 1988). The elevated catecholamine levels also make the blood more prone to clotting, thus reducing the risk of heavy bleeding in case of tissue damage, but, at the same time, increasing the risk of arterial obstruction and myocardial infarction. However, recent findings indicate that physiological stress responses are also involved in other major health problems, such as musculoskeletal disorders (Wærsted, 1997) and dysfunction of the immune system (Herbert & Cohen, 1993).

MUSCULOSKELETAL DISORDERS

The costs associated with musculoskeletal disorders increased dramatically during the 1980s, despite considerable improvements of the work environment. In British industry, for example (Symonds, Burton, Tillotson, & Main, 1995), work loss attributed to low back pain accounted for 33 million work days in 1981, 67 million in 1991, and 81 million in 1992. About 30% of the musculoskeletal costs are considered to be work-related, but it has been difficult to explain the high incidence of musculoskeletal disorders in light physical work, for example, in assembly work in the electronic industry and data entry work at video display terminals (VDTs). However, several recent studies suggest that psychosocial factors, such as low job satisfaction and lack of autonomy and variation at work, are significantly associated with back pain and shoulder problems (Bammer, 1990; Bongers, de Winter, Kompier & Hildebrandt, 1993; Johansson, 1994; Moon & Sauter, 1996).

Women generally report more musculoskeletal problems than men. The reason for this is not quite clear, but there are several possible explanations: (a) Women have less physical strength than men and gender differences in terms of muscle fiber composition may be of importance; (b) compared to men, women are overrepresented in jobs with a high incidence of muscular problems (e.g., repetitive assembly work) and they tend to stay longer in these jobs; and (c) enployed women are often exposed to more work overload and role conflicts than men, due to the combined load from paid and unpaid duties. As a consequence, women have fewer opportunities than men for off-work relaxation and their muscles may therefore remain activated too long without rest and restitution.

PHYSIOLOGICAL STRESS RESPONSES

The sympathetic adrenal medullary (SAM) system with the secretion of the catecholamines epinephrine and norepinephrine (or adrenaline and noradrenaline) has been of particular interest in the study of stress.

Research on the S AM system has its roots in the work of Walter B. Cannon in the early 20th century (Cannon, 1914). On the basis of animal experiments, he described the fight-or-flight response and the emergency function of the adrenal medulla. Psychological stress activates via the hypothalamus the sympathetic nervous system, which signals to the adrenal medulla to secrete epinephrine and norepinephrine into the blood stream. Norepinephrine is also produced by sympathetic nerve endings. This defensive reaction prepares the body for active coping efforts (Henry, 1992).

The cardiovascular and neuroendocrine functions activated by the SAM system mobilize energy to the muscles, heart, brain, and, at the same time, reduce blood flow to the internal organs and the gastrointestinal system. In response to a physical threat, this is an efficient means for survival by increasing the organism's capacity for fight or flight. Today, however, the SAM system is more often challenged by threats of a social or mental rather than physical nature. The possible health consequences of intense, repeated, and/or sustained activation (wear-and-tear) of this psychobiological program in response to psychosocial demands is a major objective for stress research (McEwen, 1998; Steptoe, 1991).

Numerous laboratory experiments, as well as studies performed in natural settings, illustrate the sensitivity of the SAM system to various psychosocial conditions, such as daily stress at work, home, school, daycare centers, hospitals, on commuter trains, buses, and so on (see reviews by Axelrod & Reisine, 1984; Frankenhaeuser, 1971, 1983; Henry & Stephens, 1977; Levi, 1972; Lundberg, 1984; Mason, 1968; Ursin, Baade, & Levine, 1978; Usdin, Kvetnansky, & Kopin, 1980).

A small but relatively constant fraction of the circulating levels of epinephrine and norepinephrine in the blood is excreted into the urine (Franken-haeuser, 1971; Levi, 1972). Urinary values provide integrated measurements for extended periods of time (usually an hour or more), which is an advantage in the study of long-term (chronic) psychosocial stress (Baum, Lundberg, Grunberg, Singer, & Gatchel, 1985). In order to reduce the influence of circadian rhythms (Åkerstedt, 1979) and differences in baseline levels, individual responses to stress are often expressed in relation to corresponding baseline levels obtained during relaxation on a different day but at the same time of the day. Thus, if reliable baseline measures can be obtained, percent change from baseline is often a more relevant measure than the absolute level.

PSYCHOPHYSIOLOGICAL STRESS ASSOCIATED WITH REPETITIVE ASSEMBLY WORK

Traditional assembly work is typically characterized by a high work pace, sometimes by machine pacing, limited opportunities for social interaction and physical mobility, and poor opportunities to influence the content of work. A list of psychosocial factors characteristic of this type of work is provided in Table 16.1. It is obvious that traditional work on the assembly line constitutes a high-strain job according to the demands-control model (Karasek & Theorell, 1990; Theorell, 1989), and this is reflected in elevated psychophysiological arousal (Lundberg, Granqvist, Hansson, Magnusson, & Wallin, 1989; Melin & Lundberg, 1997).

Timio and coworkers (Timio & Gentili, 1976; Timio, Gentili, & Pede, 1979) reported catecholamine excretion of metal workers to be significantly higher on the assembly line than in work off the line. The results were replicated 6 months later in the same group of workers. In a study of light repetitive work, Borsch-Galetke (1977) found pronounced urinary excretion of epinephrine and norepinephrine when work was paid by individual compared to group piece-rate. Johansson et al. (1978) reported self-ratings, as well as hormone output by sawmill workers in highly repetitive tasks, and found considerably larger amounts of urinary catecholamines and more irritation on the job than among fellow workers in nonrepetitive jobs. In this case, repetitiveness was associated with machine pacing, physical constraint, and higher rates of psychosomatic complaints.

In real life, repetitiveness can seldom be isolated from other psychological aspects of work. Repetitive tasks, especially in manufacturing industries, are always associated with physical constraint and often with underutilization of

TABLE 16.1
Typical Psychosocial Work Conditions Characteristic of Repetitive Work

Work Condition	Repetitive Work	Supervisory Monitoring
Opportunities for:		
Physical mobility	Restricted	Good
Social interaction	Poor	Good
Social support	Restricted	Good
Collective control	Poor	Good
Task complexity	Low	High
Predictability	High	Moderate
Control of work pace	Low to moderate	Usually high, but occasionally paced by technical system
Physical work environment	Multiple exposures	Comfortable
Physical work load	Light to moderate	Light
Night work	Unusual	Common
Payment	Piece-rate	Salary

human resources, machine pacing, and piece-rate remuneration. Therefore, laboratory studies in which repetitiveness can be isolated from other conditions, provide a useful complement to field studies. In one such laboratory experiment, Weber et al. (1980) compared repetitive discrimination and counting tasks with a simple motor task. They obtained measures of self-reports as well as neuroendocrine reactivity and indicators of cortical arousal such as critical flicker fusion and EEG recordings. They concluded that repetitive work resulted in depressed alpha activity, elevated heart rate, and elevated epinephrine excretion rates. In a laboratory experiment on machine-paced versus self-paced repetitive work (complex reaction-time task), Johansson (1981) recorded self-reports and urinary catecholamine excretion rates. Heart rate, as well as epinephrine and norepinephrine excretion, were significantly increased during performance of the machine-paced task as compared with a control session. Self-paced work resulted in a similar, but less pronounced, reaction. In addition, machine-paced, but not self-paced, repetitive work was followed by delayed recovery of epinephrine and norepinephrine excretion to baseline levels. In the same vein, Lundberg, Melin, Evans, and Holmberg (1993) found elevated physiological arousal after repetitive VDT work in a laboratory experiment, compared to more stimulating VDT work. Finally, Cox et al. (1982) designed a laboratory study simulating two repetitive tasks that were chosen to reflect various aspects of unskilled repetitive work.

Epinephrine increased considerably from night rest to work and less dramatically from day-time relaxation in the lab to work.

Although comparisons between studies should be made with great caution, it may be concluded that real-life work situations tend to give rise to larger increases in catecholamine output than laboratory settings. A reasonable explanation would be that field data reflect the totality of the work situation, including factors such as machine pacing, performance-based payment, and exposure to environmental stressors such as noise. Another circumstance to take into consideration is that subjects in a laboratory study endure the experimental condition for a limited number of hours, whereas workers on the assembly line are exposed to repetitive conditions during entire work shifts and on a regular basis. Thus, in this respect, most laboratory data can be assumed to represent under-rather than overestimations of SAM activity during repetitive work.

In order to study the physical and the psychological load associated with assembly work, a series of studies of workers at a car engine factory has been conducted (Lundberg et al., 1989; Magnusson et al., 1990; Melin & Lundberg, 1997). At the assembly line, engines were mounted on carriers that followed magnetic tracks in the floor and automatically stopped at a number of work stations where new parts were fitted to the engines by the workers. The mean work cycle was about 90 seconds. The workers were able to pace their own work, but could not influence the arrival rate of the engines; that is, if a worker kept a lower pace than the coworkers, a number of carriers would line up at the work station.

From measurements of the biomechanical load in this type of work, it was concluded (Magnusson et al., 1990) that "The large number of back problem complaints among these young assembly workers can hardly be attributed to high loads on the spine caused by poor postures or lifting of heavy burdens. It seemed more likely, therefore, that other factors like monotony, stress, and low job satisfaction could be of greater importance" (p. 778).

In a subsequent study of the psychophysiological stress levels (Lundberg et al., 1989), it was confirmed that blood pressure, heart rate, and catecholamine levels were highly elevated at work. In addition, the workers perceived very little possibility to influence their work or develop and use their abilities. They also reported lack of variety and independence at work and their job satisfaction was very low.

In a recent study (Melin, Lundberg, Söderlund, & Granqvist, 1999), stress responses of male and female assembly workers at the same car engine factory were compared in two different types of work organizations: (a) repetitive work at an assembly line, as described earlier and (b) a new and more flexible type of assembly work in autonomous groups. All participants had been working in the particular type of work organization for at least 6 months before the study. Although the actual assembly work and production demands were the same, the conditions at work differed.

Workers in the more flexible situation formed groups of 6 to 8 people, had a considerable amount of freedom and variety in their job, and could, to some extent, influence both the pace and content of their work. They also had responsibility for quality control and the organization of work. Data from questionnaires confirmed that work in the flexible situation was perceived as more varied and independent with opportunities to learn new skills. The workers were also more satisfied with their jobs in the flexible work situation. Work demands were reported to be as high, or even higher, in the flexible work situation, compared to work on the assembly line and, in keeping with this, the mean physiological arousal (epinephrine and norepinephrine) at work was about the same in both conditions (mean values over 2 work days).

However, physiological stress responses and self-reported fatigue during work on the assembly line tended to increase and reached a peak at the end of the work shift, whereas stress responses in the more flexible type of work organization were kept at a moderate and more stable level throughout the shift. In addition, epinephrine levels after assembly line work were found to be significantly higher than after the flexible work situation (see Fig. 16.1). Thus, the assembly line workers did not relax and unwind after work as quickly as workers in the flexible situation (Melin, Lundberg, Söderlund, & Granqvist, 1999). Elevated stress levels at the end of the day have also been reported in earlier studies of machine-paced repetitive work (e.g., Johansson et al., 1978) and in repetitive clerical work (Johansson & Aronsson, 1984).

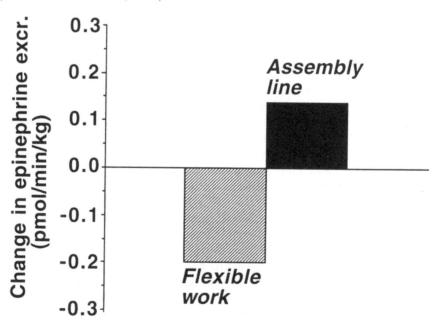

FIG. 16.1. Change in epinephrine output after work (in relation to baseline) of male and female assembly workers in two forms of work organization: assembly line and flexible work (Melin et al., 1999). Copyright © 1999 by Wiley. Reprinted with permission.

The differences between the two work organizations were particularly pronounced for women. The female workers had higher mean norepinephrine levels in repetitive assembly line work than in the flexible work situation, whereas the corresponding stress levels of men did not differ markedly. This gender difference was even more pronounced in the data collected after the work shift, that is, while unwinding at home.

A tentative explanation for the gender-differentiated data is that the female workers were stressed by having to adjust to the pace of the males at the assembly line, but were able to benefit from the more flexible work situation that, for example, gave them greater opportunities to cope with the various demands associated with their multiple roles (paid job, household chores, child care, etc.). Another possibility is that women better than men could cope with the social demands associated with the group situation (e.g., when discussing how the group should organize the assembly work, etc.).

TENTATIVE PSYCHOBIOLOGICAL MECHANISMS LINKING REPETITIVE WORK STRESS TO MUSCULOSKELETAL DISORDERS

It has generally been assumed that psychosocial stress may induce muscle tension and thus form a link to musculoskeletal disorders (Wærsted, 1997). This was confirmed in a recent laboratory experiment (Lundberg et al., 1994) where mental stress tests induced a significant increase, not only in perceived stress and blood pressure, but also in trapezius muscle tension (EMG-activity) of the women. Several other studies also support this assumption (Johansson, 1994; Svebak, Anjia, & Kårstad, 1993; Tulen, Moleman, van Steenis, & Boomsma, 1989; Wærsted, 1997), but the experimental evidence is not conclusive (Wærsted, Bjørklund, & Westgaard, 1991; Weber et al., 1980; Westgaard & Bjørklund, 1987). Significant correlations have been found between blood pressure and EMG activity and between mean norepinephrine and EMG responses in a stress condition (Lundberg et al., 1994).

Lundberg et al. (1994) found a significant increase in EMG activity when a stress test, the Stroop color-word test, was combined with a standardized physical load (test contraction), and this increase was significantly larger than that induced by the same stressor separately. This suggests an enhanced effect of psychological stress in, for example, jobs involving materials handling.

The results from the laboratory experiment have recently been followed up by a real-life study of cash register operators working in supermarkets (Lundberg et al., 1999). The aim of this study was to provide more information on the role of psychological stress and physical load in the development of muscle tension and neck-shoulder problems among women in a demanding and repetitive job with little possibility to influence the pace and content of the work. Almost 70%

of the cashiers suffered from neck-shoulder pain. In summary, the results from this study show that the increase in sympathetic arousal at work (epinephrine and norepinephrine output) among the female cashiers was quite high and that the elevation in blood pressure and muscle tension was most pronounced in women with neck-shoulder symptoms.

In a laboratory experiment, trapezius muscle tension of female cashiers was compared to that of a group of female students. It was found that the cashiers did not differ from the control group during the stress tests, but did have elevated EMG activity during rest periods (Melin, Lundberg, Kadefors, et al., 1999). This indicates that the cashiers may be at risk for muscular disorders, not necessarily due to elevated muscle tension during work, but rather due to lack of relaxation during rest periods.

In an interesting prospective study, Veiersted, Westgaard, and Andersen (1993) measured EMG activity during work and involuntary breaks at work every 10th week from the start of employment in 21 female packing workers (mean age = 25 years). Thirteen women contracted clinically diagnosed trapezius myalgia within the first year, half of them after 6 months. EMG data showed that women developing trapezius myalgia problems had higher muscle activity during breaks at work but not during actual work. It was concluded that "Sustained low-level muscle activity seems to be a risk factor for muscular pain" (Veiersted, 1995, p. 18). Additional information was obtained from analysis of the short periods of very low muscular electrical activity, that is EMG *gaps*. Female workers with a high frequency of EMG gaps seemed to have a reduced risk of developing myalgia problems compared to workers with fewer gaps (Veiersted, 1995). Similar findings have been reported by Hägg and Åström (1997).

Muscle pain associated with psychosocial factors at the workplace could be explained by a blocking of pauses in muscle activity unrelated to the actual biomechanical work being performed (Elert et al., 1992). In the modern work environment, it is possible that the lack of relaxation is an even more important health risk than the absolute level of contraction or the frequency of muscular activation.

Although knowledge about the mechanisms linking psychosocial stress to musculoskeletal disorders is still limited, new and interesting models have been proposed (Hägg, 1991; Schleifer & Ley, 1994). The model proposed by Hägg (1991), The Cinderella Hypothesis (referring to Cinderella who was first to rise and last to go to bed), is based on earlier findings by Henneman, Somjen, and Carpenter (1965), who showed that the motor units of a given muscle are recruited in a fixed order. Small, low-threshold motor units are recruited at low levels of contraction, before larger ones, and kept activated until complete relaxation of the muscle. According to this hypothesis, long-lasting activation of low-threshold motor units may cause degenerative processes, damage, and pain (Kadefors et al., 1999; Wærsted, 1997). Ongoing psychological stress may keep low-threshold motor units active more or less continuously. Wærsted et al. (1996) demonstrated continuous activation of low-threshold motor units during a 10 minute exposure to cognitive demands during VDT work in the laboratory. Although these small motor units are assumed to be fatigue resistant, there is likely to be an upper limit for continuous activation (Wærsted, 1997).

Support for the hierarchical activation of the motor units of the trapezius muscle, also during postural changes, has recently been obtained in an experimental study by Kadefors et al. (1999), using a new type of fine wire electrodes. Stress-induced muscle tension, with or without physical demands, would be enough to activate and prevent relaxation of such motor units.

A possible pathogenic mechanism for muscle pain is that nociceptors are sensitized due to local metabolic changes in fatigued low-threshold (Type I) muscle fibres (Sejersted & Vøllestad, 1993). The hypothesis of overload of certain motor units is supported by the observation of an increased number of "ragged red" Type I muscle fibers in the trapezius muscle of workers exposed to monotonous shoulder load (Bengtsson & Henriksson, 1989; S.E. Larsson et al., 1988; B. Larsson et al., 1992; Lindman et al., 1991). However, such changes have also been reported in workers without muscle pain (Lindman, 1992).

According to another model, proposed by Schleifer & Ley (1994), stress-induced hyperventilation decreases peak CO_2 levels and increases the blood pH-level (beyond 7.45=alkalosis). This contributes to elevated muscle tension and a suppression of parasympathetic activity. The sympathetic dominance may amplify the responses to catecholamines.

The role of psychosocial factors in the development of musculoskeletal disorders seems to be consistent with the allostatic load model. Cashiers working in the supermarket have an unfavorable work-rest pattern with increased muscle activity and very few periods of rest during a 2-hour work shift (Sandsjö, 1997). Jobs involving repetitive exposure to mental and physical demands (e.g., Lundberg et al., 1989), lack of relaxation on or off the job (Frankenhaeuser, Lundberg, Fredrikson, et al., 1989; Johansson et al., 1978; Melin & Lundberg, 1997; Veiersted et al., 1993), and lack of preparedness for job demands (Karasek, 1979; Theorell, 1989) are associated with elevated health risks. For example, workers in repetitive jobs (Johansson et al., 1978; Melin, Lundberg, Söderlund, & Granqvist, 1999) fail to shut down their stress responses after work. Due to multiple role responsibilities, female workers seem to be at partic-ular risk (Lundberg, 1996).

SUPERVISORY MONITORING WORK

Supervisory monitoring tasks are found in the control rooms in process industries, such as chemical plants, refineries, power stations, and paper and pulp industries. In most cases, the monitoring task includes a few fairly different components (Sandén, 1990). Supervisory control during normal operations is the most typical situation which may cover 80% to 90% of work hours in highly automated processes (Sandén & Johansson, 1990). Some typical characteristics are listed in Table 16.1.

One of the most characteristic psychosocial characteristics of this type of work is the contrast between normal operation and process disturbances, when the operator is required to shift from passive supervision to rapid information processing and decision making. Another characteristic feature is the long interval between disturbances and, thereby, the low frequency of opportunities for operators to utilize, much less develop, their process skills.

So far, supervisory monitoring has not attracted much attention in occupational stress research. One field study includes urinary measures of catecholamine excretion (Johansson & Sandén, 1997). Process operators supervising oxygen production, power generation, and the flow of cooling and heating water were studied during typical morning shifts and during work-free days spent at home. The work situation was perceived as boring, with the operators sometimes having a vague, uneasy feeling that was most pronounced during night shifts. Epinephrine excretion was significantly increased during work. Self-reported health did not differ between the process controllers and operators performing active and hectic work in control rooms.

The majority of studies on uneventful monotony that include psychoneuroendocrine measurements are laboratory experiments of fairly short duration, usually no more than a few hours. One such study was carried out by Frankenhaeuser (1976). The subjects performed a 3-hour vigilance task that caused a 50% increase in epinephrine and norepinephrine excretion. The experimental situation was associated with perceived distress and effort. In a study of simulated radar monitoring behavior of men and women, only norepinephrine excretion was significantly increased above baseline levels (O'Hanlon & Beatty, 1976). Mood and arousal were not recorded in this study. Lundberg and Frankenhaeuser (1980) studied vigilance performance as one of several experimental conditions and reported a 40% increase in epinephrine among male and female subjects. Finally, Johansson and coworkers (Johansson, Cavalini, & Pettersson, 1996) performed a study of two uneventful monitoring conditions that differed in terms of the amount of activity required by the subjects. In active as well as passive monitoring, epinephrine excretion was significantly increased above baselines, and the increase of hormonal activity was associated with an increase in perceived monotony and boredom.

To conclude, empirical evidence on psychoneuroendocrine reactions to uneventful monotony is scarce, especially with regard to real-life investigations. The data available indicate that, although prolonged uneventful monotony is associated with cortical deactivation and perceived difficulty to remain alert (Davies & Parasuraman, 1982), the excretion of epinephrine tends to increase in relation to baselines and, occasionally, the increase reaches statistical significance. The moderate activation of adrenal hormones may be interpreted to reflect the mental effort required for sustained attention and mental preparedness and does not seem to be associated with somatic complaints (Johansson & Sandén, 1997).

MENTAL AND PHYSICAL WORKLOAD AS RELATED TO STRESS HORMONES

As indicated earlier, overstimulation as well as understimulation, compared to more optimal environmental conditions, may contribute to elevated epineph-rine levels (Frankenhaeuser & Johansson, 1981; Frankenhaeuser et al., 1971; Levi, 1972). Under conditions where the individual is able to relax, catecholamine levels quickly return to baseline (Forsman, 1983). However, environmental stress may induce sustained activation of the SAM system and overexposure to stress hormones. Frankenhaeuser, Lundberg, Fredrikson, et al. (1989) demonstrated elevated urinary norepinephrine levels after work among female managers who were unable to relax due to unpaid work responsibilities at home (i.e., household chores, child care, etc.). Similarly, Rissler (1977) found that a period of overtime at work (performed on Saturdays) increased the catecholamine output in the evening at home (measured on Wednesdays) in female white-collar workers. In keeping with this, Lundberg and Palm (1989) found that overtime was associated with elevated epinephrine output during the weekend at home in full-time employed mothers of preschool children, but not in the fathers. In this context, it is also of interest to note that Alfredsson, Spetz, and Theorell (1985) found that overtime was associated with an elevated risk of myocardial infarction in women but not in men.

The magnitude of the catecholamine response is usually correlated with self-reports of stress, demands, and time pressure at the individual level (Frankenhaeuser et al., 1962; Lundberg et al., 1989). However, this pattern seems more consistent for men than for women (Frankenhaeuser, Lundberg, Augustson, et al., 1989; Lundberg, 1996) as men report more stress at work and have more elevated physiological stress responses, whereas no consistent corresponding association is found in women. By and large, data show that epineph-rine secretion mainly reflects the intensity of mental arousal induced by the environmental situation, whereas norepinephrine is more sensitive to physical demands and body posture.

Thus, measurement of these stress hormones may give information about the mental and physical load associated with different jobs. Table 16.2 summarizes findings from a series of real-life studies where catecholamine levels at work were expressed as percent increase from baseline levels for different groups of blue- and white-collar workers. Table 16.2 shows that epinephrine output is significantly elevated among blue- (assembly workers, cashiers) as well as white-collar workers (managers, clerical workers), whereas norepinephrine levels are elevated among blue-collar workers only. It is likely that the physical demands on

TABLE 16.2
Catecholamine Responses to Various Occupational Activities for Men and Women
(Increase From Nonwork Level)

Occupation/Task	Eprinephrine	Norepinephrine
Managers	(Frankenhaeuser et al., 1989)	
Men	++	+
Women	+	0
Clerical workers	(Frankenhaeuser et al., 1989)	
Men	+	0
Women	++	0
Supervisory monitoring	(Johansson & Sandén, 1997)	
Monitoring		
Men	+	+
Registered nurses	(Johansson et al., 1996)	
Women	+	++
Data entry	(Johansson & Aronsson, 1984)	
Women	+	0
Active process control	(Johansson & Sandén, 1997)	
Men	+++	0
Assembly workers	(Melin et al., 1998)	
Men	++++	++
Women	++	++
Assembly line workers	(Lundberg et al., 1989)	
Men	+++	++++
Sawmill workers	(Johansson & Aronsson, 1978)	
Men	++++	++++
Cashiers	(Lundberg et al., 1999)	
Women	+++	+++

+ = 10 < 25%; ++ = 25 < 50%; +++ = 50 < 75%; ++++ > 75%.

the white-collar workers are too low to influence their norepinephrine output. As noted earlier (e.g., Melin, Lundberg, Söderlund, & Granqvist, 1999), repetitive and monotonous blue-collar jobs are also associated with a slower physiological unwinding after work. This means that simple, monotonous, and repetitive jobs not only take a greater physiological toll at work, but also give less chance for relaxation and recovery off the job. Physical stressors such as noise may further contribute to the total load on blue-collar workers (Glass & Singer, 1972).

In view of traditional gender differences in responsibility for unpaid work at home (Hall, 1990; Kahn, 1991; Lundberg et al., 1994), with fewer possibilities for women than men to relax, the possible long-term health risks for women in repetitive work are of particular concern (Frankenhaeuser, Lundberg, & Chesney, 1991; Repetti et al., 1989; Rodin & Ickovics, 1990).

CONCLUDING COMMENTS

The main purpose of monitoring the excretion of stress hormones in occupational settings is the identification of stressors and the prevention of stress-related disease. Although such prevention of wear-and-tear of employee health ought to be an essential element in human resource management, it is often overlooked. In most Western societies, the cost of poor employee health affects the economy of industrial organizations only indirectly. There is a tendency to rely on national or private health insurance systems rather than to consider the cost in the organization's own budget.

However, the kind of psychoneuroendocrine states described here, which preceed cardiovascular and musculoskeletal disease, are associated with high arousal, feelings of tension and fatigue, and difficulties in unwinding after work. Such poor well-being is likely to affect employee motivation, performance quality, and employee loyalty in a way that may threaten organiza' tional productivity in a more direct way. The monitoring of indicators of SAM-system activity may serve as useful early warning signals.

Several psychophysiological studies suggest not only that intensity of stress during work may have damaging health effects, but also that lack of relaxation during rest is a serious problem, particularly with regard to musculoskeletal disorders. We need to learn more about the specific mechanisms linking psychosocial factors to muscular disorders. However, data consistently support the assumption that psychological stress and lack of unwinding play an important role by increasing muscle tension and preventing rest and restitution (EMG-gaps) in the absence of physical load.

As ergonomic improvements reduce the physical load at work, attention is drawn to aversive psychosocial aspects of the work environment. At the same time, the data reported earlier indicate that the influence of psychological stressors on muscular tension is enhanced in physical work. Thus, preventive actions must involve physical and psychosocial work conditions with special focus on women's stress and workload. The aim of our biopsychosocial approach is to provide data that might guide the design of such interventions. Finally, the biopsychosocial approach offers some sensitive health-related indicators that will prove useful in the evaluation of work environment intervention programs.

ACKNOWLEDGMENTS

Financial support for the research reported in this chapter has been obtained from the Swedish Council for Work Life Research (formerly the Swedish Work Environment Fund), the Swedish Council for Research in the Humanities and Social Sciences and the Bank of Sweden Tercentenary Foundation.

REFERENCES

Åkerstedt, T. (1979). Altered sleep/wake patterns and circadian rhythms. Laboratory and field studies of sympathomedullary and related variables. *Acta Physiologica Scandinavia* (Suppl. 469).

Alfredsson, L., Spetz C.-L., & Theorell, T. (1985). Type of occupation and near-future hospitalization for myocardial infarction and some other diagnoses. *International Journal of Epidemiology, 14*, 378–388.

Appley, M.H., & Trumbull, R. (Eds.). (1986). *The dynamics of stress. Physiological, psychological, and social perspectives.* New York: Plenum.

Axelrod, J., & Reisine, T.D. (1984). Stress hormones: Their interaction and regulation. *Science, 224*, 452–459.

Bammer, G. (1990). Review of current knowledge—musculoskeletal problems. In L. Berlinguet & D.Berthelette (Eds.), *Work with display units 89* (pp. 113–120). North-Holland: Elsevier.

Baum, A., Lundberg, U., Grunberg, N., Singer, J., & Gatchel, R. (1985). Urinary catecholamines in behavioral research in stress. C.R.Lake & M.G.Ziegler (Eds.), *The catecholamines in psychiatric and neurologic disorders* (pp. 55–72). Ann Arbor, MI: Butterworths.

Bengtsson, A., & Henriksson, K.G. (1989). The muscle in fibromyalgia—review of Swedish studies. *Journal of Rheumatology, 16*(Suppl. 19), 144–149.

Bongers, P.M., de Winter, C.R., Kompier, M.A.J., & Hildebrandt, V.H. (1993). Psychosocial factors at work and musculoskeletal disease. *Scandinavian Journal of Work Environment and Health, 19*, 297–312.

Borsch-Galetke, E. (1977). Katekolaminausscheidung bei Feinwerkerinnen mit und ohne Akkordarbeit [Catecholamine excretion in manual precision work with and without piece-rate remuneration]. *Zentralblatt Arbeitsmedizin, 27*, 53–58.

Bosma, H., Peter, R., Siegrist, J., & Marmot, M. (1998). Alternative job stress models and risk of coronary heart disease. The effort-reward imbalance model and the job strain model. *American Journal of Public Health, 88*, 68–74.

Cannon, W.B. (1914) The emergency function of the adrenal medulla in pain and the major emotions. *American Journal of Physiology, 33*, 356–372.

Cox, S., Cox, T., Thirlaway, M., & Mackay, C. (1982). Effects of simulated repetitive work on urinary catecholamine excretion. *Ergonomics, 25*, 1129–1141.

Davies, D.R., & Parasuraman, R. (1982). *The psychology of vigilance.* London: Academic.

Elert, J.E., Rantapää-Dahlqvist, S.B., Henriksson-Larsén, K., Lorentzon, R., & Gerdlé, B.U. C. (1992). Muscle performance, electromyography, and fibre type composition in fibromyalgia and work-related myalgia. *Scandinavian Journal of Rheumatology, 21*, 28–34.

Forsman, L. (1983). *Individual and group differences in psychophysiological responses to stress—with emphasis on sympathetic-adrenal medullary and pituitary-adrenal cortical responses.* Unpublished doctoral dissertation, Department of Psychology, Stockholm University, Sweden.

Frankenhaeuser, M. (1971). Behavior and circulating catecholamines. *Brain Research, 31*, 241–262.

Frankenhaeuser, M. (1976). The role of peripheral catecholamines in adaptation to understimulation and overstimulation. In G.Serban (Ed.), *Psychopathology of human adaptation* (pp. 173–191). New York: Plenum.

Frankenhaeuser, M. (1983). The sympathetic-adrenal and pituitary-adrenal response to challenge: Comparison between the sexes. In T.M.Dembroski, T.H.Schmidt, & G. Blümchen (Eds.) *Biobehavioral bases of coronary heart disease* (pp. 91–105). Basel: Karger.

Frankenhaeuser, M., & Johansson, G. (1981). On the psychophysiological consequences of understimulation and overstimulation. In L.Levi (Ed.), *Society, stress and disease: Vol. IV Working life* (pp. 82–89). London: Oxford University Press.

Frankenhaeuser, M., Lundberg, U., Augustson, H., Nilsson, S., Hedman, H., & Wahlström, K. (1989) *Work, stress, job satisfaction.* Swedish Work Environment Fund.

Frankenhaeuser, M., Lundberg, U., Fredrikson, M., Melin, B., Tuomisto, M., Myrsten, A.-L., Hedman, M., Bergman-Losman, B., & Wallin, L. (1989). Stress on and off the job as related to sex and occupational status in white-collar workers. *Journal of Organizational Behavior, 10*, 321–346.

Frankenhaeuser, M., Lundberg, U., & Chesney, M. (Eds.). (1991). *Women, work and stress.* New York: Plenum.

Frankenhaeuser, M., Nordheden, B., Myrsten, A.-L., & Post, B. (1971). Psychophysiological reactions to understimulation and overstimulation. *Acta Psychologica, 35*, 298–308.

Frankenhaeuser, M., Sterky, K., & Järpe, G. (1962). Psychophysiological relations in habituation to gravitational stess. *Perceptual and Motor Skills, 15*, 63–72.

Gaillard, A.W.K. (1993). Comparing the concepts of mental load and stress. *Ergonomics, 36*, 991–1005.

Glass, D.C., & Singer, J.E. (1972). *Urban stress.* New York: Academic.

Hägg, G. (1991). Static work loads and occupational myalgia—a new explanation model. In P.A.Anderson, D.J.Hobart, & J.V.Danhoff (Eds.), *Electromyographical kinesiology* (pp. 141–144). Amsterdam, The Netherlands: Elsevier.

Hägg, G., & Åström, A. (1997). Load pattern and pressure pain threshold in the upper trapezius muscle and psychosocial factors in medical secretaries with and without shoulder/neck disorders. *International Archives of Occupational and Environmental Health, 69*, 423–432.

Hall, E.M. (1990). *Womens work: An inquiry into the health effects of invisible and visible labour.* Unpublished doctoral dissertation, Karolinska Institute, Stockholm, Sweden.

Henneman, E., Somjen, G., & Carpenter, D.O. (1965). Excitability and inhibitibility of motoneurons of different sizes. *Journal of Neurophysiology, 28*, 599–620.

Henry, J.P. (1992). Biological basis of the stress response. *Integrative Physiological and Behavioral Science, 1*, 66–83.

Henry, J.P., & Stephens, P.M. (1977). *Stress, health, and the social environment. A sociobiologic approach to medicine.* New York: Springer-Verlag.

Herbert, T.B., & Cohen, S. (1993). Depression and immunity: a meta-analytic review. *Psychological Bulletin, 113,* 472–486.

Johansson, G. (1981). Psychoneuroendocrine correlated of unpaced and paced performance. In G.Salvendy & M.J.Smith (Eds.), *Machine pacing and occupational stress* (pp. 277–286). London: Taylor & Francis.

Johansson, G. (1991). Job demands and stress reactions in repetitive and uneventful monotony at work. In J.V.Johnson & G.Johansson (Eds.), *The psychosocial work environment: Work organization, democratization, and health* (pp. 61–72). Amityville, NY: Baywood.

Johansson, G., & Aronsson, G. (1984). Stress reactions in computerized administrative work. *Journal of Occupational Behavior, 5,* 159–181.

Johansson, G., Aronsson, G., & Lindström, B.O. (1978). Social psychological and neuroendocrine stress reactions in highly mechanized work. *Ergonomics, 21,* 583–599.

Johansson, G., Cavalini, P., & Pettersson, P. (1996). Psychobiological reactions to unpredictable performance stress in a monotonous situation. *Human Performance, 9,* 363–384.

Johansson, G., & Sandén, P.-O. (1997). *Biopsychosocial reactions to supervisory monitoring work.* Manuscript submitted for publication.

Johansson, J. (1994). *Psychosocial factors at work and their relation to musculoskeletal symptoms.* Unpublished doctoral dissertation, Department of Psychology, Göteborg University.

Johansson, J. (1994) . *Psychosocial factors at work and their relation to musculoskeletal symptoms.* Unpublished doctoral dissertation, Department of Psychology, Göteborg University.

Johnson, J.V., & Hall, E.M. (1988). Job strain, work place social support and cardiovascular disease. A cross-sectional study of a random sample of the Swedish working population. *American Journal of Public Health, 78,* 1336–1342.

Johnson, J.V., Hall, E., & Theorell, T. (1989). The combined effects of job strain and social isolation on the prevalence of cardiovascular disease and death in a random sample of the Swedish working male. *Scandinavian Journal of Work, Environment, and Health, 15,* 271–279.

Kadefors, R., Forsman, M., Zoéga, B., & Herberts, P. (1999). Recruitment of low threshold motor-units in the trapezius muscle in different static arm positions. *Ergonomics, 92,* 359–375.

Kahn, R.L. (1991). The forms of women's work. In M.Frankenhaeuser, U.Lundberg, & M. A.Chesney (Eds.), *Women, work and health: Stress and opportunities* (pp. 65–83). New York: Plenum.

Kahn, R.L., & Byosiere, P. (1992). Stress in organizations. In M.D.Dunette & L.M.Hough (Eds.), *Handbook of industrial and organizational psychology* (pp. 571–648). Palo Alto, CA: Consulting Psychologists Press.

Karasek, R.A. (1979). Job demands, job decision latitude and mental strain: Implications for job redesign. *Administrative Science Quarterly, 24,* 285–307.

Karasek, R.A., Russell, R.S., & Theorell, T. (1982). Physiology of stress and regeneration in job related cardiovascular illness. *Journal of Human Stress, 8,* 29–42.

Karasek, R.A., & Theorell, T. (1990). *Healthy work, stress, productivity, and the reconstruction of working life.* New York: Basic Books.

Krantz, D.S., & Manuck, S.B. (1984). Acute psychophysiologic reactivity and risk of cardiovascular disease: A review and methodologic critique. *Psychological Bulletin, 96,* 435–464.

Larsson, B., Libelius, R., & Ohlsson, K. (1992). Trapezius muscle changes unrelated to static work load. Chemical and morphologic controlled studies of 22 women with and without neck pain. *Acta Orthopaedica Scandinavica, 63,* 203–206.

Larsson, S.E., Bengtsson, A., Bodegård, L., Henriksson, K.G., & Larsson, J. (1988). Muscle changes in work related chronic myalgia. *Acta Arthopedica Scandinavia, 59,* 552–556.

Levi, L. (1972). Stress and distress in response to psychosocial stimuli. *Acta Medica Scandinavica* (Suppl. 528).

Levi, L., Frankenhaeuser, M., & Gardell, B. (1982). Reports on work stress related to social structures and processes. In G.Elliot & G.Eisdorfer (Eds.), *Stress and human health* (pp. 119–146). New York: Springer.

Lindman, R., Hagberg, M., Angqvist, K.-A., Söderlund, K., Hultman, E., & Thornell, L.E. (1991). Changes in muscle morphology in chronic trapezius myalgia. *Scandinavian Journal of Work Environmnt and Health, 17,* 347–355.

Lindman, R. (1992). Chronic trapezius myalgia—a morphological study. *Arbete och Hälsa, 34.*

Lundberg, U. (1984). Human psychobiology in Scandinavia: II. Psychoneuroendocrinology— human stress and coping processes. *Scandinavian Journal of Psychology, 25,* 214–226.

Lundberg, U. (1996). The influence of paid and unpaid work on psychophysiological stress responses of men and women. *Journal of Occupational Health Psychology, 1,* 117–130.

Lundberg, U., Elfsberg Dohns, I., Melin, B., Sansjö, L., Palmerud, G., Kadefors, R., Ekström, M., & Parr, P. (1999). Psychophysiological stress responses, muscle tension and neck and shoulder pain among supermarket cashiers. *Journal of Occupational Health Psychology, 4,* 1–11.

Lundberg, U., & Frankenhaeuser, M. (1980). Pituitary-adrenal and sympathetic-adrenal correlates of distress and effort. *Journal of Psychosomatic Research, 24,* 125–130.

Lundberg, U., Granqvist, M., Hansson, T., Magnusson, M., & Wallin, L. (1989). Psychological and physiological stress responses during repetitive work at an assembly line. *Work and Stress, 3,* 143–153.

Lundberg, U., Kadefors, R., Melin, B., Palmerud, G., Hassmén, P., Engström, M., & Elfsberg Dohns, I. (1994). Psychophysiological stress and EMG activity of the trapezius muscle. *International Journal of Behavioral Medicine, 1,* 354–370.

Lundberg, U., Melin, B., Evans, G.W., & Holmberg, L. (1993). Physiological deactivation after two contrasting tasks at a video display terminal: Learning vs. repetitive data entry. *Ergonomics, 36,* 601–611.

Lundberg, U., & Palm, K. (1989). Workload and catecholamine excretion of in parents of preschool children. *Work and Stress, 3,* 255–260.

Magnusson, M., Granqvist, M., Jonson, R., Lindell, V., Lundberg, U., Wallin, L., & Hansson, T. (1990). The loads on the lumbar spine during work at an assembly line. The risk for fatigue injuries of vertebral bodies. *Spine, 15,* 774–779.

Mason, J.W. (1968). A review of psychoendocrine research on the pituitary-adrenal cortical system. *Psychosomatic Medicine, 30,* 576–597.

McEwen, B.S. (1998). Stress, adaptation and disease: Allostasis and allostatic load. *New England Journal of Medicine, 238,* 171–179.

McEwen, B.S., & Stellar, E. (1993). Stress and the individual: Mechanisms leading to disease. *Archives of Internal Medicine, 153,* 2093–2101.

Melin, B., & Lundberg, U. (1997). A biopsychosocial approach to work-stress and musculoskeletal disorder. *Journal of Psychophysiology, 11,* 238–247.

Melin, B., Lundberg, U., Kadefors, R., Palmerud, G., Hassmén, P., Engström, M., & Elfsberg Dohns, I. (1997). *Trapezius EMG activity and psychophysiological responses of female supermarket cashiers during rest periods and experimental stress.* Unpublished manuscript.

Melin, B., Lundberg, U., Söderlund, J., & Granqvist, M. (1999). Stress reactions in assembly work: Comparison between two contrasting work organizations as related to sex. *Journal of Organizational Behavior, 20,* 47–61.

Moon, S.D., & Sauter, S.L. (Eds.). (1996). *Psychosocial aspects of musculoskeletal disorders in office work.* London: Taylor & Francis.

O'Hanlon, J.F., & Beatty, J. (1976). Catecholamine correlates of radar monitoring performance. *Biological Psychology, 4,* 293–303.

Repetti, R., Matthews, K.A., & Waldron, I. (1989). Employment and women's health: Effects of paid employment on women's mental and physical health. *American Psychologist, 44,* 1394–1401.

Rissler, A. (1977) Stress reactions at work and after work during a period of quantitative overload. *Ergonamics, 20,* 13–16.

Rodin, J., & Ickovics, J.R. (1990). Women's health. Review and research agenda as we approach the 21st century. *American Psychologist, 45,* 1018–1034.

Rozanski, A., Bairey, C.N., Krantz, D.S., Friedman, J., Resser, K.J., Morell, M., Hilton-Chalfen, S., Hestrin, L., Bietendorf, J., & Berman, D.S. (1988). Mental stress and the induction of silent myocardial ischemia in patients with coronary artery disease. *The New England Journal of Medicine, 318,* 1005–1011.

Sandén, P.-O. (1990). *Work in the control room. Studies of sociotecnical systems, job satisfaction, mental load and stress reactions.* Unpublished doctoral dissertation, Stockholm University, Sweden.

Sandén, P.-O., & Johansson, G. (1990). *Job content and technology in process control: Consequences for mental load and job involvement* (Report No. 725). Stockholm, Sweden: Stockholm University, Department of Psychology.

Sandsjö, L. (1997). *Long term trapezius EMG monitoring in different occupations. Proceedings from the 13th Triennal Congress of the International Ergonomics Association* (pp. 216–218). Tampere, Finland: Finnish Institute of Occupational Health, Helsinki.

Schleifer, L.M., & Ley, R. (1994). End-tidal PCO_2 as an index of psychophysiological activity during VDT data-entry work and relaxation. *Ergonomics, 37,* 245–254.

Seeman, T.E., McEwen, B.S., Singer, B.H., Albert, M.S., & Rowe, J.W. (1997). Increase in urinary cortisol excretion and memory declines. Mac Arthur studies of successful aging. *Journal of Clincal Endocrinology and Metabolism, 82,* 2458–2465.

Seeman, T.E., Singer, B.H., Rowe, J.W., Horwitz, R.I., & McEwen, B.S. (1997). The price of adaptation—Allostatic load and its health consequences: Mac Arthur studies of successful aging. *Archives of Internal Medicine, 157,* 2259–2268.

Sejersted, O.M., & Vøllestad, N.K. (1993). Physiology of muscle fatigue and associated pain. In H.Værøy & H.Merskey (Eds.), *Progress in fibromyalgia and myofascial pain* (pp. 41–51). Amsterdam, The Netherlands: Elsevier.

Selye, H. (1956). *The stress of life.* New York: McGraw-Hill.

Siegrist, J. (1996). Adverse health effects of high-effort/low-reward conditions. *Journal of Occupational Health Psychology, 1,* 27–41.

Steptoe, A. (1991) The links between stress and illness. *Journal of Psychosomatic Research*, *35*, 633–644.

Svebak, S., Anjia, R., & Kårstad, S.I. (1993). Task-induced electromyographic activation in fibromyalgia subjects and controls. *Scandinavian Journal of Rheumatology, 22,* 124–130.

Symonds, T.L., Burton, A.K., Tillotson, K.M., & Main, C.J. (1995). Absence resulting from low back trouble can be reduced by psychosocial intervention at the work place. *Spine, 20,* 2738–2745.

Taylor, F.W. (1923). *The principles of scientific management.* New York: Harper.

Theorell, T. (1989). Personal control at work and health. In A.Steptoe & A.Appels (Eds.), *Stress, personal control, and health.* New York: Wiley.

Timio, M., & Gentili, S. (1976). Adrenosympathetic overactivity under conditions of work stress. *British Journal of Preventive and Social Medicine, 30,* 262–265.

Timio, M., Gentili, S., & Pede, S. (1979). Free adrenaline and noradrenaline excretion related to occupational stress. *British Heart Journal, 42,* 471–474.

Tulen, J.H.M., Moleman, P., van Steenis, H.G., & Boomsma, F. (1989). Characterization of stress reactions to the stroop color word test. *Pharmacology Biochemistry, and Behavior, 32,* 9–15.

Ursin, H., Baade, E., & Levine, S. (1978). *Psychobiology of stress. A study of coping men.* New York: Academic.

Usdin, E., Kvetnansky, R., & Kopin, I.J. (Eds.). (1980). *Catecholamines and stress: Recent advances.* New York: Elsevier North-Holland.

Veiersted, B. (1995). *Stereotyped light manual work, individual factors and trapezius myalgia.* Unpublished doctoral dissertation, University of Oslo, Norway.

Veiersted, K.B., Westgaard, R.H., & Andersen, P. (1993). Electromyographic evaluation of muscular work pattern as a predictor of traplezius myalgia. *Scandinavian Journal of Work and Environmental Health, 19,* 284–290.

Wall, T.D., & Martin, R. (1987). Job and work design. In C.L.Cooper & I.T.Robertson (Eds.), *International review of industrial and organizatonal psychology* (pp. 61–91).

Wærsted, M. (1997). *Attention-related muscle activity—a contributor to sustained occupational muscle load.* Unpublished doctoral dissertation, National Institute of Occupational Health, Oslo, Norway.

Wærsted, M., Bjørklund, R., & Westgaard, R. (1991). Shoulder muscle tension induced by two VDU-based tasks of different complexity. *Ergonomics, 34,* 137–150.

Wærsted, M., Eken, T., & Westgaard, R.H. (1996). Activity of single motor units in attention-demanding tasks: Firing pattern in the human trapezius muscle. *European Journal of Applied Physiology, 72,* 323–329.

Weber, A., Fussler, C., O'Hanlon, J.F., Gierer, R., & Grandjean, E. (1980). Psychophysiological effects of repetitive tasks. *Ergonomics, 23,* 1033–1046.

Westgaard, R., & Bjørklund, R. (1987). Generation of muscle tension additional to posture muscle load. *Ergonomics, 30,* 911–923.

Chapter 17

Engineering Psychophysiology in Japan

Akihiro Yagi
Kwansei Gakuin University, Japan

For a number of years, I have been proposing the concept of *psychological engineering,* which is best defined as engineering relating to human psychological activities (Yagi, 1981, 1995a). Psychological engineering is a broad concept that includes engineering psychophysiology. It consists of several subordinate themes. The first theme is the development of new systems between the human mind and machines. Some systems are being developed in industrial settings. The reason for the need for developing such new approaches is the fact that the purpose of development of a machine is gradually shifting from convenience to comfort.

The second theme is the development of the technology to measure psychological effects in industrial settings. This includes the development of new types of questionnaires and the application of multidimensional analysis to the psychological activities. In engineering areas, it also includes the development of new psychophysiological methods, such as neural imaging techniques including fMRI and MEG, and the application of analysis, such as procedures attempting to identify electric dipoles in the brain.

The third theme is the development of new types of human-machine systems incorporating concepts and procedures utilizing virtual reality. The fourth is the development of the concept of an artificial mind. This is an attempt to develop a computer that operates similarly to the way humans think. Artificial in-telligence is a well-known example of this development. Further, artificial emotion and motivation may be implemented in computers or artificial neural systems of the future. The fifth theme is engineering psychophysiology. The purpose of this chapter is to introduce the reader to trends in recent engineering psychophysiological efforts in Japan.

PSYCHOPHYSIOLOGY IN ERGONOMICS

The field that traditionally has had a close relation to engineering psychophysiology is ergonomics. Many studies have been conducted involving the use of psychophysiological techniques to ergonomic issues in Japan. Because most of the studies are published in Japanese, people in other countries may be unfamiliar with this research. Many such studies involving the use of psychophysiology are reported at the annual meetings of several academic societies in Japan. Two societies that deal with psychophysiology and ergonomics are the Japanese Society for Physiological Psychology and Psychophysiology (JSPP) and the Japan Ergonomics Research Society (JES).

The research theme in JSPP is quite varied, ranging from the study of basic problems to applied research (e.g., clinical and engineering). On the other hand, the important research themes exemplifying papers presented at the meetings of JES deal with issues such as mental workload, human-machine interaction, fatigue assessment, and so on. Not only are scientists trained in psychology and physiology engaged in such efforts, but many engineers are also conducting psychophysiological research in both laboratory and practical field settings. Methods and theories utilized in these studies are similar to those in other countries. Because some of the studies are presented at the conference of IEA, I do not discuss them in detail here.

CHARACTERISTICS OF ENGINEERING
PSYCHOPHYSIOLOGY IN JAPAN

The assessment of mental workload and fatigue have been very important themes in traditional psychophysiological ergonomic studies in Japan. These efforts are similar to those reported from Western countries. *Kansei* and neural imaging technologies are becoming current topics in engineering psychophysiology in Japan. The first characteristic of engineering psychophysiology is the study of *kansei*. *Kansei* is an ordinal word that means sensitivity and feeling. The measurement of *kansei* is described in detail in the next section.

The second characteristic of engineering psychophysiology is the study of the neural images of the mind. Recently, several new types of devices have been used to study the relationship between mental activity and neural systems. MEG, functional MRI, and the electric dipole estimation procedure have been implemented as noninvasive methods. They have been implemented in a number of engineering laboratories, in some national institutes, and in some private industrial settings.

At present, the main theme of research in such settings focuses on cognitive neuroscience. For instance, topics for research include sensation, attention mechanism, and *kansei*. However, the final aim of studies in these engineering laboratories is to apply the results to the development of better human-machine interfaces and new types of information-processing machines or computers. The investigators are principally engineers. However, the use of highly technological devices does not invariably guarantee the finding of new and useful information concerning mental activities. Some psychologists and physiologists who are members of the laboratories work collaboratively with each other and engineering members of the organization. Psychologists and physiologists who are members of other institutions and universities engage in joint research with these organizations.

ENGINEERING RELATING TO *KANSEI*

The term *kansei* is often used in the engineering area in Japan and Korea. The lexical meaning of *kansei* is intuitional mental activities related to feeling and desire. Because the ordinal meaning of *kansei* is very vague, it is difficult to translate the term *kansei* into English. There are a few descriptions in the dictionary of psychology, although *kansei* has even been used as a translation of the term *sensory* in psychology.

Kansei originally referred to a personal ability to discriminate and evaluate the high quality of art or the purity of the natural environment. For instance, it is a common expression that an excellent artist shows sharp *kansei.* Therefore, it may be psychologically explained that *kansei* refers to the intelligence associated with the ability to make fine discriminations and evaluations.

In the engineering field, *kansei* is used to refer to sensitivity and feeling when a person looks at or has contact with a product. Engineers wish to control such qualities as beauty, desirability, comfort, and pleasantness manifested in an industrial product.

Such good quality is called *kaitekisei.* A new product that shows *kaitekisei* may have a good influence upon a person as an end user. In other words, the product with *kaitekisei* will stimulate the *kansei* of a person. Therefore, engineers wish to measure a person's *kansei* by using psychophysiological methods. The engineer's goal is to make use of the measurement of *kansei* in designing the products. The studies of not only usability, but also *kansei* are required in engineering field in Japan.

Some of the most useful measures for determining *kansei* involve the use of psychological methods. Psychophysical methods, the semantic differential method, questionnaires, and multidimensional scaling are used in the engineering field. Nagamachi (1988) proposed *kansei* engineering that makes use of psychological techniques in the design of a product. As mentioned later, however, engineers in Japan prefer to use objective physiological methods rather than the more subjective psychological procedures.

RESEARCH PROJECT ON *KANSEI*
AND STRESS

A research project supported by the Japanese government started in 1990. The title of the project is "Human Sensory Measurement Application Technology." The two major themes of the project are the measurement of *kansei* and stress. Six national research institutes for engineering and 22 laboratories that belong to private companies are supported by the Japanese government (Ministry of International Trade and Industry, Agency of Industrial Science and Technology) to conduct this research effort. The Research Institute of Human Engineering for Quality Life (HQL) coordinates the project. In Korea, a new research project whose title is "*Kansei* Engineering" has recently been initiated. It is modeled on the Japanese project. Many laboratories are involved in the development of this new technology for the assessment of stress, fatigue, and activation. Some laboratories are attempting to develop tools for the measurement of *kansei*.

Some of the laboratories in private companies in Japan had initiated such psychophysiological research before the new project started. The head of the laboratory in a manufacturing company is usually an engineer. Most of these leaders are not educated in experimental psychology. Frankly speaking, they frequently dislike questionnaires and psychological tests since these measures deal with subjective impression. They prefer physiological measures to psychological measures for the assessment of mental activities. It is most likely one of the reasons why many laboratories in private companies have initiated psychophysiological investigations. In some excellent companies (e.g., Panasonic (Matsushita), Mitsubishi, Toyota, etc.), psychologists and engineers have been working together in the laboratory. They report that psychological methods in conjunction with psychophysiological procedures are the most useful measure for *kansei*.

EXAMPLES OF PSYCHOPHYSIOLOGICAL
STUDIES ON *KANSEI*

One of the main themes of the project is the assessment of *kaitekisei,* feelings of comfort and pleasantness. Another theme focuses on stress. *Kaitekisei* in relation to *kansei* refers to an agreeable quality, one generated by aspects of the environment. As mentioned before, the engineer wishes to identify conditions producing comfort and feelings of pleasantness associated with specific products. For that purpose, they study the psychological factors (e.g., feelings of comfort and pleasantness) of a person as a user.

Alpha activity in the EEG is used as an index of relaxation and feelings of comfort in many laboratories. However, some of the conclusions remain questionable, especially those suggesting that subjects felt comfortable only when alpha activity appeared.

Yoshida and Kaneko (1989) proposed a new method for analyzing alpha wave activity. They developed a system to estimate the fluctuation of the frequencies in the alpha band (8–13 Hz). Because the algorithm for the analysis is too complicated to present here, the interested reader is referred to their paper. This group (Yoshida et al., 1991) reported that the coefficient of fluctuation was related to the feeling of comfort of the subject rather than the appearance of alpha activity. Kato and Yagi (1993) have attempted to discriminate between pleasant and unpleasant feelings using facial muscle activities. When the subject perceived an unpleasant smell, most showed a uniform pattern in the facial muscle activities. However, these authors could not obtain unique muscle patterns associated with pleasant smells. It is still difficult to assess feelings of pleas-antness using psychophysiological technique. Psychophysiological studies on *kansei* were initiated fairly recently in Japan. Thus, the study of *kansei* is still in the early stages of development.

RESEARCH PROJECTS DEALING WITH THE ASSESSMENT OF STRESS

The assessment of stress may be easier than that of *kaitekisei*. Some methods developed in Western countries have already been used by Japanese investigators to assess stress. Heart rate and heart rate variability (HRV) are often used in industrial settings. Because such studies are similar to those conducted in Western countries, only examples of such research being conducted in private companies and supported by the research project are introduced here. Some investigators in engineering laboratories are proposing to develop new techniques of assessment for transient stress as well as long-lasting stress. A group at Mitsubishi Electric is conducting research on the assessment of stress (e.g., Ohsuga et al., 1995), especially with subjects working at VDT tasks. They proposed a model for the identification of stress and are developing a system to assess stress using neural network methodologies. A group at Shiseido Cosmetics is measuring stress using endocrinological assessment procedures (Abe, 1996). Engineers at Sanyo Electric have developed a unique method to assess stress with a noncontact measurement procedure. They found a relationship between stress levels and the temperature of the bridge of the nose using an improved infrared camera (Genno et al., 1997). Omuron Life Science manufactured, for trial, a portable measurement device for recording a number of physiological variables in industrial settings (Nishio et al., 1997). The project will be succeeded by a new research project that includes cognitive and behavoral processes.

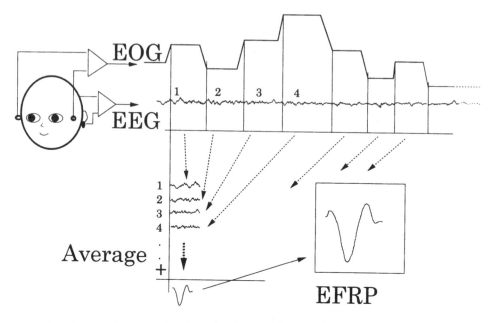

FIG. 17.1. Detection of eye fixation related potentials (EFRP). EEG epochs time-locked to onset of eye fixations were collected and averaged.

EVENT-RELATED POTENTIALS APPLIED TO ENGINEERING

Finally, I would like to introduce unique event-related brain potentials(ERP) as a useful index for psychophysiological assessment. The so-called ERP changes with visual attention or workload. However, it is difficult to apply visual ERP technology to the assessment of visual tasks because eye movements produce major artifacts in the EEG. During visual task performance, the eyes usually move. Eye movement records demonstrate a step-like pattern consisting of saccades and fixation pauses. Saccades occur about 3 to 4 times per second. Information concerning the nature of the visual object is sent from the retina to the brain during the fixation pause. When EEGs time-locked to fixation pause onset (i.e., offset of saccades) are averaged, the eye fixation related potential (EFRP) is obtained (see Fig. 17.1). The author has developed a system to detect these EFRP.

The EFRP consists of several components. Some components change as a function of stimulus properties (e.g., spatial frequency, contour and brightness of the stimulus) and subjective factors (e.g., attention, signal detection, and language processing; Yagi, 1995b). There is thus the possibility to apply the EFRP to studies in engineering fields.

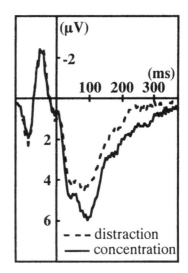

FIG. 17.2. Grand averaged EFRPs in concentration (solid line) and distraction (dotted line). The vertical line indicated the moment of onset of the eye fixation. Upward negative.

APPLICATION OF EYE FIXATION
RELATED POTENTIAL

In the first series of such studies, EFRP was measured while subjects worked at a VDT task. The wave form of the EFRP changed with the type of screen (positive or negative) on the CRT (Yagi & Ogata, 1995). When subjects enjoyed perform- ing a computer graphic task, the amplitude of EFRP increased (Yagi et al., 1997). On the other hand, the amplitude decreased when the subject was tired or the arousal level decreased (see Fig. 17.2). We came to the conclusion that variations in EFRP are related to the attention and arousal level during task performance.

In the second series of experiments, we examined the relationship between the lighting environment and EFRP. We measured EFRP under five types of lighting systems ranging from working under a spotlight to general lighting conditions. The stability (variability) of the wave form of EFRP reflected the level of concentration of attention to the tasks under the lighting conditions (Sakaue et al., 1997; Yagi et al., 1998).

In the third series of experiments, we measured EFRP when subjects observed patterns of textiles. The amplitudes of early components in the EFRP changed with the complexity of the textile pattern. The result suggests that EFRP may be applicable as a psychophysiological index to evaluate textile patterns (Sakamoto & Yagi, 1992).

Most of these studies were laboratory experimental studies. Of course, there may be some difficulties in applying EFRP technology to practical ergonomic problems. However, it is our conviction that EFRP will be a more useful measure in assessing visual environments and visual processing than the so called

ERP. Although we are still in an early stage of developing and evaluating this technology, we will hopefully be able to apply EFRP to the assessment of *kansei* in the future.

REFERENCES

Abe, T. (1996). Evaluation of bath salts by cortisone. *Fragrance Journal, 24,* 74–83 (in Japanese).

Genno, H., Ishikawa, K., Kanbara, O., & Kikumoto, M. (1997). Using facial skin temperature to objectively evaluate sensations. *Industrial Journal of Industrial Ergonomics, 19,* 161–171.

Kato, M., & Yagi, A. (1993). The effects of pleasantness of odors on facial EMG. In K. Kurihara, N.Suzuki, & H.Ogawa (Eds.), *Olfaction and taste* XI (pp. 676–677). Tokyo: Springer-Verlag.

Nagamachi, M. (1988). Image technology based on knowledge engineering and its application to design consultation. *Ergonomics International, 88,* 72–74.

Nishio, Y., Suzuki, M., & Tanimura, Y. (1997). Application of small physiological measuring devices to the study of human responses. *Proceedings of the 13th Triennial Congress of the International Ergonomics Association, Vol. 7,* 270–272.

Ohsuga, M., Terashita, H., Shimono, F., & Toda, M. (1995). Assessment of mental workload based on a model of autonomic regulations on the cardiovascular system. In Y.Anzai, K. Ogawa, & H.Mori (Eds.), *Symbiosis of human and artifact* (pp. 771–776). Tokyo: Elsevier.

Sakamoto, K., & Yagi, A. (1992). The effect of textile patterns on eye fixation related potentials. *Proceedings of Annual Meeting of the Japan Research Association for Textile End-Uses,* 74–75 (in Japanese).

Sakaue, M., Akashi, Y., Umeno, C., & Yagi, A. (1997). Relationship between mental concerntration of workers and ambient illuminations. *Journal of the Illuminating Engineering Institute of Japan, 81,* 385–391 (in Japanese).

Yagi, A. (1981). Psychological engineering. *Japanese Journal of Biofeedback Research, 9,* 8–60 (in Japanese).

Yagi, A. (1995a).The role of psychological engineering in study of high quality of human life. *Proceedings of Summit Meeting for Quality of Life, Vol 2* (pp. 1–5). Osaka, Japan: HQL.

Yagi, A. (1995b). Eye fixation related potential as an index of visual function. In S.Tsutsui & K.Shirakura, (Eds.), *Biobehavioral self-regulation in the east and the west* (pp. 177–181). Tokyo: Springer Verlag.

Yagi, A., Imanishi, S., Konishi, H., Akashi, Y., & Kanaya, S. (1998). Brain potentials associated with eye fixations during visual tasks under different lighting systems. *Ergonomics.*

Yagi, A., & Ogata, M. (1995). Measurement of work load using brain potentials during VDT tasks. In Y.Anzai, K.Ogawa, & H.Mori (Eds.), *Symbiosis of human & artifact* (pp. 823–826). Tokyo: Elsevier.

Yagi, A., Sakamaki, E., & Takeda, Y. (1997). Psychophysiological measurement of attention in a computer graphic task. *Proceedings of the 5th International Scientific Conference of Work With Display Unit,* (pp. 203–204).

Yoshida, T., & Kaneko, T. (1989). The 1/f frequency-fluctuation of background EEG during children's TV watching. In N.W.Bond & D.A.T.Siddle (Eds.), *Psychology: Issues and applications* (pp. 315–323). Amsterdam, The Netherlands: Elsevier.

Yoshida, T., Ohmoto, S., & Kanamura, S. (1991). 1/f frequency-fluctuation of human EEG and emotional changes. In T.Musha, S.Sato, & M.Yamamoto (Eds.), *Proceedings of the international conference on noise in physical systems and 1/f fluctuations* (pp. 719–722). Tokyo: Ohmsha.

Author Index

Subject Index